Plant-Derived Anti-inflammatory Agents: Molecular Pharmacology and Further Studies

Plant-Derived Anti-inflammatory Agents: Molecular Pharmacology and Further Studies

Editor

Diana Roxana Pelinescu

MDPI

Basel • Beijing • Wuhan • Barcelona • Belgrade • Novi Sad • Cluj • Manchester

Editor
Diana Roxana Pelinescu
Department of Genetics
University of Bucharest
Bucharest
Romania

Editorial Office
MDPI
St. Alban-Anlage 66
4052 Basel, Switzerland

This is a reprint of articles from the Special Issue published online in the open access journal *Pharmaceuticals* (ISSN 1424-8247) (available at: www.mdpi.com/journal/pharmaceuticals/special_issues/Plant_Anti_inflammatory_Agents).

For citation purposes, cite each article independently as indicated on the article page online and as indicated below:

Lastname, A.A.; Lastname, B.B. Article Title. *Journal Name* **Year**, *Volume Number*, Page Range.

ISBN 978-3-0365-9189-6 (Hbk)
ISBN 978-3-0365-9188-9 (PDF)
doi.org/10.3390/books978-3-0365-9188-9

Contents

About the Editor

Diana Roxana Pelinescu

Associate Prof. Pelinescu Diana, PhD, has a research activity spanning over 20 years, which has mainly been carried out at the University of Bucharest, Faculty of Biology. Her studies were mainly focused on: microorganisms (lactic acid bacteria and yeasts) with biotechnological potential; analysis of microbial biodiversity from human and animal gastrointestinal tracts and from fermented foods; the gut microbiota diversity and dynamics in the presence of some bioactive compounds like mushroom extracts, tea, probiotics, synbiotics, and nutraceuticals; mechanisms involved in the beneficial effects of food supplements and plant extracts; biotechnological studies on obtaining microbial biomass and bioactive compounds; and in vitro analysis of nanomaterials influence on human and microbial cells.

Preface

Mankind has used medicinal plants for thousands of years due to their healing properties. In the days when there were no synthetic drugs, plants represented the only alternative to treating health problems. Many plant compounds have anti-inflammatory properties, and they are an essential source of pharmaceutical drugs. Some modern pharmaceuticals are either directly extracted from plants or synthesized based on compounds found in plants. The most well-known compounds are aspirin, morphine, quinine, ephedrine, and atropine.

Since the technological advances achieved during the last decades in all fields, chemical extraction has become more accurate, and the characterization of plant compounds has become more rigorous. Moreover, numerous in vitro and in vivo studies have demonstrated the beneficial effects of plant compounds on human health.

Due to their great diversity and huge number of compounds, plants remain an important source of compounds and inspiration for the pharmaceutical industry and modern medicine.

Diana Roxana Pelinescu
Editor

pharmaceuticals

Article

Malva parviflora Leaves and Fruits Mucilage as Natural Sources of Anti-Inflammatory, Antitussive and Gastro-Protective Agents: A Comparative Study Using Rat Models and Gas Chromatography

Ahmed E. Altyar [1], Ans Munir [2], Saiqa Ishtiaq [2], Muhammad Rizwan [3], Khizar Abbas [4], Osama Kensara [5], Sameh S. Elhady [6], Waleed Y. Rizg [7,8], Fadia S. Youssef [9] and Mohamed L. Ashour [9,10,*]

1 Department of Pharmacy Practice, Faculty of Pharmacy, King Abdulaziz University, P.O. Box 80260, Jeddah 21589, Saudi Arabia; aealtyar@kau.edu.sa
2 Department of Pharmacognosy, College of Pharmacy, University of the Punjab, Lahore 54000, Pakistan; ansmunir92@gmail.com (A.M.); saiqa.pharmacy@pu.edu.pk (S.I.)
3 Department of Pathology, Lahore Medical and Dental College, University of Health Sciences, Lahore 54600, Pakistan; muhammad.rizwan@lmdc.edu.pk
4 Department of Pharmacognosy, Faculty of Pharmacy, Bahauddin Zakariya University, Multan 60800, Pakistan; khizarabbas@bzu.edu.pk
5 Department of Clinical Nutrition, Faculty of Applied Medical Sciences, Umm Al-Qura University, P.O. Box 7067, Makkah 21955, Saudi Arabia; oakensara@uqu.edu.sa
6 Department of Natural Products, Faculty of Pharmacy, King Abdulaziz University, P.O. Box 80260, Jeddah 21589, Saudi Arabia; ssahmed@kau.edu.sa
7 Department of Pharmaceutics, Faculty of Pharmacy, King Abdulaziz University, P.O. Box 80260, Jeddah 21589, Saudi Arabia; wrizq@kau.edu.sa
8 Center of Excellence for Drug Research and Pharmaceutical Industries, King Abdulaziz University, P.O. Box 80200, Jeddah 21589, Saudi Arabia
9 Department of Pharmacognosy, Faculty of Pharmacy, Ain-Shams University, Abbasia, Cairo 11566, Egypt; fadiayoussef@pharma.asu.edu.eg
10 Department of Pharmaceutical Sciences, Pharmacy Program, Batterjee Medical College, P.O. Box 6231, Jeddah 21442, Saudi Arabia
* Correspondence: ashour@pharma.asu.edu.eg

Abstract: *Malva parviflora* L., Little mallow, has been traditionally used as an alternative food source. It acts as a medicinal herb containing a potential source of mucilage thus herein; we aimed to assess the toxicity, anti-inflammatory, antitussive and gastro-protective actions of *M. parviflora* mucilage extracted from its leaves (MLM) and fruit (MFM). Toxicity studies were investigated by in vitro hemolytic assay whereas acute anti-inflammatory and antitussive activities were assessed by carrageenan-induced paw edema and sulphur dioxide induced cough model in rats, respectively. Gastro-protective effects were studied using ethanol induced acute and chronic gastric ulcer rat models. Their metabolic profiles were determined using gas chromatography. The results revealed that MLM and MFM were non-toxic towards human erythrocytes and their lethal doses were found to be greater than 5 g/kg. Pretreatment with MLM (500 mg/kg) and MFM (500 mg/kg) significantly reduced the carrageenan-induced paw thickness ($p < 0.001$). Maximum edema inhibition (%) was observed at 4 h in diclofenac sodium (39.31%) followed by MLM (27.35%) and MFM (15.68%). Animals pretreated with MLM (500 mg/kg) significantly lower the cough frequency in SO_2 gas induced cough models in contrast to control. Moreover, MLM at doses of 250 and 500 mg/kg reduced the ethanol induced gastric mucosal injuries in acute gastric ulcer models presenting ulcer inhibition of 23.04 and 38.74%, respectively. The chronic gastric ulcer model MFM (500 mg/kg) demonstrated a remarkable gastro-protective effect showing 63.52% ulcer inhibition and results were closely related to standard drug sucralfate. In both models, MLM and MFM decreased gastric juice volume and total acidity in addition to an increased gastric juice pH and gastric mucous content justifying an anti-secretary role of this mucilage that was further confirmed by histopathological examination. Meanwhile, GC analyses of the mucilage revealed their richness with natural as well as acidic monosaccharides. It is

check for updates

Citation: Altyar, A.E.; Munir, A.; Ishtiaq, S.; Rizwan, M.; Abbas, K.; Kensara, O.; Elhady, S.S.; Rizg, W.Y.; Youssef, F.S.; Ashour, M.L. *Malva parviflora* Leaves and Fruits Mucilage as Natural Sources of Anti-Inflammatory, Antitussive and Gastro-Protective Agents: A Comparative Study Using Rat Models and Gas Chromatography. *Pharmaceuticals* **2022**, *15*, 427. https://doi.org/10.3390/ph15040427

Academic Editor: Diana Roxana Pelinescu

Received: 7 March 2022
Accepted: 28 March 2022
Published: 31 March 2022

concluded that MLM and MFM can be used therapeutically for the management of inflammation, cough and gastric ulcer.

Keywords: anti-inflammatory; antitussive; cytoprotection; gastric ulcer; *M. parviflora*; mucilage

---

1. Introduction

Demulcents form a protective coating over mucous membranes alleviating pain and inflammation becoming slimy upon contact water [1]. They are indicated to heal hot, irritated, dry and inflamed mucous membranes and for ulcers, sore throat, inflammation in bowels and upper respiratory tract, and for bladder infections [2]. They can also act as prebiotics and normalize intestinal flora population. They repair the mucosa by reducing the irritation of bowel and decreasing sensitivity to gastric acid [3]. Histological studies have revealed that consumption of demulcent herbs results in the presence of polysaccharides layers on the membrane surface [4].

Demulcents are medicinally important as they contain high amounts of mucilage. Plant mucilage is a gelatinous substance chiefly comprising polysaccharides and uronic acid, as well as glycoproteins with other biologically active substances including tannins, alkaloids and steroids [5,6]. Mucilage is a renewable and inexpensive source of non-toxic, bioactive and eco-friendly compounds and have extensive applications in pharmaceutical, food and nutraceutical, cosmetics, textile, paper and paint industries [7]. The biologically active and nutraceutical characteristics, along with their possible health blessings, mark mucilage as an important ingredient in a healthy diet. They have been used as an antioxidant, antidiabetic, anticancer, antifungal, antimicrobial, anti-inflammatory, wound healer, ACE enzyme inhibitor, hypolipidemic agent and immune-stimulator. Further, they are used to treat skin, gastrointestinal, respiratory and urinary disorders [8,9]. It is worthy to highlight that a number of plant mucilages are available in the literature that have been approved by FDA for their additive roles in food, cosmetics and pharmaceutical industries i.e., Fenugreek seed mucilage, *Hibiscus rosa-sinensis* mucilage *Lepidium sativum* mucilage, *Aloe vera*, *Plantago ovata* seed mucilage [10].

Moreover, inflammation is the first response of body's immune system to stress, injury and infection. Prolonged and persistent inflammation could be harmful leading to the development of diseases such as fever, asthma, arthritis, atherosclerosis, auto-immune disorders and cancer [11–13]. The treatment of inflammation usually depends on steroidal and non-steroidal anti-inflammatory drugs that reveal side effects such as gastrointestinal ulcer, hypertension, osteoporosis, hepatotoxicity, kidney disorders as well as allergies [14]. Gastric ulcers are a long-lasting illness affecting thousands of people globally. This not only disturbs everyday life of affected individuals but is also sometimes accompanied by fatal complications i.e., gastric bleeding and perforations [15].

Traditionally, natural products were utilized to cure and manage stomach ulcers [16–18]. They proposed effective, inexpensive and easily available forms of treatment for individuals affected by gastric ulcers [19]. Thus, unveiling of the most active and safe ulcer healing and protective agent from natural products is of great value [20]. This suggests that seeking new natural anti-inflammatory and gastro-protective agents with limited side effects is needed for human health [21]. Furthermore, mucilage polysaccharides obtained from different plants are used in children with acute cough due to their emollient and demulcent properties. Their mucus-protective and bio-adhesive properties are useful in oral and gastric disorders [22].

Little mallow plant (*Malva parviflora* L.) are among the edible crops and have been part of the Mediterranean diet for a long time [23]. It is generally recognized as cheese-weed, small whorl mallow and in Pakistan as Sonchal. Traditionally decoction of the entire plant has been used as a remedy for fever, cold and cough meanwhile its leaves are used as a vegetable and an emollient. They are used to treat inflammation, wounds, gastritis, bladder ulcers, diuretic, constipation, abdominal pain, diarrhea, anthelmintic, hair loss, ocular disease, scorpion sting and profuse menstruation [24]. Immature fruits are used as snacks. Mallows have a long history of medicinal use due to their high antioxidant activity and anti-inflammatory potential. Pharmacologically, *M. parviflora* have been evaluated for its antioxidant, anti-inflammatory, antimicrobial, antiulcer, antidiabetic, anti-irritant, hepatoprotective, neuroprotective, analgesic and wound healing properties [25]. Although anti-inflammatory and gastro-protective effects of crude methanol and ethanol extracts of *M. parviflora* leaves have been reported, no research was found in the literature regarding such activities of *M. parviflora* leaves and fruit mucilage [26–29].

Hence, our previous study reported the potent antioxidant, DNA damage and skin protection activity of *M. parviflora* leaves mucilage [25], thus, herein our main objective was to compare both *M. parviflora* leaves and fruit mucilage in virtue of their toxicity, anti-inflammatory, antitussive and gastro-protective activities aiming to use them as potential food sources to alleviate many health disorders. In vitro and in vivo toxicity was studied by hemolytic assay and following the OECD 425 guidelines, respectively. Acute antitussive and anti-inflammatory activities were investigated using sulphur dioxide induced cough models and carrageenan-induced paw edema models in rats, respectively. In addition, ethanol induced acute and chronic gastric ulcer models were used to study the gastro-protective effects of *M. parviflora* leaves (MLM) and fruit (MFM) mucilage. Moreover, their comparative metabolic profiles were determined using gas chromatography. Meanwhile, it is noteworthy to mention that this is the first study to report the toxicity and antitussive activity of *M. parviflora* leaves and fruit mucilage. Previously, no data were available regarding *M. parviflora* mucilage, meanwhile the clinical role of mucilage is an emerging research area and remains still under investigation.

2. Results

2.1. In Vitro Hemolytic Activity

Hemolysis assay is suitable to investigate whether cytotoxicity is associated with direct damage to cell membrane or not. Hemolytic activity of MLM and MFM was screened against human erythrocytes at four different concentrations which are 125, 250, 500 and 1000 µg/mL. Full hemolysis (100%) was attained using 0.1% Triton X-100 whereas no hemolysis was observed with Phosphate-buffered saline (PBS). Both mucilages displayed little hemolytic result towards human RBCs. Very weak hemolysis was observed for MLM and MFM even at high concentrations showing 3.81 and 2.43% hemolysis at 1000 µg/mL respectively. Results for hemolytic activity as depicted in Figure 1 supports the safety of MLM and MFM towards human erythrocytes.

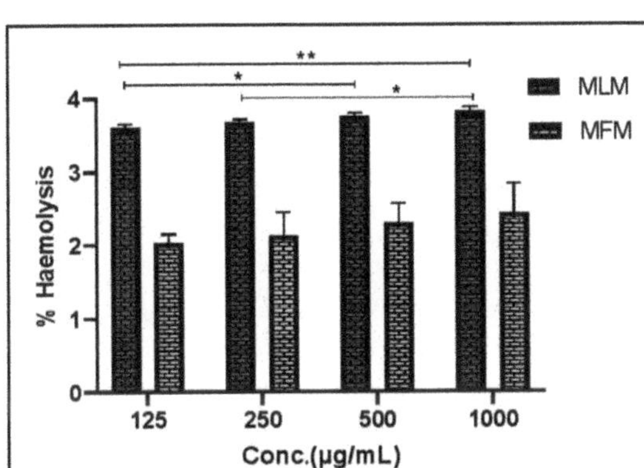

Figure 1. Hemolytic activity of MLM and MFM at different concentrations, 125, 250, 500 and 1000 µg/mL; PBS (negative control) showed no hemolysis whereas Triton-X (positive control) showed 100% hemolysis; Significant at $p < 0.05$ * and 0.01 **.

2.2. Acute Toxicity Study

In acute toxicity studies, all rats remained alive showing no signs of toxicity or abnormalities observed at a dose of 5 g/kg of MLM and MFM, administered orally. For the period of fourteen days, there were no signs of abnormalities evidenced by no behavioral changes or alteration in body weight. There was a non-significant difference in serum biochemical parameters of treated group and control group (Table 1). Histological analysis showed no signs of hepatic or renal toxicity as compared to control group (Figure 2). Thus, it was concluded that the oral lethal dose for MLM and MFM would be greater than 5 g/kg. These results support the safety of MLM and MFM even at high dose.

Table 1. Results of biochemical parameters from acute toxicity studies performed using MLM and MFM (5 g/Kg) treated rats.

Groups	Urea (mg/dL)	Creatinine (mg/dL)	Bilirubin (mg/dL)	SGPT (u/L)	SGOT (u/L)	ALP (u/L)
Normal	42.83 ± 1.05	0.9 ± 0.05	0.89 ± 0.039	107.5 ± 9.98	118 ± 8.88	139 ± 6.96
MLM	38.83 ± 1.8 ^{ns}	0.82 ± 0.06 ^{ns}	0.76 ± 0.05 ^{ns}	105.83 ± 11 ^{ns}	114.67 ± 7.8 ^{ns}	128.5 ± 5.5 ^{ns}
MFM	40.5 ± 1.57 ^{ns}	0.73 ± 0.05 ^{ns}	0.79 ± 0.04 ^{ns}	86.67 0± 6.66 ^{ns}	114.67 ± 7.8 ^{ns}	133 ± 3.21 ^{ns}

All values are represented as (Mean ± SEM), ns = not significant.

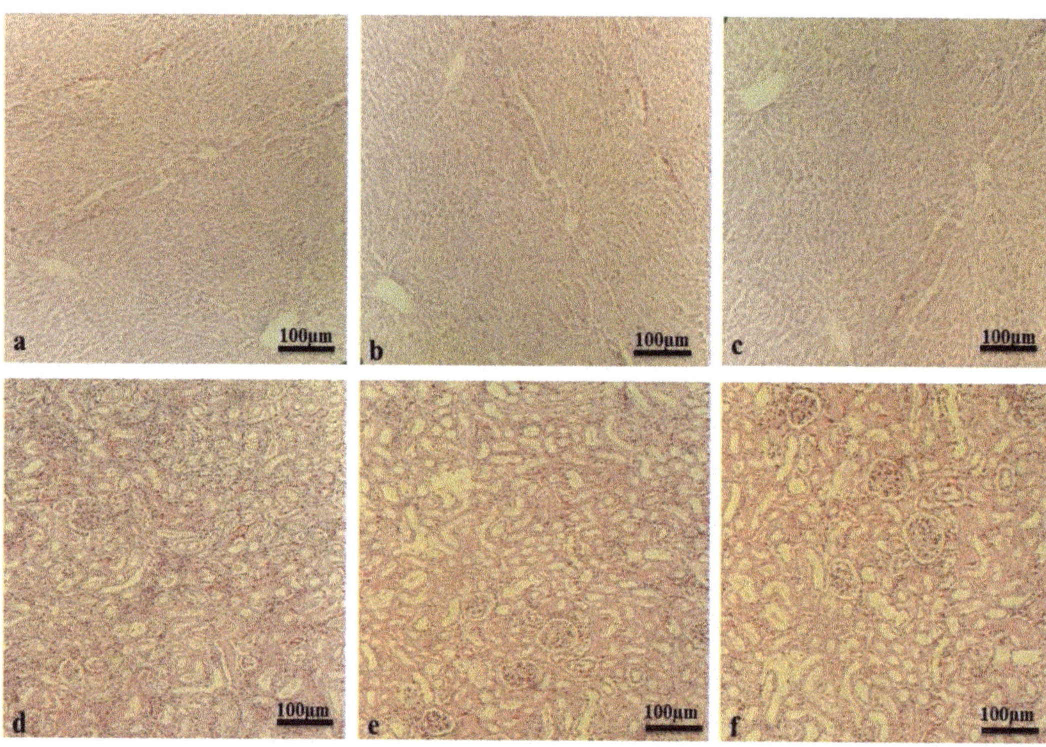

Figure 2. Liver and kidney histology of acute toxicity assay (**a**) Liver of vehicle control group (**b**) Liver of MLM treated group (**c**) Liver of MFM treated group (**d**) Kidney of vehicle control group (**e**) kidney of MLM treated group (**f**) Kidney of MFM treated group.

2.3. Acute Anti-Inflammatory Activity

The carrageenan-induced paw edema model was carried out to investigate the anti-inflammatory potential of mucilage. Sub plantar injection of 1% carrageenan (0.1 mL) to rat hind paw steadily increased the paw thickness that extends to maximum after 4 h. Pretreatment with MLM (500 mg/kg) and MFM (500 mg/kg) significantly reduced the paw thickness with maximum inhibition at 4 h as compared to the disease control group ($p < 0.001$). MLM exhibited better inhibition of paw thickness as compared to MFM but not as much as diclofenac sodium (Table 2). The maximum percentage of anti-inflammatory activity was observed in diclofenac sodium (39.31%) at 4 h followed by MLM (27.35%) and MFM (15.68%) as depicted in Figure 3. Percentage inhibition of edema observed by MLM was significantly comparable with diclofenac sodium ($p < 0.05$).

Table 2. Paw thickness at different time intervals in rats treated with MLM and MFM at a dose of 500 mg/kg.

Group Name	Paw Thickness (mm)				
	0 h	1 h	2 h	3 h	4 h
Control	2.09 ± 0.19	3.0 ± 0.16	3.40 ± 0.16	3.52 ± 0.14	3.55 ± 0.15
Standard	2.10 ± 0.26 ^{ns}	2.30 ± 0.25 ***	2.38 ± 0.17 ***	2.28 ± 0.20 ***	2.15 ± 0.25 ***
MLM	2.07 ± 0.25 ^{ns}	2.68 ± 0.18 *	2.80 ± 0.09 ***	2.71 ± 0.1 ***	2.58 ± 0.17 ***
MFM	2.05 ± 0.16 ^{ns}	2.73 ± 0.18 ^{ns}	2.99 ± 0.20 **	2.97 ± 0.18 ***	2.99 ± 0.16 ***

Results are expressed as mean ± S.D; Control is distilled water (10 mL/kg); the standard is diclofenac sodium (10 mg/kg). Significant at $p < 0.05$ *, 0.01 ** and 0.001 ***, ns = not significant.

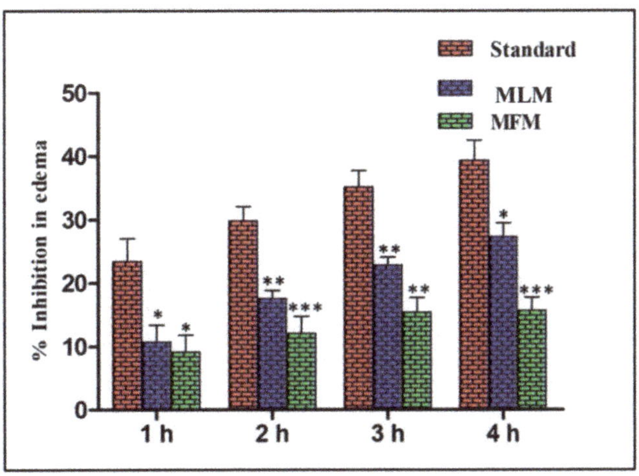

Figure 3. Percentage inhibition of edema by MLM and MFM (500 mg/kg) at different time intervals; Significant at $p < 0.05$ *, 0.01 ** and 0.001 ***.

2.4. Acute Antitussive Activity

Codeine (10 mg/kg) reduced cough frequency from 127.6 to 53.8; marking a 57.8% inhibition in the incidence of cough. Oral administration of 500 mg/kg of MLM significantly decreased the number of coughs induced by SO_2 gas from 127.4 to 66.2; marking a 47.99 % inhibition in the incidence of cough. The least antitussive activity was observed in animals treated with 500 mg/kg of MFM that reduced the cough frequency from 133 to 91.66 with 31.37% inhibition of cough (Figure 4). There was a non-significant difference in percentage inhibition of cough in MLM treated rats when compared with standard drug.

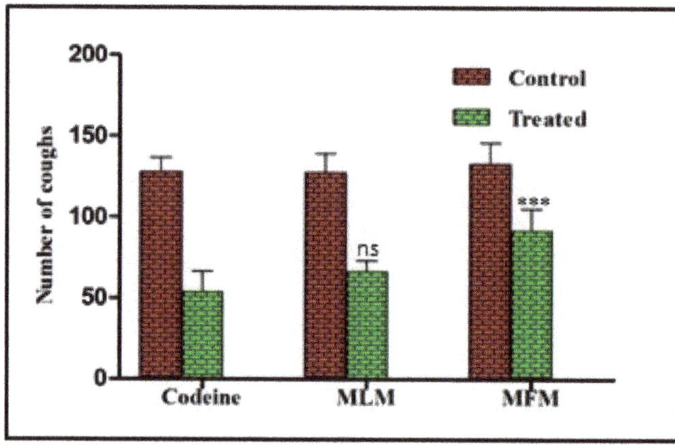

Figure 4. Number of coughs in control and treated groups with codeine phosphate (10 mg/kg), MLM (500 mg/kg) and MFM (500 mg/kg); Significant at $p < 0.001$ ***, ns = not significant.

2.5. Antiulcer Activity

2.5.1. Macroscopic Analysis of Stomach Mucosa

The stomach walls of each rat was observed by the naked eye and then by hand lens. Macroscopic view of gastric walls of normal group showed smooth surface without noticeable scars. In acute gastric ulcer models, severe lesions with extensive visible hemorrhagic streaks were observed in the ulcer control group. Moderate lesions were seen in rats pretreated with MLM (250 mg/kg) and MFM (250 mg/kg and 500 mg/kg). Low moderate lesions were seen in rats pretreated with ranitidine, sucralfate and MLM (500 mg/kg). However, mild lesions were observed in omeprazole treated group which directs high protection against gastric ulcer (Figure 5).

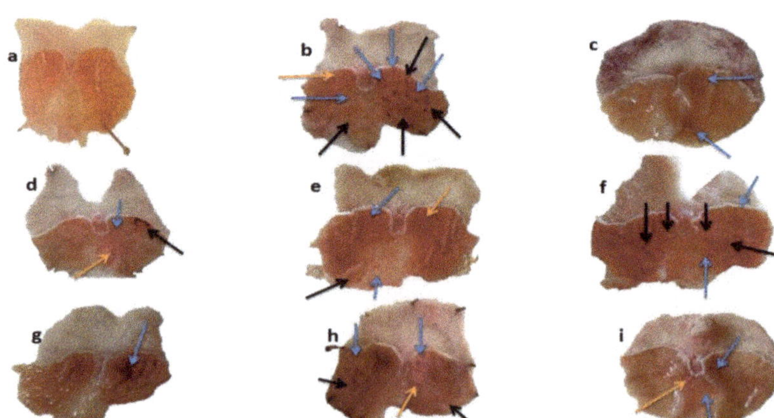

Figure 5. Macroscopic view of gastric mucosa in acute gastric ulcer model of (**a**) Normal (**b**) Ethanol (**c**) Omeprazole (**d**) Ranitidine (**e**) Sucralfate (**f**) MLM 250 mg/kg (**g**) MLM 500 mg/kg (**h**) MFM 250 mg/kg (**i**) MFM 500 mg/kg groups. Black arrows = hemorrhagic streaks, Blue arrows = Ulcer spots, Orange arrows = reddish mucosa.

In chronic gastric ulcer models, ethanol induced gastric ulcer scars and lesions that were visible, but the nature of lesions was different as compared to the acute ulcer models. The lesions are generally categorized by the existence of white rounded scars with rare hemorrhage. Moderate scars were present in gastric mucosa of rats pretreated with ranitidine, omeprazole and MLM (250 mg/kg). However, rare scars were observed in rates pretreated with sucralfate, MLM (500 mg/kg) and MFM (Figure 6).

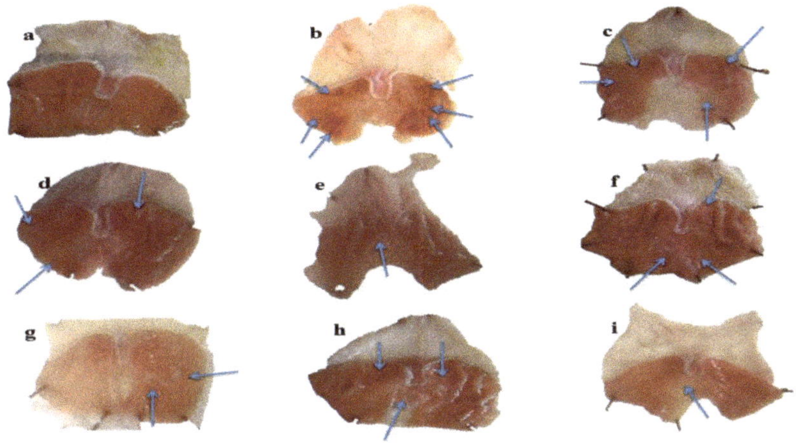

Figure 6. Macroscopic view of gastric mucosa in chronic gastric ulcer model of (**a**) Normal (**b**) Ethanol (**c**) Omeprazole (**d**) Ranitidine (**e**) Sucralfate (**f**) MLM 250 mg/kg (**g**) MLM 500 mg/kg (**h**) MFM 250 mg/kg (**i**) MFM 500 mg/kg groups. Blue arrows = Ulcer spots.

2.5.2. Determination of Ulcer Score, Ulcer Index and Percentage Inhibition

The ulcer index (UI) and % ulcer inhibition of experimental and control groups in acute gastric ulcer are given in Table 3. UI for the ethanol group was 11.36 meanwhile omeprazole, ranitidine and sucralfate considerably reduced the UI to 3.46, 5.21 and 5.27 with ulcer protection of 69.56%, 54.21% and 53.63%, respectively, where omeprazole showed maximum ulcer protection. Pretreatment with MLM at dosage of 250 mg/kg and 500 mg/kg decreased the UI to 8.74 and 6.96 and gave ulcer protection of 23.04% and 38.74%, respectively. Pretreatment with MFM at dosage of 250 mg/kg and 500 mg/kg decreased the UI to 10.47 and 7.03 and gave ulcer protection 7.85% and 38.08%, respectively. Thus, it was clear that pretreatment with MFM and MLM resulted in an increased ulcer inhibition activity in a dose–dependent manner (Table 3).

Table 3. Effect of MLM and MFM on ulcer score, ulcer index and percentage inhibition in acute ulcer.

Group Name	Ulcer No.	Ulcer Score	Incidence of Ulcer (%)	Ulcer Index	Inhibition of Ulcer (%)
Normal (10 mL/kg p. o)	-	-	-	-	-
Ethanol (10 mL/kg p. o)	7.83 ± 0.60	5.75 ± 0.57	100	11.36	-
Ranitidine (50 mg/kg p. o)	1 ± 0.52 ***	1.0 ± 0.47 ***	50	5.21	54.21
Omeprazole (20 mg/kg p. o)	0.67 ± 0.42 ***	0.58 ± 0.37 ***	33.33	3.46	69.56
Sucralfate (100 mg/kg p. o)	1.16 ± 0.54 ***	1.5 ± 0.67 ***	50	5.27	53.63
MLM (250 mg/kg p. o)	2.0 ± 0.73 ***	2.08 ± 0.69 ***	83.33	8.74	23.04
MLM (500 mg/kg p. o)	1.16 ± 0.60 ***	1.75 ± 0.70 ***	66.66	6.96	38.74
MFM (250 mg/kg p. o)	2.17 ± 0.48 ***	2.5 ± 0.56 **	100	10.47	7.85
MFM (500 mg/kg p. o)	1.5 ± 0.5 ***	2.17 ± 0.70 ***	66.66	7.03	38.08

The results are shown in the form of Mean ± SEM. Significant at $p < 0.01$ ** and 0.001 ***.

Meanwhile, in chronic gastric ulcer model, UI of ulcer control group, ethanol group, was 10.8. Pretreatment with omeprazole, ranitidine and sucralfate considerably reduced the UI to 5.22, 3.5 and 3.51 showing ulcer protection of 51.66, 67.59 and 67.5%, respectively. As compared to acute ulcer model, in chronic gastric ulcer model ranitidine and sucralfate showed maximum ulcer protection estimated by 3.5 and 3.51 accounting for 67.59 and 67.50%. Pretreatment with MLM at dosages of 250 mg/kg and 500 mg/kg decreased the UI to 7.08 and 5.67 and gave ulcer protection of 34.44 and 47.44%, respectively. Pretreatment with MFM at dosages of 250 mg/kg and 500 mg/kg decreased the UI to 6.95 and 3.95 and gave ulcer protection of 35.65 and 63.52%, respectively (Table 4). As observed in acute gastric ulcer models, dose–dependent ulcer inhibition was also observed in chronic gastric ulcer model; however, in chronic ulcer MFM was found to be more effective in contrast to acute gastric ulcer models where MLM revealed higher efficacy.

Table 4. Effect of MLM and MFM on ulcer score, ulcer index and percentage inhibition in chronic ulcer.

Group Name	Ulcer No.	Ulcer Score	Incidence of Ulcer (%)	Ulcer Index	Inhibition of Ulcer (%)
Normal (10 mL/kg p. o)	-	-	-	-	-
Ethanol (10 mL/kg p. o)	3.67 ± 0.71	4.33 ± 0.79	100	10.8	-
Ranitidine (50 mg/kg p. o)	0.67 ± 0.49 ***	1 ± 0.74 ***	33.33	3.5	67.59
Omeprazole (20 mg/kg p. o)	0.83 ± 0.40 ***	1.42 ± 0.70 ***	50	5.22	51.66
Sucralfate (100 mg/kg p. o)	0.83 ± 0.54 ***	0.92 ± 0.58 ***	33.33	3.51	67.5
MLM (250 mg/kg p. o)	1.83 ± 0.79 ***	2.33 ± 1.08 ***	66.67	7.08	34.44
MLM (500 mg/kg p. o)	0.83 ± 0.40 ***	1.58 ± 0.78 ***	50	5.67	47.44
MFM (250 mg/kg p. o)	1.00 ± 3.7 ***	1.83 ± 0.69 **	66.67	6.95	35.65
MFM (500 mg/kg p. o)	0.83 ± 0.54 ***	1.08 ± 0.76 ***	33.33	3.94	63.52

The results are shown in the form of Mean± SEM. Significant at $p < 0.01$ ** and 0.001 ***.

2.5.3. Histology of Stomach Wall

Histological examination of gastric mucosa revealed that ethanol caused severe disruption in the gastric mucosa that pierced deeply, accompanied by mucosal and sub-mucosal edema and leukocyte infiltration. Thickness of mucosal layer decreased extensively in chronic ulcer models due to the daily interruption of ethanol with gastric mucosa. The standard drugs and mucilage established enhanced protection of gastric mucosa in a dose–dependent manner accompanied by a pronounced reduction in edema and leukocytes infiltration of sub-mucosal layers. Histological observations in acute and chronic ulcer model are shown in Figures 7 and 8.

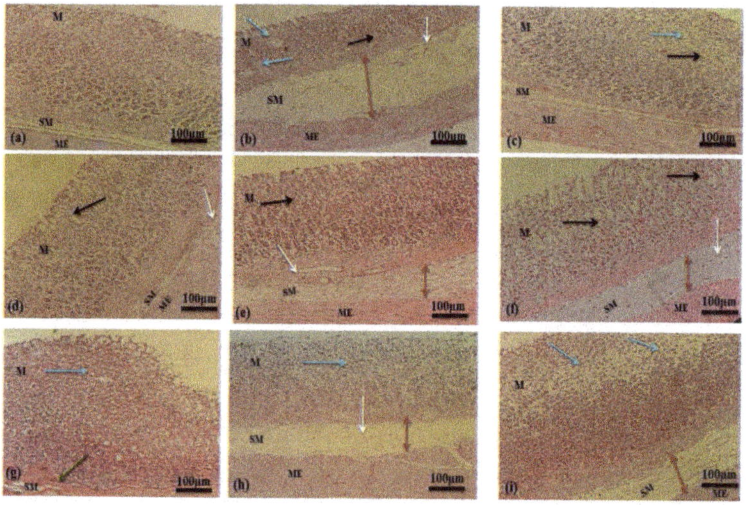

Figure 7. Histological evaluations in acute ulcer model (**a**) Normal control group (**b**) Ethanol (Ulcer control group) (**c**) Omeprazole (**d**) Ranitidine (**e**) Sucralfate (**f**) MLM 250 mg/kg (**g**) MLM 500 mg/kg (**h**) MFM 250 mg/kg (**i**) MFM 500 mg/kg. M represents mucosa, SM represents submucosa and ME represent muscularis externa. Blue arrows indicate areas of ulceration and focal erosion. Black arrows indicate the mucosal edema. Dark red double headed arrows indicate submucosal edema. White arrows indicate leukocyte infiltration. Green arrows indicate congestions.

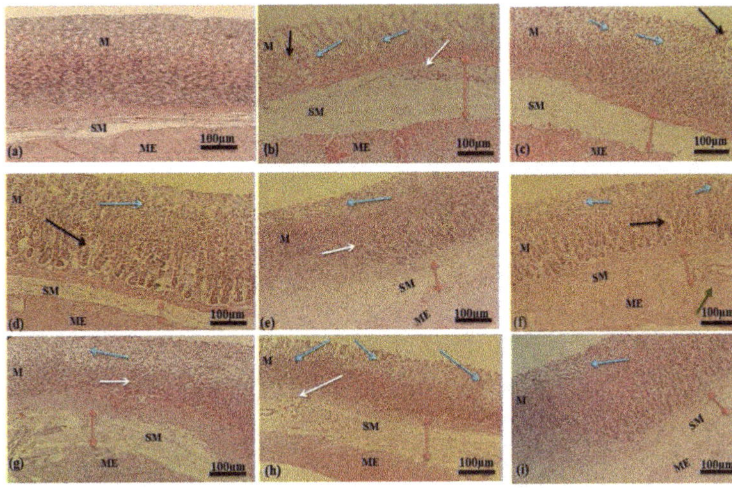

Figure 8. Histological evaluations in chronic ulcer model (**a**) Normal control group (**b**) Ethanol (Ulcer control group) (**c**) Omeprazole (**d**) Ranitidine (**e**) Sucralfate (**f**) MLM 250 mg/kg (**g**) MLM 500 mg/kg (**h**) MFM 250 mg/kg (**i**) MFM 500 mg/kg. M represents mucosa, SM represents submucosa and ME represent muscularis externa. Blue arrows indicate areas of ulceration and focal erosion. Black arrows indicate the mucosal edema. Dark red double headed arrows indicate submucosal edema. White arrows indicate leukocyte infiltration. Green arrows indicate congestions.

2.5.4. Mechanism of Action

Effect on Gastric Juice Parameters

Gastric juice of ulcer control group displayed higher gastric juice volume, lower pH and high total acidity in contrast to normal animals. Pretreatment with MLM (500 mg/kg) and MFM (500 mg/kg) significantly ($p < 0.001$) reduced the gastric juice volume and total acidity but increased pH when compared with ulcer control group as illustrated in Figure 9. Similar changes were also observed in chronic gastric ulcer models (Table 5).

Table 5. Effect of MLM and MFM on mucous content and protein content in acute and chronic gastric ulcer.

	Acute Gastric Ulcer		Chronic Gastric Ulcer	
Group Name	**Mucous Content**	**Total Protein**	**Mucous Content**	**Total Protein**
Normal (10 mL/kg p.o)	482.5 ± 15.59	52.16 ± 3.66	504.17 ± 14.52	74.64 ± 3.71
Ethanol (10 mL/kg p. o)	411.17 ± 9.13	19.55 ± 3.44	295 ± 37.13	38.68 ± 3.26
Ranitidine (50 mg/kg p. o)	452.5 ± 13.14 ns	44.27 ± 3.66 **	445 ± 21.29 ns	72.72± 4.62 ***
Omeprazole (20 mg/kg p. o)	501.67 ± 13.52 ***	38.58 ± 6.37 ns	637.67 ± 32.44 ***	68.41 ± 3.55 ***
Sucralfate (100 mg/kg p. o)	432.5 ± 12.63 ns	28.52 ± 5.93 ns	947.17 ± 65.21 ***	70.46 ± 4.09 ***
MLM (250 mg/kg p. o)	440.5 ± 12.63 ns	23.54 ± 4.46 ns	440.5 ± 12.63 ns	68.58 ± 4.37 ***
MLM (500 mg/kg p. o)	520 ± 17.89 ***	47.47 ± 6.50 **	960.33 ± 58.24 ***	69.35 ± 3.40 ***
MFM (250 mg/kg p. o)	430 ± 12.11 ns	26.73 ± 5.03 ns	840.17 ± 26.24 ***	48.64 ± 3.31 ns
MFM (500 mg/kg p. o)	488.16 ± 17.09 **	30.28 ± 3.95 ns	1239.33 ± 54.29 ***	70.05 ± 4.02 ***

Results are expressed in the form of Mean ± SEM.; Significant at $p < 0.01$ ** and 0.001 ***, ns = not significant; Mucous content are expressed as μg of Alcian blue/g wet tissue; Total protein is expressed as μg/mL.

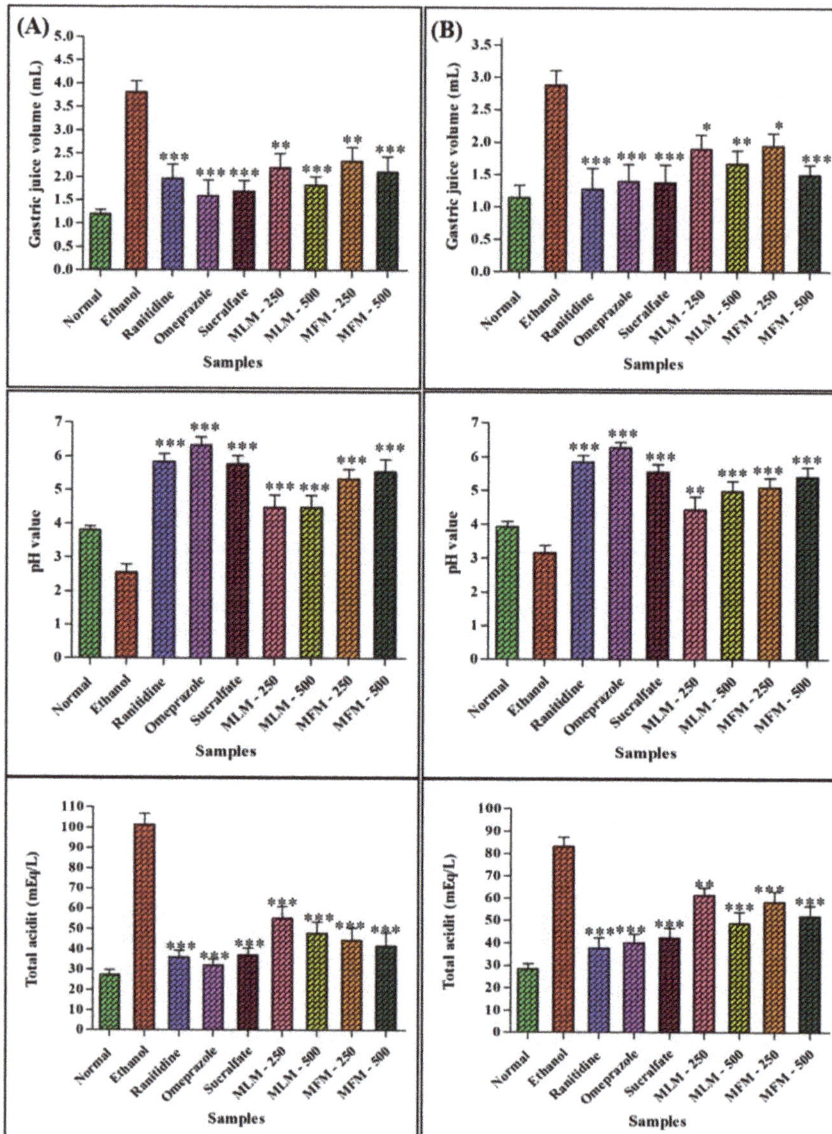

Figure 9. Effect of MLM and MFM on gastric juice volume, pH and total acidity in acute (**A**) and chronic ulcer (**B**); The results are expressed in the form of Mean ± SEM. Significant at $p < 0.05$ *, 0.01 ** and 0.001 ***.

Effect on Total Protein and Total Mucus Content

The Alcian blue binding capacity was used as a marker for quantification of gastric mucus content. Ethanol treatment reduced the total gastric mucus content and ultimately Alcian blue binding capacity. Pretreatment with MLM (500 mg/kg) and MFM (500 mg/kg) in acute gastric ulcer models significantly ($p < 0.01$) protected the gastric mucosa increasing the mucus content in contrast to the ulcer control group. However, at low dosages (250 mg/kg), a non-significant difference was observed in total mucus content of MLM and

MFM with ulcer control group. Ranitidine and MLM (500 mg/kg) had total protein content significantly dissimilar from ulcer control group ($p < 0.01$). However, in all other treated group a non-significant difference was observed in comparison to ulcer control group. In chronic gastric ulcers, there was a dramatic rise of total mucous content in animals pretreated with MLM (500 mg/kg), MFM (500 mg/kg), MFM (250 mg/kg) and sucralfate (100 mg/kg) significantly different from ulcer control group ($p < 0.001$). A non-significant difference was observed between total mucus content of animals pretreated with MLM (250 mg/kg) and ranitidine when compared with ulcer control group. Total protein content was also significantly different in the treated group as compared to the ulcer control group (Table 5).

2.6. Gas Chromatography Analyses of MLM and MFM

GC/MS analysis of MFM revealed the existence of four neutral monosaccharides which are galactose, rhamnose, arabinose as well as glucose constituting about 50.21, 7.61, 6.18 1.90 mg/g of MFM, respectively, whereas galacturonic acid constitutes the identified acidic monosaccharide showing 16.02 mg/g of MFM (Figure S1). Regarding the main monosaccharides detected in MLM, they are previously analyzed by the authors and reported [25] where GC/MS analysis of MLM showed the presence of five neutral monosaccharides together with one acidic monosaccharide which are galactose, rhamnose, arabinose, mannose as well as glucose and galacturonic acid accounting for 51.09, 10.24, 8.90, 1.80, 0.90 and 15.06 mg/g of MLM, respectively. Results showed that the leaves mucilage showed slightly higher levels of neutral monosaccharides compared to the fruit mucilage in contrast to the acidic monosaccharide represented by galacturonic acid that displayed a higher level in the fruit mucilage.

3. Discussion

Plant derived natural therapeutic agents have become part of primary healthcare in developing countries. Because of their natural origin some of them are mistakenly considered as absolute safe drugs. Thus, there is a great need of scientific studies on toxicities of drugs obtained from natural sources [30]. Erythrocytes are abundantly present in the human body, and they show vast biological and structural attributes playing an important role in drug transport. Breakdown of erythrocytes is known as hemolysis that occurs upon exposure of erythrocytes to toxicants. Some phytochemicals can cause hemolysis that in turn reflects that given substance is cytotoxic to RBCs [31,32].

Erythrocyte membranes are made up of polyunsaturated fatty acids and proteins, therefore, more vulnerable to peroxidation. Hemoglobin present in RBCs also catalyzes the oxidation process. Oxidation of RBCs represents oxidative damage of other biological membranes. Thus, the chemicals which produced free radicals can damage the erythrocyte membrane and serve as hemolytic agent [33]. In vitro hemolytic activity of MLM and MFM was assessed in an effort to check their toxicity to red blood cells. The results clearly justified the safety of MLM and MFM for RBCs and hence, can be used for further in vivo studies. Furthermore, acute oral toxicity of extracted mucilage was also performed to identify the dose that can be safely used for *in-vivo* studies. It was concluded that MLM and MFM are totally safe for systemic use and lethal dose is above 5 g/kg. As 5 g/kg showed no signs of toxicities therefore 1/20 and 1/10 of this lethal dose were selected in the current study (250 mg/kg and 500 mg/kg, respectively) for in vivo studies.

Regarding inflammation, it is defined as a localized reaction accompanied by soreness, warmth, swelling, tenderness and loss of function. Inflammatory mediators are released upon exposure to inflammatory agents. These inflammatory mediators widen the blood vessels and cause chemotaxis and ultimately lead to many disorders [34]. Red seaweeds (Rhodophyceae) are the source of carrageenan which is a sulphated polyglactan with esterified sulphate group that is responsible for its chemical activity. Carrageenan-induced paw edema is the most often practiced model for acute anti-inflammatory activity [34].

Carrageenan generates oxygen free radicals by changing the neutrophil membrane. Neutrophils are responsible for acute external inflammation and rapidly move towards the site of inflammation. They engulfed the pathogens and trashes of damaged tissues and consequently produce the reactive oxygen species and proteolytic enzyme to tear down the engulfed particles. Host tissues are also damaged during this process. As a result of these activities, a deficiency of ATP occurs resulting in the loss of cell function ultimately leading to necrosis [35]. The current study determined the anti-inflammatory activity of mucilage. The results showed that maximum edema inhibition occurred at the fourth hour by diclofenac sodium followed by MLM and MFM where MLM showed better anti-inflammatory activity when compared to MFM. Anti-inflammatory activity of MLM increased with increase in time with maximum results at the fourth hour. This might be due to the radical scavenging effects of MLM and MFM against free radicals produced at site of inflammation. The polysaccharides present in this mucilage may be the active ingredient against inflammation. Previously, mucilage extracted from lemon demonstrated pronounced anti-inflammatory activity in carrageenan-induced paw edema model in rats [35]. Okra and Baobab belongs to the family Malvaceae and their mucilage polysaccharides have been reported to possess anti-inflammatory properties [36].

Concerning the antitussive activity, there are various causes of cough that can be induced by any type of irritation to bronchi and trachea. This irritation can be the result of exposure to light, allergens, chemicals or any other foreign matter. As a result, airway sensory nerves are activated followed by increased mucous secretion, release of inflammatory mediators and damage to airway epithelium. Antitussives suppress cough either by acting centrally or peripherally. Centrally acting antitussive agents suppress the cough center in the brain and peripherally acting agents work as a demulcent or local anesthetic [37]. In this study, SO_2 gas was used as an irritant to induce cough in rats. SO_2 was produced by chemical reaction between sodium hydrogen sulphite and concentrated sulphuric acid. As all conditions were kept the same, it was therefore supposed that the quantity and saturation of SO_2 would remain same during each exposure. Therefore, quantification of SO_2 was not performed. The results showed that MLM exhibited 47.99% inhibition against chemical-induced cough. MLM was a better cough suppressant than MFM thus supporting the traditional use of leaves of *M. parviflora* as an antitussive. MLM and MFM might act peripherally due to the demulcent properties of mucilage. Folk use of *M. parviflora* is as a demulcent with mucilage responsible for this demulcent action [38]. Hence, this study consolidated the demulcent action of mucilage of *M. parviflora* based on scientific evidence. Rhamnogalacturonan was previously isolated from Marshmallow (*Althea officinalis*) mucilage that also belongs to family Malvaceae and showed high cough suppressant action in guinea pigs [39]. Similarly in an open trial, patients suffering from cough were treated successfully with herbal cough syrup containing marshmallow root mucilage [40].

Gastric ulcer is the chief health issue of GIT where gastric acid, pepsin and *H. pylori* are among the destructive factors that aggravate gastric ulcers, meanwhile the protective factors include mucin, NO, bicarbonates, blood flow, prostaglandins and growth factors. When a disturbance occurs between destructive and defensive factors, gastric ulcers occur [41]. The mucosal layer of stomach releases mucous that contains glycoprotein and lipids. Aside from suppressing the oxygen-derived free radicals it stimulates smooth movement of food and protects the stomach against the damaging effects of HCl and pepsin by preventing their penetration in the mucosal membrane [42,43]. Ethanol causes gastric ulcers as it solubilizes the mucous and diffuses into gastric mucosa. As a result, HCl and pepsin come in contact with gastric mucosa and injure it, in addition, ethanol increases gastric acid secretion. Moreover, it ruptures vascular endothelium, increases vascular permeability, increases oxidative stress in tissue and decreases gastric mucous secretion and thus ethanol induced gastric ulcer models were found to be suitable [44].

Hence, the antiulcer activities of MLM and MFM were evaluated in ethanol induced acute and chronic gastric ulcer models. It has been reported that antacids, anti-secretory agents and cytoprotective agents are the main classes of drugs used for the management of ulcers [45]. Therefore, in this study omeprazole, ranitidine and sucralfate were used as standard drugs. The results showed that *M. parviflora* mucilage and standard drugs significantly reduced the ulcer count and ulcer index with increased cytoprotective effects in both acute and chronic models, whereas the ulcer control group presented severe hemorrhagic streaks. The necrotic and petechial lesions in gastric mucosa leading to gastric ulcers are attributed to ethanol treatment [43]. Extreme stomach mucosa disruption leads to lower release of bicarbonate ions that in turn lowers the pH via increased production of gastric contents. Hence, ethanol increases gastric acid secretion that may result in increased gastric volume, decreased pH, increased total acidity and increased ulcer index [46]. A significant increase in pH and reduction in gastric volume and acidity was noticed in rats pretreated with MLM, MFM and standard drugs when compared with ulcer control groups as represented in Figure 9. This predicted the anti-secretory and antacid mechanism of the gastroprotective effects of MLM and MFM.

In addition, a reduced amount of protein content in stomach tissue homogenate is marked as a sign of damage to normal cellular functions. Therefore, treatments that cause an increase in protein content may be considered to contain auto-healing agents that support the mucosal regeneration process [40]. The low protein content in the disease control group of acute and chronic models was a sign of cellular dysfunction. The protein content in experimental and standard groups was higher compared with the disease control group thus supporting the presence of auto-healing effects. Treatment with MLM and MFM augmented the regeneration of epithelial cells and thus considerably amplified the protein concentration. Meanwhile, gastric mucus also acts as a shield for the protection of gastric walls against aggressive factors and as a first line of mucosal protection from luminal acid [47]. It contains viscid, flexible and translucent gel composed of glycoproteins (5%) and water (95%) and can be identified by the extent of Alcian blue binding. The defensive role of the mucus shield relies on the structure as well as the density of the layer protecting the mucosal surface. Furthermore, mucus can also act as an antioxidant and can diminish free radical-induced mucosal damage [48]. Administration of MLM and MFM significantly raised the amount of mucous in contrast to the disease control group. MLM and MFM treatment increased the gastric mucous content directly dependent on dosage. Interestingly, production of mucus in chronic ulcer models have remarkably been increased. The highest mucus content was observed in rats pretreated with MFM (500 mg/kg) followed by MLM (500 mg/kg) and results were comparable with rats pretreated with sucralfate. This unusual increase in gastric mucus suggested that MLM and MFM may trigger the discharge of chemical mediators responsible for the gastric mucus production i.e., gastrin, prostaglandin, secretin and acetylcholine [49]. Hence, it can be predicted that an increased secretion of mucus by administering MLM and MFM may be one of the potential mechanisms of their gastroprotective action. Mucilage contains polysaccharides where previous studies reported that polysaccharides possess antiulcer activity. These mucilaginous polysaccharides may form a protective covering on gastric mucosa or regenerate it [19]. This can be justified by the high mucus content in both acute and chronic gastric ulcer models.

4. Materials and Methods

4.1. Plant Material

Malva parviflora was collected in February 2018 from Lahore, Punjab, Pakistan. Plant Taxonomist, Dr. Zaheer-ud-din khan at Department of Botany, Government College University, Lahore authenticated the plant and issued a voucher specimen number GC.Herb.Bot.3533. Dusty material was removed by washing the leaves and unripe fruits with tap water. Leaves were dried in shade and concomitantly powdered in a mechanical grinder. The powdered leaves and fresh unripe fruits were used for the extraction. The mucilage was dried in an oven at 45 °C till it was completely dried. The dried mucilage was then powdered by mortar and pestle. The powdered mucilage was passed through a sieve#80. Particle size of the obtained powder was uniform [50–52].

4.2. Drugs and Chemicals

Omeprazole, sucralfate and ranitidine were used as standard anti-ulcer drugs and were kindly provided by Schazoo Zaka (Pvt.) Ltd. (Sheikhupura, Punjab, Pakistan), Highnoon Laboratories (Pvt.) Ltd. (Lahore, Punjab, Pakistan) and Surge Laboratories (Pvt.) Ltd. (Sheikhupura, Punjab, Pakistan), respectively. Alcian blue dye was obtained from Unichem, china, bovine serum albumin from Bioshop (Burlington, Ontario, Canada) and normal saline from Otsuka (Karachi, Pakistan. Diethyl ether and sucrose were acquired from Labscan (Bangkok, Thailand). Sodium hydrogen sulphite, sodium hydroxide, Sodium carbonate, potassium chloride, Tris HCl and phenolphthalein were purchased from BDH chemicals (Poole, UK). All other chemicals used in this study were obtained from Sigma-Aldrich (Steinheim, Germany). All the reagents were prepared freshly and were of pharmaceutical grade.

4.3. Animals

Wistar albino rats of both sexes (180–200 g) were maintained under standard conditions at (22–24 °C). The relative humidity of 50–60% and photoperiod of 12 h light and 12 h dark cycle was retained. Animals were fed on commercial pellet food with unrestricted supply of water. The experiments were performed following the guidelines set by the Institutional Ethical Committee for animal care and experimentation, College of Pharmacy, University of the Punjab, Lahore, Pakistan (AEC/PUCP/1094 dated 11 February 2019).

4.4. Preparation of the Mucilage

Mucilage was extracted from dried powdered leaves and fresh fruits following the previously described method by Munir et al. [25]. *Malva parviflora* leaves and fruit mucilage was labelled as MLM and MFM respectively.

4.5. Evaluation of the Biological Activities of the Mucilage Extracts

4.5.1. In Vitro Hemolytic Activity

Blood (3 mL) was collected in an anticoagulant tube from a volunteer human (male, 26 years, Blood group O+) by a Phlebotomist in Punjab University Healthcentre lab (Allama Iqbal campus, University of Punjab, Lahore, Pakistan). The sample was centrifuged at 850 rpm for 5 min. The clear supernatant was poured off followed by washing of residue pellets with 5 mL of chilled (4 °C) phosphate buffer saline (PBS) solution (pH 7.4). Washed cells suspension was made in 20 mL cool sterilized PBS. Cells were calculated using hemocytometer and for each assay 7.068×10^8 cell/mL were used. 20 µL of plant mucilage were taken in an Eppendorf tube and PBS and 0.1% TritonX-100 were used as negative and positive control respectively. 180 µL of diluted blood cell suspension were added into each tube followed by incubation at 37 °C. After incubation for 35 min, suspension was allowed to cool for 5 min followed by centrifugation at 1500 rpm for 5 min. 100 µL of supernatant were collected and diluted with 900 µL chilled sterile phosphate buffer saline. An aliquot of 200 µL of all these samples including positive and negative control were transferred to 96-well plate. ELISA microplate reader was used to measure the absorbance at 630 nm [53].

$$\% \text{ Hemolysis} = \frac{Abs(Sample\ absorbance)}{Abs\ (control\ absorbance)} \times 100$$

4.5.2. Acute Toxicity Study

Acute oral toxicity of MLM and MFM was determined in Wistar albino rats according to OECD guidelines 425 [54]. Limit test was performed at 5000 mg/kg where the control group was given only distilled water. The animals were observed for death or any noxious outcome in earliest four hours after the dosing and regularly for fourteen days. During this period, parameters such as weight, physical appearance, behavioral changes, injury, illness signs and mortality were observed. On the 15th day animals were anesthetized and sacrificed with an overdose of xylazine and ketamine anaesthetic drugs and blood samples were obtained by cardiac puncture. Animals were dissected and organs such as liver and kidney were obtained for histopathological examination.

4.5.3. Anti-Inflammatory Activity

Acute anti-inflammatory activity of MLM and MFM was evaluated by carrageenan-induced paw edema in rats [55]. The rats were arbitrarily separated in four groups ($n = 5$). In the first group, the animals received a dose of 10 mL/kg of distilled water (carrageenan control); meanwhile in the second group, the animals were orally administered with diclofenac sodium in a dose of 10 mg/kg (standard group) [56]. In the third and fourth groups, the animals were orally given 500 mg/kg of MLM (MLM experimental group) and MFM (MFM experimental group), respectively. After 1 h, all the animals were injected with 0.1 mL of carrageenan in normal saline (1% (*w*/*v*)) into subplantar region of left hind paw of each rat. Thickness of paw was measured using digital vernier caliper at 0 h, 1 h, 2 h, 3 hand 4 h intervals. Edema inhibition (%) was calculated using the following formula:

$$\%\ \text{inhibition of edema} = \frac{T_c - T_t}{T_c} \times 100$$

where T_c = Paw thickness of control group and T_t = Paw thickness of experimental group.

4.5.4. Antitussive Activity

Acute antitussive activity of MLM and MFM was evaluated by SO_2 induced cough model in rats [57]. Each animal served his own control due to variation in number of coughs in each individual animal. Groups were treated in the following manner where in the first group, the animals were orally given codeine phosphate in a dose of 10 mg/kg. In the second and third groups, the animals were orally given 500 mg/kg of MLM and MFM respectively. The animals were exposed to SO_2 for 60 s after an hour of receiving the treatment and the number of coughs was counted. The number of coughs was compared before and after the treatment to determine the reduction in cough. The percentage inhibition frequency of cough was determined using the following formula:

$$\%\ \text{inhibition of frequency of cough} = \left(\frac{C_C - C_T}{C_C}\right) \times 100$$

where C_C is the cough frequency in control animal and C_T is the cough frequency in treated animal [57].

4.5.5. Antiulcer Activity

Study Design

The animals were arbitrarily separated into 9 groups ($n = 6$) where the first group (normal group) was orally treated with distilled water (DW) 1 mL/100g [58]. The second group was orally treated with 1 mL/100 g of ethanol (ethanol group) meanwhile from 3–5 groups, the animals received orally standard drugs which are ranitidine (50 mg/kg) [59], omeprazole (20 mg/kg) [60] and sucralfate (100 mg/kg) [61], respectively. Groups 6–7 were orally administered MLM in doses of 250 and 500 mg/kg, respectively whereas groups 8–9 were orally administered MFM in doses of 250 and 500 mg/kg, respectively.

Ethanol Induced Acute Gastric Ulcer

Animals were prevented from food for 24 h with unrestricted excess toward water. They were deprived of water just 2 h prior starting the experimental procedure. Animals of all groups were pretreated with drugs and extract as mentioned above. After 30 min all groups were orally administered 90% ethanol (1 mL/100 g) except normal group. Animals were euthanized with an overdose of xylazine and ketamine anaesthetic drugs after 1 h of ethanol administration [62]. Animals were dissected, removed the stomachs and incised along the larger curve. Washed with ice cold normal saline and gastric mucosa was examined [63].

Ethanol Induced Chronic Gastric Ulcer

Animals were prevented from food for 24 h with unrestricted excess toward water. They were deprived of water just 2 h prior to starting the experimental procedure. All animals were treated with drug and extract for fourteen days as mentioned above. On the first day of experiment after 30 min above treatment animals were orally administered with 90% ethanol (1 mL/100 g) except normal group. From the 2nd day 30% (*v*/*v*) ethanol was orally given to all groups for fourteen days except normal group. On 14th day, all the animals were anesthetized and sacrificed with an overdose of xylazine and ketamine anaesthetic drugs after 1 h of ethanol administration [62]. Animals were dissected, removed stomachs and incised along larger curve and washed with ice cold normal saline then gastric mucosa was examined [64].

Parameters of Gastric Ulcer Evaluation

Macroscopic and Microscopic Evaluation

Normal saline was used to wash the stomach and the apparent changes in inner walls of stomach were carefully examined macroscopically by using magnifying glass and microscope. The number of lesions was carefully noted and was used for the determination of ulcer index. Photos were taken using mobile camera (Oppo A57, Guangdong Oppo Mobile Telecommunications Corp., Ltd., Dongguan, Guangdong, China) [65].

Ulcer Scoring

Based on severity, the ulcers were given scores as previously reported by Gupta et al. [66] as follows: 0 = no ulcer; 0.5 = reddish mucosa; 1 = red spots; 1.5 = hemorrhagic streaks; 2 = deep ulcers and 3 = perforations.

Ulcer Index

Ulcer index can be calculated as previously described by Gul et al. using the following formula [67].

$$\text{Ulcer index} = (\text{UN} + \text{US} + \text{UP}) \times 10^{-1}$$

where UN = average number of ulcers per animal; US = average of severity score; UP = percentage of animals with ulcer.

Ulcer Protection (%)

Ulcer protection (%) was calculated described by Gul et al. using the following formula [67].

$$\%\ \text{Protection} = \frac{\text{Ulcer index of ethanol treated group} - \text{ulcer index of treated group}}{\text{ulcer index of ethanol treated group}} \times 100$$

Histological Analysis

Samples of stomach walls were fixed in formalin solution (10%) for 48 h and then dehydrated by washing with ascending grades of ethanol. Samples were cleaned by xylene and embedded in paraffin wax. The rotary microtome was used to cut sections of 5–6 mm thickness followed by staining with hematoxylin and eosin. The sections were examined under a microscope for histopathological variations such as ulceration, congestion, edema, necrosis and leukocyte infiltration [68].

Mode of Gastro-Protective Activity

Estimation of the Gastric Volume

After opening the stomach gastric content was squeezed out in a falcon tube followed by centrifugation at 1000 rpm) for 15 min, then the supernatant was collected, and its volume was determined [69].

Determination of the pH Value and Total Acidity

Gastric juice pH was determined using a digital pH meter as previously described by Sen et.al [68]. A 1 mL volume of gastric juice was obtained in a vial then 2 drops of phenolphthalein were added as indicator. Then, it was titrated against 0.1 N NaOH until the end point that is the change from colorless to light pink. The volume of NaOH used was determined then the following formula was used to calculate total acidity where results were expressed in terms of clinical unit mEq/L [70].

$$\text{Total acidity} = \frac{\text{Volume of NaOH used} \times \text{Normality of NaOH used} \times 100}{0.1}$$

Determination of Gastric Mucin Content

The glandular portion of stomach was excised and weighed, and this portion was soaked for 2 h in 10 mL of 0.1% Alcian blue solution in 0.16 M sucrose buffered with 0.05 M sodium acetate adjusted to pH = 5. Excess uncomplexed dye was removed by rinsing with 0.25 M sucrose solution at an interval of 15 min and 45 min respectively. Dye forms a complex with mucus that was extracted with 10 mL of 0.5 M magnesium chloride solution for 2 h with consecutive shaking for one minute after 30 min interval. A 4 mL volume of this solution was shaken with an equal volume of diethyl ether. The resulting emulsion was centrifuged at 3000 rpm and the absorbance of aqueous layer was taken at 580 nm. The mucin content of the sample was determined from the standard curve, which was expressed in microgram of Alcian blue extracted per gram of wet gland tissue [71].

Determination of the Total Protein Content

Total protein of glandular tissues homogenate was estimated by Lowry method [72]. The tissue homogenate of glandular portion of stomach was mixed with 1 mL of 0.1 M Tris-HCl buffer (pH 7.4) by a homogenizer followed by centrifugation at 1500 rpm for 15 min at 4 °C. The homogenate was diluted with 0.1 M tris-HCl buffer to reach 10 mL. Bovine serum albumin (BSA) was used as standard and 0.1 N of NaOH was used as blank. A 1 mL volume of the sample was mixed with 4.5 mL of reagent I that is formed by mixing 48 mL of 2% Na_2CO_3 in 0.1 N NaOH with 1 mL of 1% NaK Tartrate in H_2O and 1 mL of 0.5% $CuSO_4 \cdot 5\,H_2O$ in H_2O followed by addition of 0.5 mL of reagent II (1 part Folin-Phenol [2 N]: 1 part water) after 10 min. The test tubes were placed in the dark for 30 min and the absorbance was taken at 750 nm. BSA standard calibration curve was used to interpret the results.

4.6. *Gas Chromatography Coupled with Mass Spectrometry Analysis*

Monosaccharide composition of MLM and MFM was analyzed using a modified GC-MS analytical procedure previously adopted by Xia et al. [73] depending upon trimethylsilyl dithioacetal (TMSD) derivatization. A 1 µL volume of the sample was applied to Agilent 7890A Gas Chromatography coupled to Agilent 5975C Mass spectrometer (Agilent Technologies, Santa Clara, CA, USA) with HP-5MS column using a temperature range of 80 °C for 0 min, 80–90 °C at 2.5 °C/min, 190–252 °C at 2 °C/min, 252–300 °C at 25 °C/min, 300–310 °C at 25 °C/min and held for 15 min. Mass spectra were recorded employing total ion chromatogram (TIC) mode and interpreted using NIST 5 software [25].

4.7. *Statistical Analysis*

GraphPad prism 8.4.3 was used for statistical analysis. Statistical significance difference was calculated using one way ANOVA followed by Dunnett's test. The values $p < 0.05$ *, 0.01 ** and 0.001 *** were considered as statistically significant.

5. Conclusions

In conclusion, the current findings suggest that mucilage extracted from the leaves and fruit of *M. parviflora* are safe for in vivo studies and could be incorporated in many significant therapeutic applications after clinical trials. Treatment with mucilage reduced cough, inflammation and ulcers in animal models. The underlying mechanism of gastro protection is in the anti-secretory and mucus protective potential of mucilage. However, further in vivo studies are recommended to comprehensively understand its exact mechanism of action. This study clearly highlights the application of *M. parviflora* mucilage in adjuvant therapy of gastric ulcers. Further studies are in progress to isolate bioactive polysaccharides and glycoprotein from this mucilage.

Supplementary Materials: The following supporting information can be downloaded at: https://www.mdpi.com/article/10.3390/ph15040427/s1, Figure S1: GC/MS chromatograms of neutral (A) and acidic (B) polysaccharides in *M. parviflora mucilage* obtained from the fruits (MFM).

Author Contributions: Conceptualization, A.M. and S.I.; methodology A.M., S.I., M.R., S.S.E. and K.A.; software, F.S.Y.; resources, A.E.A., O.K., W.Y.R. and S.S.E.; writing—original draft preparation, A.M. and S.I.; writing—review and editing, F.S.Y., S.S.E. and M.L.A.; supervision, F.S.Y., S.S.E. and M.L.A.; funding acquisition, W.Y.R., A.E.A., O.K. and S.S.E. All authors have read and agreed to the published version of the manuscript.

Funding: This research was funded by the Deanship of Scientific Research (DSR) at King Abdulaziz University, Jeddah, Saudi Arabia under grant no. (G-68-166-38).

Institutional Review Board Statement: The animal study protocol was approved by the Punjab University College of Pharmacy, (AEC/PUCP/1094 dated 11 February 2019).

Informed Consent Statement: Informed consent was obtained from all subjects involved in the study.

Data Availability Statement: Data is contained within the article and Supplementary Materials.

Acknowledgments: This project was funded by the Deanship of Scientific Research (DSR) at King Abdulaziz University, Jeddah, Saudi Arabia under grant no. (G-68-166-38). All the authors, therefore, acknowledge with thanks DSR for technical and financial support.

Conflicts of Interest: The authors declare no conflict of interest.

References

1. Eccles, R.; Mallefet, P. Soothing properties of glycerol in cough syrups for acute cough due to common cold. *Pharmacy* **2017**, *5*, 4. [CrossRef] [PubMed]
2. Romm, A.; Ganora, L.; Hoffmann, D.; Yarnell, E.; Abascal, K.; Coven, M. Chapter 3—Fundamental Principles of Herbal Medicine. In *Botanical Medicine for Women's Health*; Room, A., Hardy, M.L., Mills, S., Eds.; Churchill Livingstone: Saint Louis, France, 2010; pp. 24–74.
3. Czibulka, A. Probiotics and Herbal Therapies. In *Laryngopharyngeal Reflux Disease*; Springer: Berlin/Heidelberg, Germany, 2019; pp. 103–113.
4. Schmidgall, J.; Schnetz, E.; Hensel, A. Evidence for bioadhesive effects of polysaccharides and polysaccharide-containing herbs in an ex vivo bioadhesion assay on buccal membranes. *Planta Med.* **2000**, *66*, 48–53. [CrossRef] [PubMed]
5. Tosif, M.M.; Najda, A.; Bains, A.; Kaushik, R.; Dhull, S.B.; Chawla, P.; Walasek-Janusz, M. A Comprehensive review on plant-derived mucilage: Characterization, functional properties, applications, and its utilization for nanocarrier fabrication. *Polymers* **2021**, *13*, 1066. [CrossRef] [PubMed]
6. Nazari, M.; Riebeling, S.; Banfield, C.C.; Akale, A.; Crosta, M.; Mason-Jones, K.; Dippold, M.A.; Ahmed, M.A. Mucilage polysaccharide composition and exudation in maize from contrasting climatic regions. *Front. Plant Sci.* **2020**, *11*, 1968. [CrossRef] [PubMed]
7. Dybka-Stępień, K.; Otlewska, A.; Góźdź, P.; Piotrowska, M. The renaissance of plant mucilage in health promotion and industrial applications: A review. *Nutrients* **2021**, *13*, 3354. [CrossRef]
8. El-Shiekh, R.A.; Salama, A.; Al-Mokaddem, A.K.; Abdel-Sattar, E.A. Gastroprotective effect of mucilage fraction from Solenostemma argel via cytoprotection and attenuation of oxidative stress, inflammation and apoptosis. *J. Herb. Pharmacol.* **2021**, *10*, 232–240. [CrossRef]
9. Wadhwa, J.; Nair, A.; Kumria, R. Potential of plant mucilages in pharmaceuticals and therapy. *Curr. Drug Deliv.* **2013**, *10*, 198–207. [CrossRef]
10. Alam, M.T.; Parvez, N.; Sharma, P.K. FDA-approved natural polymers for fast dissolving tablets. *J. Pharm.* **2014**, *2014*, 952970. [CrossRef]
11. Hou, C.; Chen, L.; Yang, L.; Ji, X. An insight into anti-inflammatory effects of natural polysaccharides. *Int. J. Biol. Macromol.* **2020**, *153*, 248–255. [CrossRef]

12. Aboulwafa, M.M.; Youssef, F.S.; Gad, H.A.; Altyar, A.E.; Al-Azizi, M.M.; Ashour, M.L. A comprehensive insight on the health benefits and phytoconstituents of *Camellia sinensis* and recent approaches for its quality control. *Antioxidants* **2019**, *8*, 455. [CrossRef]
13. Ashour, M.L.; Youssef, F.S.; Gad, H.A.; El-Readi, M.Z.; Bouzabata, A.; Abuzeid, R.M.; Sobeh, M.; Wink, M. Evidence for the anti-inflammatory activity of *Bupleurum marginatum* (Apiaceae) extracts using in vitro and in vivo experiments supported by virtual screening. *J. Pharm. Pharmacol.* **2018**, *70*, 952–963. [CrossRef] [PubMed]
14. El Senousy, A. Immunomodulatory and anti-inflammatory activities of the defatted alcoholic extract and mucilage of *Hibiscus sabdariffa* L. leaves, and their chemical characterization. *J. Pharmacog. Phytochem.* **2019**, *8*, 982–990.
15. Carlotto, J.; Maria-Ferreira, D.; de Souza, L.M.; da Luz, B.B.; Dallazen, J.L.; de Paula Werner, M.F.; Cipriani, T.R. A polysaccharide fraction from "ipê-roxo" (*Handroanthus heptaphyllus*) leaves with gastroprotective activity. *Carbohydr. Polym.* **2019**, *226*, 115239. [CrossRef]
16. Nascimento, A.M.; Maria-Ferreira, D.; de Souza, E.F.J.; de Souza, L.M.; Sassaki, G.L.; Iacomini, M.; Werner, M.F.d.P.; Cipriani, T.R. Gastroprotective effect and chemical characterization of a polysaccharide fraction from leaves of Croton cajucara. *Int. J. Biol. Macromol.* **2017**, *95*, 153–159. [CrossRef]
17. El-Din, M.I.G.; Youssef, F.S.; Said, R.S.; Ashour, M.L.; Eldahshan, O.A.; Singab, A.N.B. Chemical constituents and gastro-protective potential of *Pachira glabra* leaves against ethanol-induced gastric ulcer in experimental rat model. *Inflammopharmacology* **2021**, *29*, 317–332. [CrossRef] [PubMed]
18. El-Din, M.I.G.; Youssef, F.S.; Ashour, M.L.; Eldahshan, O.A.; Singab, A.N.B. New γ-pyrone glycoside from Pachira glabra and assessment of its gastroprotective activity using an alcohol-induced gastric ulcer model in rats. *Food Funct.* **2020**, *11*, 1958–1965. [CrossRef] [PubMed]
19. Khan, M.S.A.; Khundmiri, S.U.K.; Khundmiri, S.R.; Al-Sanea, M.M.; Mok, P.L. Fruit-derived polysaccharides and terpenoids: Recent update on the gastroprotective effects and mechanisms. *Front. Pharmacol.* **2018**, *9*, 569. [CrossRef]
20. Lajili, S.; Ammar, H.H.; Mzoughi, Z.; Amor, H.B.H.; Muller, C.D.; Majdoub, H.; Bouraoui, A. Characterization of sulfated polysaccharide from *Laurencia obtusa* and its apoptotic, gastroprotective and antioxidant activities. *Int. J. Biol. Macromol.* **2019**, *126*, 326–336. [CrossRef]
21. Fathy, S.; Emam, M.; Agwa, S.A.; Zahra, F.A.; Youssef, F.; Sami, R. The antiproliferative effect of *Origanum majorana* on human hepatocarcinoma cell line: Suppression of NF-kB. *Cell. Mol. Biol.* **2016**, *62*, 80–84.
22. Murgia, V.; Ciprandi, G.; Votto, M.; De Filippo, M.; Tosca, M.A.; Marseglia, G.L. Natural remedies for acute post-viral cough in children. *Allergol. Immunopathol.* **2021**, *49*, 173–184. [CrossRef]
23. Kahramanoğlu, İ.; Wan, C. Determination and improvement of the postharvest storability of Little Mallow (*Malva parviflora* L.): A novel crop for a sustainable diet. *HortScience* **2020**, *55*, 1378–1386.
24. Ododo, M.M.; Choudhury, M.K.; Dekebo, A.H. Structure elucidation of β-sitosterol with antibacterial activity from the root bark of *Malva parviflora*. *SpringerPlus* **2016**, *5*, 1210. [CrossRef] [PubMed]
25. Munir, A.; Youssef, F.S.; Ishtiaq, S.; Kamran, S.H.; Sirwi, A.; Ahmed, S.A.; Ashour, M.L.; Elhady, S.S. *Malva parviflora* leaves mucilage: An eco-friendly and sustainable biopolymer with antioxidant properties. *Polymers* **2021**, *13*, 4251. [CrossRef] [PubMed]
26. Dugani, A.; Dakhil, B.; Treesh, S. Protective effect of the methanolic extract of *Malva parviflora* L. leaves on acetic acid-induced ulcerative colitis in rats. *Saudi J. Gastroenterol.* **2016**, *22*, 226–233.
27. El-Naggar, M.E.; Hussein, J.; El-sayed, S.M.; Youssef, A.M.; El Bana, M.; Latif, Y.A.; Medhat, D. Protective effect of the functional yogurt based on *Malva parviflora* leaves extract nanoemulsion on acetic acid-induced ulcerative colitis in rats. *J. Mat. Res. Technol.* **2020**, *9*, 14500–14508. [CrossRef]
28. Afolayan, A.J.; Aboyade, O.M.; Adedapo, A.A.; Sofidiya, M.O. Anti-inflammatory and analgesic activity of the methanol extract of *Malva parviflora* Linn (Malvaceae) in rats. *Afr. J. Biotechnol.* **2010**, *9*, 1225–1229.
29. Bouriche, H.; Meziti, H.; Senator, A.; Arnhold, J. Anti-inflammatory, free radical-scavenging, and metal-chelating activities of *Malva parviflora*. *Pharm. Biol.* **2011**, *49*, 942–946. [CrossRef]
30. Jothy, S.L.; Zakaria, Z.; Chen, Y.; Lau, Y.L.; Latha, L.Y.; Sasidharan, S. Acute oral toxicity of methanolic seed extract of *Cassia fistula* in Mice. *Molecules* **2011**, *16*, 5268–5282. [CrossRef]
31. Afsar, T.; Razak, S.; Khan, M.R.; Mawash, S.; Almajwal, A.; Shabir, M.; Haq, I.U. Evaluation of antioxidant, anti-hemolytic and anticancer activity of various solvent extracts of *Acacia hydaspica* R. Parker aerial parts. *BMC Complement. Altern. Med.* **2016**, *16*, 258.
32. Lakshmi, G.; Smitha, N.; Ammu, S.; Priya, C.; Bhaskara Rao, K. Phytochemical profile, in vitro antioxidant and hemolytic activities of various leaf extract of *Nymphaea Nouchali* Linn: An in vitro study. *Int. J. Pharm. Pharm. Sci.* **2014**, *6*, 548–552.
33. Vidhya, R.; Udayakumar, R. Phytochemical screening and evaluation of in vitro haemolytic, thrombolytic and antiinflammatory activities of *Aerva lanata* (L.). *IAJPS* **2016**, *6*, 6–7.
34. Elmajdoub, A.A.; Awidat, S.K.; El-Mahmoudy, A.M. Anti-inflammatory potential of Agaricus in carrageenan-induced model of local inflammation in rats. *Int. J. Basic Clin. Pharmacol.* **2015**, *4*, 497. [CrossRef]
35. Galati, E.; Cavallaro, A.; Ainis, T.; Tripodo, M.; Bonaccorsi, I.; Contartese, G.; Taviano, M.F.; Fimiani, V. Anti-Inflammatory effect of lemon mucilage: In vivo and in vitro studies. *Immunopharmacol. Immunotoxicol.* **2005**, *27*, 661–670. [CrossRef] [PubMed]
36. Alba, K.; Nguyen, P.T.; Kontogiorgos, V. Sustainable polysaccharides from Malvaceae family: Structure and functionality. *Food Hydrocolloid.* **2021**, *118*, 106749. [CrossRef]

37. Kaur, H.B.; Ruknuddin, G.; Nariya, M.; Patgiri, B.; Bedarkar, P.; Prajapati, P.K. Anti-tussive activity of Ashtangavaleha prepared by two different methods against sulphur dioxide induced cough in mice. *Med. J.* **2018**, *11*, 471–475. [CrossRef]
38. Ridh, D.A.A.M.; Sarheed, N.M.; Kokaz, O. Activity of ethanolic extraction of *Malva parviflora* and liqourice as antifungal and antioxidant in male rats. *J. Pharm. Sci. Res.* **2018**, *10*, 777–781.
39. Sutovska, M.; Capek, P.; Franova, S.; Joskova, M.; Sutovsky, J.; Marcinek, J.; Kalman, M. Antitussive activity of *Althaea officinalis* L. polysaccharide rhamnogalacturonan and its changes in guinea pigs with ovalbumine-induced airways inflammation. *Bratisl. Lek. Listy* **2011**, *112*, 670–675.
40. Büechi, S.; Vögelin, R.; Eiff, M.; Ramos, M.; Melzer, J. Open trial to assess aspects of safety and efficacy of a combined herbal cough syrup with Ivy and Thyme. *Res. Complement. Nat. Class. Med.* **2006**, *12*, 328–332. [CrossRef]
41. Adefisayo, M.A.; Akomolafe, R.O.; Akinsomisoye, S.O.; Alabi, Q.K.; Ogundipe, O.L.; Omole, J.G.; Olamilosoye, K.P. Gastro-protective effect of methanol extract of *Vernonia amygdalina* (del.) leaf on aspirin-induced gastric ulcer in Wistar rats. *Toxicol. Rep.* **2017**, *4*, 625–633. [CrossRef]
42. Chen, S.H.; Liang, Y.C.; Chao, J.C.; Tsai, L.H.; Chang, C.C.; Wang, C.C.; Pan, S. Protective effects of *Ginkgo biloba* extract on the ethanol-induced gastric ulcer in rats. *World J. Gastroenterol.* **2005**, *11*, 3746–3750. [CrossRef]
43. Rahim, N.A.; Hassandarvish, P.; Golbabapour, S.; Ismail, S.; Tayyab, S.; Abdulla, M.A. Gastroprotective effect of ethanolic extract of *Curcuma xanthorrhiza* leaf against ethanol-induced gastric mucosal lesions in Sprague-Dawley rats. *BioMed Res. Int.* **2014**, *2014*, 416409. [CrossRef] [PubMed]
44. Adinortey, M.; Ansah, C.; Galyuon, I.; Nyarko, A. In vivo models used for evaluation of potential antigastroduodenal Ulcer Agents. *Ulcers* **2013**, *2013*, 489–512. [CrossRef]
45. Sharifi-Rad, M.; Fokou, P.V.T.; Sharopov, F.; Martorell, M.; Ademiluyi, A.O.; Rajkovic, J.; Salehi, B.; Martins, N.; Iriti, M.; Sharifi-Rad, J. Antiulcer agents: From plant extracts to phytochemicals in healing promotion. *Molecules* **2018**, *23*, 1751. [CrossRef] [PubMed]
46. Balogun, M.E.; Besong, E.E.; Obimma, J.N.; Djobissie, S.F.; Mbamalu, O.S. Gastroprotective effect of ethanolic extract of *Vigna subterranea* in ethanol-induced gastric mucosal ulceration in Rats. *Ind. J. Physiol. Pharmacol.* **2018**, *62*, 347–358.
47. Zakaria, Z.A.; Balan, T.; Azemi, A.K.; Omar, M.H.; Mohtarrudin, N.; Ahmad, Z.; Abdullah, M.N.H.; Desa, M.N.M.; Teh, L.K.; Salleh, M.Z. Mechanism (s) of action underlying the gastroprotective effect of ethyl acetate fraction obtained from the crude methanolic leaves extract of *Muntingia calabura*. *BMC Complement. Altern. Med.* **2016**, *16*, 78. [CrossRef] [PubMed]
48. Arawwawala, L.; Arambewela, L.; Ratnasooriya, W. Gastroprotective effect of *Piper betle* Linn. leaves grown in Sri Lanka. *J. Ayurveda Integ. Med.* **2014**, *5*, 38. [CrossRef]
49. Onoja, S.O.; Chinagorom, O.; Ikpa, C.B.; Madubuike, K.G.; Ezeigbo, I.I.; Ijioma, S.N.; Anaga, A.O.; Ezeja, M.I. Gastroprotective effects of methanol extract of *Eremomastax speciosa* leaf harvested in Southern part of Nigeria in rat. *EuroBiotech J.* **2018**, *2*, 200–208. [CrossRef]
50. Monrroy, M.; García, E.; Ríos, K.; García, J.R. Extraction and physicochemical characterization of mucilage from *Opuntia cochenillifera* (L.) Miller. *J. Chem.* **2017**, *2017*, 4301901. [CrossRef]
51. Patel, D.; Prajapati, D.; Patel, N. Seed mucilage from *Ocimum americanum* linn. as disintegrant in tablets: Separation and evaluation. *Ind. J. Pharm. Sci.* **2007**, *69*, 431. [CrossRef]
52. Kamble, M.S. Studies on isolation and evaluation of *Ocimum tenuiflorum* linn seed mucilage. *J. Drug Del. Therap.* **2012**, *2*, 6. [CrossRef]
53. Zubair, M.; Rizwan, K.; Rashid, U.; Saeed, R.; Saeed, A.A.; Rasool, N.; Riaz, M. GC/MS profiling, in vitro antioxidant, antimicrobial and haemolytic activities of *Smilax macrophylla* leaves. *Arab. J. Chem.* **2017**, *10*, S1460–S1468. [CrossRef]
54. OECD. Test No. 425: Acute oral toxicity—Up-and-down procedure. In *The OECD Guidelines for the Testing of Chemicals*; OECD: Paris, France, 2001; ISBN 9789264071049.
55. Igbe, I.; Ching, F.P.; Eromon, A. Anti-inflammatory activity of aqueous fruit pulp extract of *Hunteria umbellata* K. schum in acute and chronic inflammation. *Acta Pol. Pharm. Drug Res.* **2010**, *67*, 81–85.
56. Singh, M.; Kumar, V.; Singh, I.; Gauttam, V.; Kalia, A.N. Anti-inflammatory activity of aqueous extract of *Mirabilis jalapa* Linn. leaves. *Pharmacog. Res.* **2010**, *2*, 364. [CrossRef] [PubMed]
57. Gupta, R.; Gupta, M.; Bhandari, A.; Gupta, J. Evaluation of antitussive activity of polyherbomineral formulation on cough reflex induced by different cough induced models in mice. *Int. J. Drug Dev. Res.* **2014**, *6*, 280–289.
58. Ateufack, G.; Domgnim Mokam, E.C.; Mbiantcha, M.; Dongmo Feudjio, R.B.; David, N.; Kamanyi, A. Gastroprotective and ulcer healing effects of piptadeniastrum Africanum on experimentally induced gastric ulcers in rats. *BMC Complement. Altern. Med.* **2015**, *15*, 214. [CrossRef] [PubMed]
59. Katary, M.; Salahuddin, A. Gastroprotective Effect of punicalagin against ethanol-induced gastric Ulcer: The possible underlying Mechanisms. *Biomark. J.* **2017**, *3*, 1. [CrossRef]
60. Rafiei, M.; Namazi, F.; Rajaian, H.; Nazifi, S. Gastroprotective effects of various *Scrophularia striata* extracts on ethanol-induced gastric ulcer in Rats. *Turk. J. Pharm. Sci.* **2016**, *13*, 328–334. [CrossRef]
61. Arab, H.H.; Salama, S.A.; Omar, H.A.; Arafa, E.-S.A.; Maghrabi, I.A. Diosmin protects against ethanol-induced gastric injury in rats: Novel anti-ulcer actions. *PLoS ONE* **2015**, *10*, e0122417. [CrossRef]

62. Tamaddonfard, E.; Erfanparast, A.; Farshid, A.A.; Imani, M.; Mirzakhani, N.; Salighedar, R.; Tamaddonfard, S. Safranal, a constituent of saffron, exerts gastro-protective effects against indomethacin-induced gastric ulcer. *Life Sci.* **2019**, *224*, 88–94. [CrossRef]
63. Sharma, A.L.; Bhot, M.A.; Chandra, N. Gastroprotective effect of aqueous extract and mucilage from *Bryophyllum pinnatum* (Lam.) Kurz. *Anci. Sci. Life.* **2014**, *33*, 252–258. [CrossRef]
64. Das, A.K.; Bigoniya, P.; Verma, N.K.; Rana, A. Gastroprotective effect of *Achyranthes aspera* Linn. leaf on rats. *Asian Pac. J. Trop. Med.* **2012**, *5*, 197–201. [CrossRef]
65. Dashputre, N.; Naikwade, N. Evaluation of anti-ulcer activity of methanolic extract of *Abutilon indicum* Linn. leaves in experimental rats. *Int. J. Pharm. Sci. Drug Res.* **2011**, *3*, 97–100.
66. Gupta, J.; Kumar, D.; Gupta, A. Evaluation of gastric anti–ulcer activity of methanolic extract of *Cayratia trifolia* in experimental animals. *Asian Pac. J. Trop. Dis.* **2012**, *2*, 99–102. [CrossRef]
67. Gul, H.; Abbas, K.; Qadir, M.I. Gastro-protective effect of ethanolic extract of *Mentha longifolia* in alcohol-and aspirin-induced gastric ulcer models. *Bang. J. Pharmacol.* **2015**, *10*, 241–245. [CrossRef]
68. Sen, S.; Asokkumar, K.; Umamaheswari, M.; Sivashanmugam, A.; Subhadradevi, V. Antiulcerogenic effect of gallic acid in rats and its effect on oxidant and antioxidant parameters in stomach tissue. *Ind. J. Pharm. Sci.* **2013**, *75*, 149.
69. Kore Kakasaheb, J.; Shete Rajkumar, V.; Patel Apsari, J.; Kulkarni Jitendra, B. Antiulcer activity of aqueous extract of *Spinacia oleracia* in rats. *IJRPC* **2011**, *13*, 29–31.
70. Shukla, P.; Porwal, A.; Roy, S.; Chaturvedi, S.; Tripathi, S.; Arya, N. Preliminary study on antiulcer effect of agomelatine and its potentiation with pyridoxine. *Int. J. Basic Clin. Pharmacol.* **2017**, *6*, 2566–2570. [CrossRef]
71. AlRashdi, A.S.; Salama, S.M.; Alkiyumi, S.S.; Abdulla, M.A.; Hadi, A.H.A.; Abdelwahab, S.I.; Taha, M.M.; Hussiani, J.; Asykin, N. Mechanisms of gastroprotective effects of ethanolic leaf extract of *Jasminum sambac* against HCl/ethanol-induced gastric mucosal injury in rats. *Evid. Based Complement. Altern. Med.* **2012**, *2012*, 786426. [CrossRef]
72. Lowry, O.H.; Rosebrough, N.J.; Farr, A.L.; Randall, R.J. Protein measurement with the Folin phenol reagent. *J. Biol. Chem.* **1951**, *193*, 265–275. [CrossRef]
73. Xia, Y.G.; Sun, H.M.; Wang, T.L.; Liang, J.; Yang, B.Y.; Kuang, H.X. A modified GC-MS analytical procedure for separation and detection of multiple classes of carbohydrates. *Molecules* **2018**, *23*, 1284. [CrossRef]

pharmaceuticals

Article

Multidirectional Effects of Terpenoids from *Sorbus intermedia* (EHRH.) PERS Fruits in Cellular Model of Benign Prostate Hyperplasia

Agnieszka Sołtys [1], Agnieszka Galanty [1,*], Karolina Grabowska [1], Paweł Paśko [2], Paweł Zagrodzki [2] and Irma Podolak [1]

[1] Department of Pharmacognosy, Faculty of Pharmacy, Jagiellonian University Medical College, Medyczna 9, 30-688 Kraków, Poland; agnieszka.soltys@doctoral.uj.edu.pl (A.S.); karolina1.grabowska@uj.edu.pl (K.G.); irma.podolak@uj.edu.pl (I.P.)

[2] Department of Food Chemistry and Nutrition, Faculty of Pharmacy, Jagiellonian University Medical College, Medyczna 9, 30-688 Kraków, Poland; p.pasko@uj.edu.pl (P.P.); pawel.zagrodzki@uj.edu.pl (P.Z.)

* Correspondence: agnieszka.galanty@uj.edu.pl

Abstract: Benign prostatic hyperplasia (BPH) is a common urological disease affecting aging men. Its pathogenesis is regarded as complex and multifactorial, with sex hormones and inflammation as key contributory factors. In the current study, we investigated the anti-BPH potential of terpenoids present in the fruits of *Sorbus intermedia* (EHRH.) PERS. Not only the effects on testosterone-stimulated normal prostate epithelial PNT2 cells, namely suppression of 5-α-reductase activity, PSA secretion, and cell proliferation, were determined but also the inhibitory activity on heat-induced protein denaturation, hyaluronidase, as well as IL-6, TNF-α, and NO release in LPS-treated macrophages. *Sorbus* terpenoids significantly inhibited 5-α-reductase activity and reduced PSA secretion in PNT2 cells, reversing the stimulatory effect of testosterone. PNT2 cell proliferation was also found to be attenuated. Subsequently, all compounds reduced the release of pro-inflammatory mediators in RAW 264.7 cells. In addition, ursolic acid (UA) and its aldehyde (UAL) were the most potent hyaluronidase inhibitors of all compounds, with IC_{50} values of 225.75 μg/mL and 369.77 μg/mL, respectively. For better understanding and interpretation of the overall effect of *Sorbus* terpenoids on different aspects of BPH pathogenesis and development, cluster analysis was applied.

Keywords: *Sorbus intermedia*; terpenoids; benign prostatic hyperplasia; antiproliferative activity; antiandrogenic activity; anti-inflammatory activity; fruits

check for updates

Citation: Sołtys, A.; Galanty, A.; Grabowska, K.; Paśko, P.; Zagrodzki, P.; Podolak, I. Multidirectional Effects of Terpenoids from *Sorbus intermedia* (EHRH.) PERS Fruits in Cellular Model of Benign Prostate Hyperplasia. *Pharmaceuticals* **2023**, *16*, 965. https://doi.org/10.3390/ph16070965

Academic Editor: Diana Roxana Pelinescu

Received: 26 May 2023
Revised: 3 July 2023
Accepted: 4 July 2023
Published: 5 July 2023

1. Introduction

Benign prostatic hyperplasia (BPH) is a non-malignant enlargement of the prostate gland. It is a very common urological disease affecting aging men. Histopathologically, it is manifested by an increase in the number of cells of the epithelial and fibromuscular tissues of the transition zone of the prostate gland and the periurethral region [1].

The aetiology and pathophysiology of BPH are still not fully understood. Several theories of its origin and development have been proposed [2,3], and various permissive factors have been indicated to date [1,4]. The role of sex hormones and chronic inflammation are highly disputed. Animal and human studies suggest a significant role of dihydrotestosterone (DHT), which is the main androgen of the prostate [5]. As for chronic inflammation, it is also thought to play an important role in the development of BPH, particularly in the occurrence, progression, and severity of clinical symptoms associated with BPH [1,5–7].

Although BPH is not life-threatening, pathological changes within the prostate gland often affect urinary function, leading to clinical symptoms known as LUTS (lower urinary tract symptoms), which are mainly related to urination dysfunction. However, it should be noted that not all men experience LUTS, and a relatively low correlation between prostate volume, urinary tract symptoms, and urinary flow rate has been observed [8,9]. Recent studies have shown that the lifetime prevalence of BPH-associated LUTS is estimated at 26.2% worldwide and increases from 14.8% to 36.8% in men aged 40 and ≥80, respectively [10]. Nevertheless, urinary symptoms cause discomfort and inconvenience in performing daily activities and significantly reduce quality of life [5]. Furthermore, as BPH is closely associated with age, the incidence and prevalence rates of this disorder are expected to increase in an aging population, and this expected burden will challenge health care systems [8,11].

Current treatments of BPH include watchful waiting, pharmacotherapy (mainly 5-α-reductase inhibitors and α1-blockers), phytotherapy, and several forms of active intervention, such as various surgical and non-surgical procedures [8,9]. The appropriate therapy is selected on the basis of the clinical manifestation of the disease as well as the patients' health status and individual preferences [8].

Herbal medicines are very popular among patients and administered in mild to moderate BPH. It is also not without significance that they are easily available and usually well tolerated. They also appear to be suitable for prevention, as BPH is characterized by high morbidity and a long latency period [3]. Several modes of action of natural products used in BPH have been proposed, including antiandrogenic, antiproliferative, and anti-inflammatory activity, but most studies refer to the activity of herbal extracts [2,3]. Some studies point to triterpenes as potential anti-BPH agents and, although the literature is scarce, provide encouraging results [12–17].

Recently, we reported on isolation of ursane terpenoids, i.e., ursolic acid (UA), ursolic aldehyde (UAL), 3-*O*-β-acetoxy-ursolic acid (AUA), 3-*O*-β-acetoxy-19α-hydroxy-ursolic acid (AHUA), and uvaol (UO), as well as β-sitosterol (βSIT), from fruits of *Sorbus intermedia* (EHRH.) PERS (Rosaceae), which is a popular ornamental tree native to northern Europe [18]. Some of these compounds were found to exhibit cytotoxic activity against, among others, human prostate cancer cell lines DU145 and PC3, with almost null toxic effects on normal prostate epithelial PNT2 cells. Satisfactory results against PC3 cells after 48 h of incubation were obtained for UA, AUA, and UAL, with IC_{50} values of 4.45 µg/mL, 16.40 µg/mL, and 22.45 µg/mL, respectively [18]. Previously, chloroform extracts also prepared from *S. intermedia* fruits at different developmental stages were found to reduce the viability of prostate cell lines [19]. This prompted us to extend our investigations to another prostate-related health problem, namely BPH. Therefore, the aim of the current study was to further investigate terpenoids isolated from *S. intermedia* with regard to their anti-BPH potential. Given the complexity of BPH pathogenesis, these compounds have been subjected to a number of in vitro studies to assess their multidirectional effects on different aspects of BPH. We designed and proposed an in vitro model based on testosterone-stimulated PNT2 cells and assessed whether these compounds could affect 5-α-reductase activity, PSA release, and cell hyperproliferation. We then determined the anti-inflammatory properties of all compounds. Finally, we applied hierarchical agglomeration cluster analysis to investigate the similarities between the activities of these compounds in the different experiments performed in this work to better interpret their overall effect in the human BPH model and inflammation and possibly identify the most promising multi-target compounds.

2. Results and Discussion

2.1. Antiandrogenic Effects of S. intermedia Terpenoids on Testosterone-Stimulated PNT2 Cells

Dihydrotestosterone (DHT) is the main androgen of the prostate. More than 90% of testosterone is converted in stromal cells to the more active DHT by 5-α-reductase. Moreover, the prostate remains sensitive to androgens throughout life, and simultaneously, intraprostatic DHT levels remain high despite aging [20]. Animal and human studies suggest its significant involvement in the development of BPH, as DHT is believed to play a key role in maintaining homeostasis between cell proliferation and apoptosis [5,20]. DHT binds, with higher affinity than testosterone, to the androgen receptor, which ultimately promotes transcription of androgen-dependent genes and further protein synthesis, differentiation, as well as cell growth. Moreover, androgens act indirectly by stimulation of production of several growth factors, e.g., EGF (epidermal growth factor), KGF (keratinocyte growth factor), and IGFs (insulin-like growth factors), involved in cell proliferation [3,5,20]. The prostatic stroma and epithelium interact through cell signalling mechanisms mediated by DHT and DHT-stimulated growth factors. DHT influences prostate cells in both an autocrine and paracrine manner [5]. For this reason, 5-α-reductase has become a pharmacological target in the treatment of BPH, and subsequently, inhibitors of this enzyme, such as dutasteride and finasteride, were introduced to the therapeutics. These drugs effectively reduce DHT levels and prostate volume; however, they are burdened with some bothersome side effects, such as decreased libido and impotence.

On this basis, we investigated the effect of natural terpenoids (Figure 1) from *S. intermedia* on 5-α-reductase activity in testosterone-stimulated PNT2 cells. Although the post hoc analysis showed no significance between the control groups in this study, there is a strong and obvious trend that clearly indicates a stimulating effect of the hormone on prostate cells. We also measured PSA (prostate-specific antigen) secretion, as the PSA gene (*KLK3*) is regulated by androgens, and PSA serves as biomarker for prostate diseases such as prostate cancer and BPH [21]. The results of our experiments are shown in Figure 2. In this study, we used dutasteride as a reference drug. It is one of the 5-α-reductase inhibitors that is indicated in moderate to severe BPH.

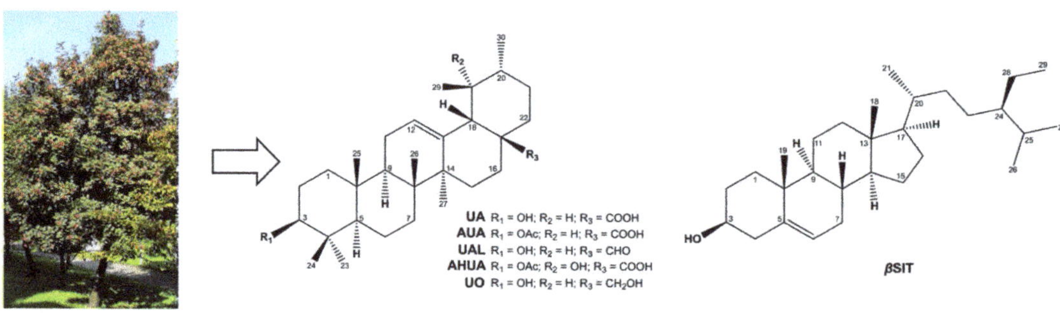

Figure 1. Chemical structures of terpenoids isolated from *S. intermedia* fruits [18].

All *Sorbus* compounds tested, including ursane derivatives and β-sitosterol, significantly inhibited 5-α-reductase activity and reduced PSA secretion in testosterone-stimulated PNT2 cells. UA, AUA, and βSIT showed suppression of the enzyme activity by approximately 20% at a concentration of 40 µg/mL, which was comparable to the reference drug. Both dutasteride and *Sorbus* compounds reversed the stimulatory effect of testosterone on 5-α-reductase activity. In addition, all compounds reduced PSA release in PNT2 cells to 76–86% at the highest concentrations, and there were no statistically significant differences between tested compounds and dutasteride. Although we did not observe a strict dose–effect relationship, some compounds showed a tendency to act in dose-dependent manner, especially UA, UAL, and AUA.

Of the *Sorbus* compounds tested in the current work, only βSIT has been studied more extensively, including by in vivo human studies, as a potential anti-BPH agent. Systematic reviews indicate that βSIT improves urinary symptoms but does not reduce prostate size [22,23]. Previously, βSIT was found to reduce 5-α-reductase activity in the hamster prostate in a dose-dependent manner [24]. In addition, there is evidence for the antiandrogenic activity of several herbs used in the treatment of BPH, such as *Serenoa repens*, *Pygeum africanum*, and *Urtica dioica*, whose main active constituents are sterols [25,26]. Interestingly, a standardized saw palmetto extract enriched in βSIT was recently studied, and no significant inhibitory activity on 5-α-reductase was observed [27].

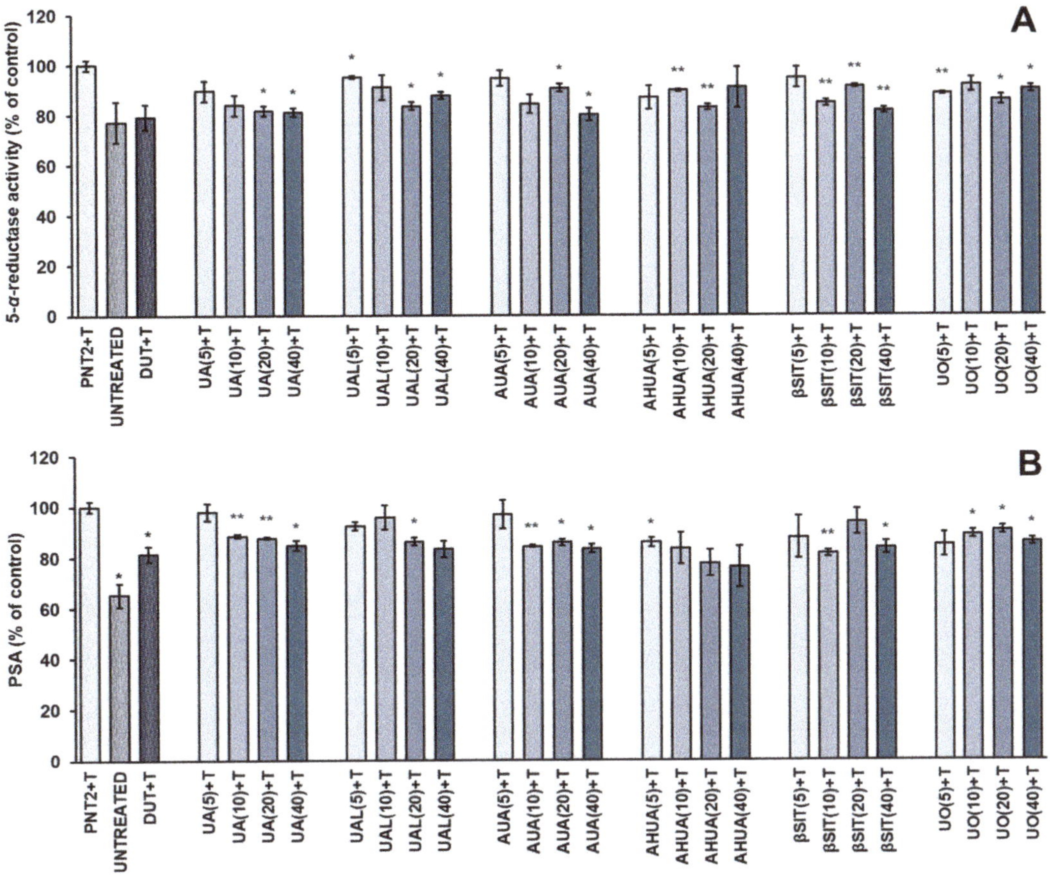

Figure 2. The effects of *S intermedia* terpenoids on 5-α-reductase activity (**A**) and PSA secretion (**B**) in PNT2 cells stimulated by testosterone. PNT2 cells were incubated with terpenoids at different concentrations (the numbers in parentheses indicate concentrations in µg/mL) in the presence of testosterone (T). The results are expressed as the mean ± SD of three experiments. The results are set together with untreated PNT2 cells (UNTREATED) and cells treated with testosterone and dutasteride as the reference drug (DUT+T). Statistical analysis was performed using one-way ANOVA and T3 Dunnett post hoc test with * $p < 0.05$, ** $p < 0.01$, against the testosterone-stimulated cells. Abbreviations: UA, ursolic acid; UAL, ursolic aldehyde; AUA, 3-*O*-β-acetoxy-ursolic acid; AHUA, 3-*O*-β-acetoxy-19α-hydroxy-ursolic acid; βSIT, β-sitosterol; UO, uvaol.

To the best of our knowledge, literature reports on the anti-BPH potential of ursane-based triterpenes are extremely scarce. One study in a rat model of testosterone-induced BPH showed that treatment with ursolic acid resulted in a reduction of prostate volume as well as in serum and prostate tissue DHT levels [12]. Interestingly, an isomer of ursolic acid–oleanolic acid downregulated 5-α-reductase II expression in a rat model of BPH while inhibiting prostate growth and serum DHT levels, and the results were comparable to finasteride [17].

2.2. Antiproliferative Activities of S. intermedia Terpenoids

The development of BPH is related to excessive and abnormal cell proliferation. Pathological hypertrophy of prostate gland is considered to be associated with the imbalance between cell proliferation and apoptosis, which is modulated predominantly by prostatic androgens. Therefore, we investigated the antiproliferative activity of *S. intermedia* terpenoids and βSIT towards testosterone-treated PNT2 cells. Effects were observed after 24 h, 48 h, and 72 h and were compared with dutasteride, which was used as a reference drug. The results are shown in Figure 3.

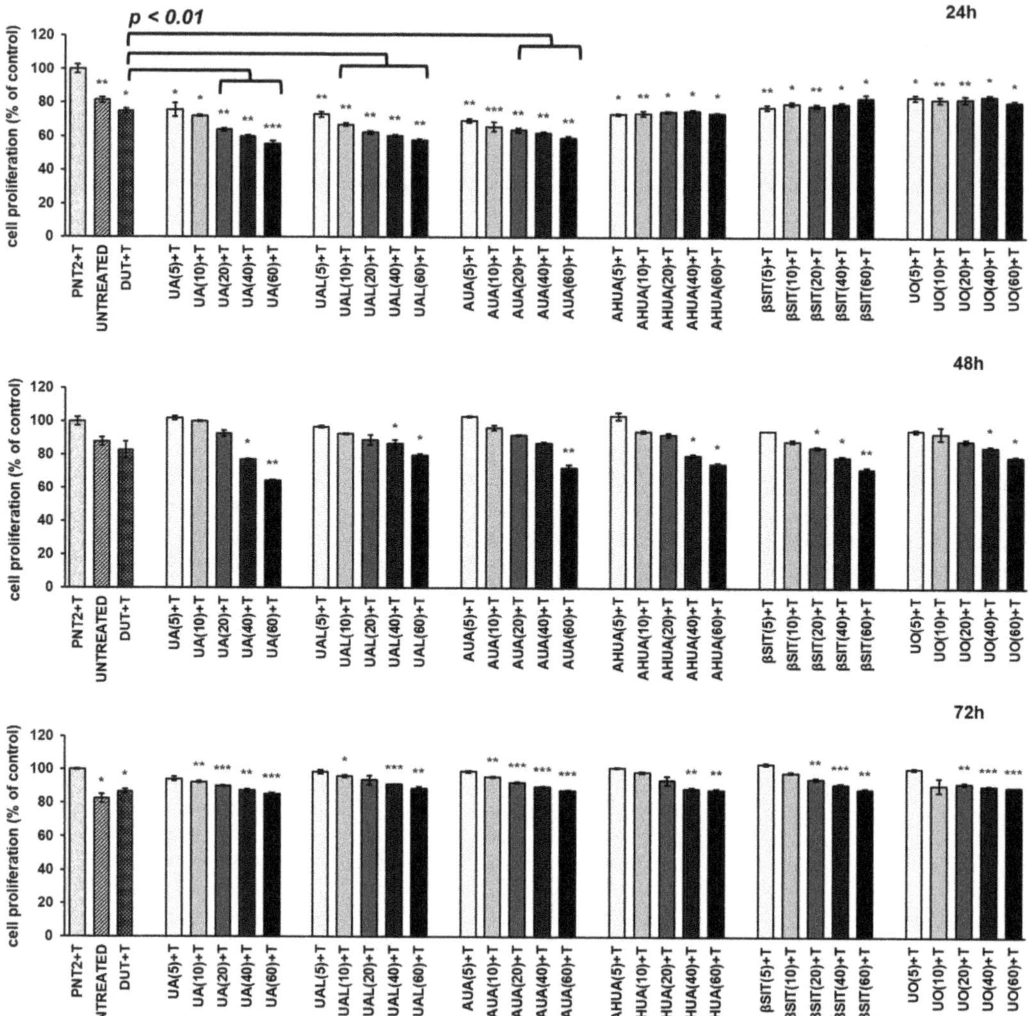

Figure 3. The anti-proliferative effects of *S intermedia* terpenoids on testosterone-stimulated PNT2 cells. PNT2 cells were incubated with terpenoids at different concentrations (the numbers in parentheses indicate concentrations in µg/mL) in the presence of testosterone (T). The results obtained after 24 h, 48 h, and 72 h of incubation are presented separately on each bar chart and expressed as the mean ± SD of three experiments. The results are set together with untreated PNT2 cells (UNTREATED) and cells treated with testosterone and dutasteride as the reference drug (DUT+T). Statistical analysis was performed using one-way ANOVA and T3 Dunnett post hoc test with * $p < 0.05$, ** $p < 0.01$, and *** $p < 0.001$ against the testosterone-stimulated cells. Significant differences ($p < 0.01$) between tested compounds and reference drug are marked by black line. Abbreviations: UA, ursolic acid; UAL, ursolic aldehyde; AUA, 3-*O*-β-acetoxy-ursolic acid; AHUA, 3-*O*-β-acetoxy-19α-hydroxy-ursolic acid; βSIT, β-sitosterol; UO, uvaol.

All compounds reduced cell proliferation, with better results obtained after 24 h and 48 h of incubation. With the exception of the results obtained after 24 h for AHUA, βSIT, and UO, all other compounds supressed or tended to supress cell proliferation in a dose-dependent manner. UA, UAL, and AUA were more potent than the reference drug, causing an approximately 45% decrease in cell proliferation at a concentration of 60 µg/mL after 24 h. Although their cytostatic effect was less pronounced after 72 h of incubation, they were still as active as the reference drug.

UA was previously reported to exert cytotoxic [18,28–30] or cytostatic [31] activities against various prostate cell lines. For example, a 24 h incubation of PC3 cells with 40 µM of UA resulted in a significant increase in the number of cells in the G1 phase [31]. Its combination at a low dose (4.1 µM) with oleanolic acid (5.47 µM) induced cytotoxic autophagy and inhibited the growth of BPH-1 cells by 50%, which proved more effective than administration of UA alone (50% growth inhibition at a concentration of 67 µM) [16]. In addition, UA arrested the cell cycle in several other cells of non-prostatic origin [32–35]. Similarly, the antiproliferative activity of βSIT against cells of various cancer types is quite well documented [36], including its activity towards PC3 cells, where a 1-day incubation with 16 µM of βSIT led to a 35% growth inhibition [37].

For other *Sorbus* terpenoids, i.e., UAL, AUA, AHUA, and UO, there is a lack of information in the literature, especially with regard to their effects on prostate cells. In a previous work, we already determined their cytotoxic effects against DU145, PC3, and PNT2 cells using the lactate dehydrogenase assay, which indicates a cell death. AHUA and UO showed no effects on all cells tested, while UAL and AUA selectively affected the viability of DU145 and PC3 cells without toxicity towards normal PNT2 cells [18]. Nevertheless, to the best of our knowledge, there is no other report relating to the antiproliferative activity of these compounds on any cells of prostatic origin. With regard to non-prostatic cell lines, AUA and UO inhibited the growth of A375 [38] as well as HepG2 and MCF-7 cells [39,40], respectively, while UAL, in a mixture with oleanolic aldehyde, exerted an antiproliferative effect on MCF-7 cells with GI_{50} value equal to 202 µM [41]. To the best of our knowledge, no similar studies on in vitro hyperproliferation of prostate cells were found for any of the terpenoids tested.

2.3. *Anti-Inflammatory Activities of S. intermedia Terpenoids*

The impact of inflammation on the development of BPH appears to be complex and reciprocal, as inflammation can not only promote the development of BPH alone or in interaction with sex hormones, but it can also be induced during BPH, thus leading to expansion of its severity [7]. Several studies have found that chronic inflammation is associated with higher prostate volume, IPSS (international prostate symptom score), acute urinary retention, and/or higher risk of prostatectomy [42]. Therefore, we decided to assess the anti-inflammatory potential of *S. intermedia* compounds.

2.3.1. Inhibition of Albumin Heat-Induced Denaturation

As part of anti-inflammatory experiments, *Sorbus* ursane terpenoids and βSIT were first screened for their ability to stabilize serum albumin against heat-induced denaturation, as protein denaturation is considered part of the inflammation process. Non-steroidal anti-inflammatory drugs, e.g., indomethacin or diclofenac sodium, were found to exhibit anti-denaturation effect and stabilize serum albumin. In our study, however, most of the compounds tested failed to inhibit heat-induced denaturation of albumin, and only AHUA and UO (Table 1) slightly protected the protein, showing approximately 30% inhibition in the concentration range of 25–1000 µg/mL and 100–1000 µg/mL, respectively. The results were compared to diclofenac sodium, which was used as a reference drug.

Table 1. The effects of *S. intermedia* terpenoids against heat-induced albumin denaturation.

Concentration	Albumin Denaturation Inhibition (%)		
(μg/mL)	AHUA	UO	DS
5	NE	NE	34.29 ± 2.07
10	19.61 ± 6.76	9.35 ± 2.50	52.45 ± 2.84
25	28.85 ± 2.50 *	11.17 ± 1.69 *	81.06 ± 3.04
50	31.46 ± 3.07 *	18.35 ± 5.37	95.75 ± 1.71
100	30.63 ± 3.34 *	28.49 ± 3.82 *	99.06 ± 0.26
250	22.79 ± 5.00	31.98 ± 2.34 *	99.76 ± 0.26
500	28.67 ± 2.72 *	30.78 ± 1.56 *	100.00 ± 0.00
750	30.24 ± 2.72 *	29.30 ± 1.28 *	100.00 ± 0.00
1000	32.81 ± 4.75 *	30.37 ± 3.36 *	100.00 ± 0.00
IC_{50}	>Cmax	>Cmax	8.61

Inhibition of albumin denaturation was measured using turbidimetric assay. Values are presented as the mean ± SD of three experiments. Statistical analysis was performed using one-way ANOVA and T3 Dunnett post hoc test against positive control. * $p < 0.05$ indicates significant denaturation inhibition. UA, UAL, AUA, and βSIT were not active. Abbreviations: AHUA, 3-*O*-β-acetoxy-19α-hydroxy-ursolic acid; UO, uvaol; DS, diclofenac sodium; NE, not examined.

2.3.2. The Effects of *S. intermedia* Terpenoids on the Release of Pro-Inflammatory Mediators in LPS-Stimulated RAW 264.7 Macrophages

Robert et al. (2009) showed that more than 80% of surgically treated patients had CD8 T-lymphocyte and macrophage infiltrates in prostate tissues [43]. Some have identified BPH as an autoimmune disease and TNF-α as a potential therapeutic target, pointing to the ability of TNF antagonists to reduce epithelial hyperplasia, NFκB activation, and macrophage-mediated inflammation [44,45]. A retrospective study indicated that prostate tissues from TNF-antagonist-treated patients had a lower level of inflammation when compared to untreated group [44]. Tong et al. (2022) attempted to investigate the relationship between DHT and inflammation in BPH development and progression. The authors revealed that DHT stimulates the prostate stromal cells' proliferation by increasing TNF-α expression in LPS-stressed M1 macrophages. Moreover, they analysed tissues from BPH patients and found that TNF-α expression was increased in patients with larger prostate volume [46]. Subsequently, elevated levels of other pro-inflammatory cytokines, including IL-6, IL-8, and IL-17, were found in BPH tissues [16]. A higher risk of BPH is also linked to metabolic syndrome, in which increased levels of C-reactive protein, IL-1β, IL-6, IL-8, and TNF-α were observed [42].

Hence, we decided to test whether compounds isolated from fruits of *S. intermedia* are able to influence the release of selected pro-inflammatory mediators in LPS-stimulated RAW 264.7 macrophages. The results are shown in Figure 4. In this assay, dexamethasone, a glucocorticoid with an anti-inflammatory properties, was used as a reference drug.

All compounds at sub-cytotoxic concentrations significantly reduced the release of all tested mediators in LPS-stimulated RAW 264.7 macrophages. The strongest suppression of IL-6 release was observed with UA (about 30% at concentration of 10 μg/mL), and this effect was dose-dependent ($p < 0.001$). UAL was most effective towards TNF-α and reduced its secretion up to 72% at a concentration of 20 μg/mL. *Sorbus* terpenoids reduced NO release in LPS-stimulated RAW cells comparably by approximately 20–25% at their most effective concentrations. A strong dose–effect relationship ($p < 0.001$) was observed only for UA.

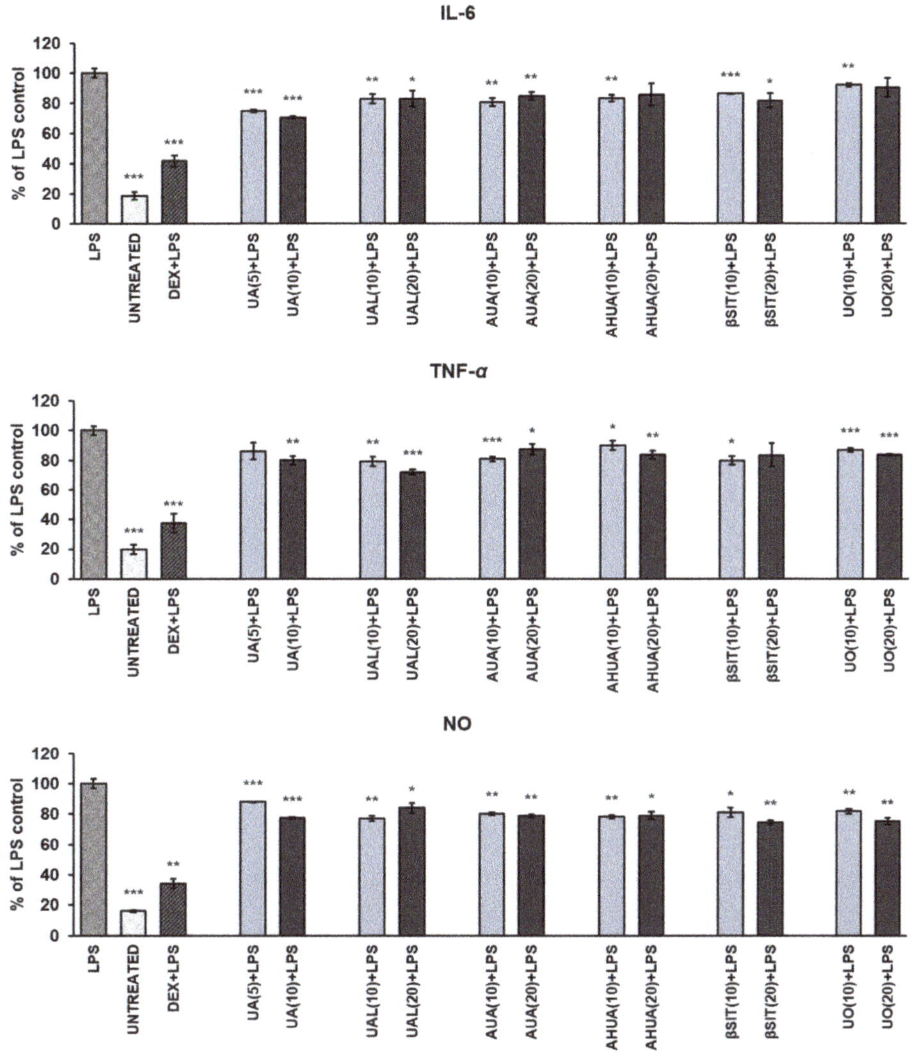

Figure 4. The impact of *S intermedia* terpenoids on the release of IL-6, TNF-α, and NO in LPS-stimulated RAW 264.7 macrophages. RAW cells were pre-treated with terpenoids at different concentrations (the numbers in parentheses indicate concentrations in μg/mL) for 1 h, followed by the addition of 10 ng/mL of LPS to induce inflammation. Exceptionally, due to high cytotoxicity towards RAW cells, UA were analysed at lower concentrations. Values are presented as the mean ± SD of three experiments. The results are set together with untreated RAW cells (UNTREATED) and cells treated with LPS and dexamethasone as the reference drug (DEX+LPS). Statistical analysis was performed using one-way ANOVA and T3 Dunnett post hoc test with * $p < 0.05$, ** $p < 0.01$, and *** $p < 0.001$ against the LPS-stimulated cells. Abbreviations: UA, ursolic acid; UAL, ursolic aldehyde; AUA, 3-*O*-*β*-acetoxy-ursolic acid; AHUA, 3-*O*-*β*-acetoxy-19*α*-hydroxy-ursolic acid; *β*SIT, *β*-sitosterol; UO, uvaol.

Biological studies on the anti-inflammatory activity of UAL and AHUA are extremely unaddressed. Only UA and UO are reasonably well examined. Du et al. (2020) showed that in LPS-treated macrophages, UO significantly suppressed NO production as well as mRNA expression and secretion of several mediators, including IL-6 and TNF-α [47]. Significant reductions in NO production at concentrations of 3 µM, 10 µM, and 30 µM were also reported by Wang et al. (2020), but 50% inhibitory activity was not achieved [48]. UO was also considered to lack activity against NO production in LPS and IFN-γ -treated [49] and LPS-stressed RAW 264.7 cells, with IC_{50} above 100 µM [50]. With regard to UA, several studies indicate its inhibitory activity against NO production in stimulated macrophages [49,51,52]. In contrast, in the study of Zhou and Wink (2019), UA did not cause a statistically significant decrease in NO level in RAW264.7 cells at the concentration range of 3–9 µM; however, it downregulated the TNF-α expression [53]. In one study, UA and AUA suppressed the expression and production of TNF-α and IL-6 in TNF-α-stimulated RA synovial fibroblasts [54]. βSIT was also previously reported to reduce TNF-α and IL-6 production in LPS-stimulated RAW 264.7 cells [55]. However, it was also regarded as inactive against NO secretion in LPS and IFN-γ-treated RAW cells with an IC_{50} above 100 µM [48].

2.3.3. The Effects of *S. intermedia* Terpenoids on Hyaluronidase Activity

The extracellular matrix is a network composed of macromolecules such as collagens, elastin, proteoglycans/glycosaminoglycans (e.g., hyaluronic acid), and different glycoproteins [56]. It not only serves as a structural scaffold but also contributes to several cellular processes, such as proliferation, migration, differentiation, autophagy, and angiogenesis. For example, hyaluronic acid and its degradation products interact with CD44 in a size-dependent manner. Long-chain hyaluronic acid induces receptor clustering, while low-mass molecules induce various signalling pathways involved in the regulation of cytoskeletal organization, cell growth, and proliferation [57].

Because matrix components have such important regulatory functions, the enzymes involved in their catabolism are essential in maintaining homeostasis. Indeed, increased hyaluronic acid degradation, catalysed by hyaluronidases, is observed in several pathophysiological processes [58]. The role of hyaluronidase in BPH is unknown, but there are reports linking the activity of this enzyme to the development and progression of prostate cancer [59,60]. It is also worth noting that hyaluronidases leading to increased tissue permeability are recognized as factors that spread inflammation.

Therefore, we examined the hyaluronidase inhibitory potential of compounds isolated from *S. intermedia* using the turbidimetric method and quercetin as a reference substance. The results are shown in Table 2. βSIT and UO were found to be practically inactive with IC_{50} > 1000 µg/mL, but UA and UAL were more potent than the reference (quercetin, IC_{50} = 517.05 µg/mL), with IC_{50} values 225.75 µg/mL and 369.77 µg/mL, respectively. AHUA exerted similar activity to quercetin. Interestingly, AUA and βSIT slightly inhibited the enzyme at lower concentrations.

Table 2. Anti-hyaluronidase activity of *S. intermedia* terpenoids.

Concentration (µg/mL)	Hyaluronidase Inhibition (%)						
	UA	UAL	AUA	AHUA	βSIT	UO	QUERCETIN
10	0.00 ± 0.00	NE	7.87 ± 0.95 *	0.00 ± 0.00	3.72 ± 0.78 *	0.00 ± 0.00	0.61 ± 0.53
25	2.52 ± 2.18	0.00 ± 0.00	11.70 ± 1.55 *	0.17 ± 0.30	12.09 ± 0.28 *	0.00 ± 0.00	1.17 ± 0.41
50	3.23 ± 1.71	1.62 ± 0.35	16.76 ± 4.36	0.35 ± 0.30	20.53 ± 0.75 *	0.00 ± 0.00	1.18 ± 0.95
100	11.65 ± 0.92 *	3.91 ± 0.56	19.27 ± 4.85	2.32 ± 1.88	23.32 ± 0.75 *	0.00 ± 0.00	4.36 ± 0.88
250	60.31 ± 0.88 *	15.45 ± 1.67	30.57 ± 0.68 *	18.54 ± 3.44	25.79 ± 0.45 *	0.52 ± 0.52	18.08 ± 1.00
500	82.38 ± 3.77 *	78.79 ± 1.63 *	45.63 ± 1.37 *	42.07 ± 1.72	27.78 ± 3.83	4.01 ± 1.14	38.84 ± 1.33
750	85.35 ± 2.30	NE	52.13 ± 4.94	93.41 ± 2.75	30.56 ± 3.89	13.71 ± 1.53	87.27 ± 1.16
1000	96.01 ± 3.54	96.25 ± 0.86	56.73 ± 4.26	99.22 ± 0.68 *	36.75 ± 4.61	35.13 ± 4.85	90.94 ± 1.72
IC_{50}	225.75	369.77	705.74	519.87	>Cmax	>Cmax	517.05

Hyaluronidase inhibition was measured using turbidimetric assay. Values are presented as the mean ± SD of three experiments. Statistical analysis was performed using one-way ANOVA and T3 Dunnett post hoc test with * $p < 0.05$ vs quercetin. Abbreviations: UA, ursolic acid; UAL, ursolic aldehyde; AUA, 3-*O*-β-acetoxy-ursolic acid; AHUA, 3-*O*-β-acetoxy-19α-hydroxy-ursolic acid; βSIT, β-sitosterol; UO, uvaol; NE, not examined.

To the best of our knowledge, this is the first report on the anti-hyaluronidase activity of UAL, AHUA, and UO. It was previously reported that βSIT inhibits hyaluronidase, with IC_{50} = 888.5 ± 44.9 μg/mL [61]. In another study, it suppressed the enzyme activity by approximately 32% at a concentration of 100 μg/mL [62]. These results are in agreement with ours. Meanwhile, the literature data on ursolic acid activity are divergent, ranging from 40–60% inhibition at a concentration of 1000 μg/mL [63,64] to about 50% suppression of the enzyme at a concentration range from 25 to 50 μg/mL [65]. Our results best match those obtained by Michel et al. (2017), who determined the IC_{50} value for this compound equals to 380.14 ± 10.92 μg/mL [66]. In contrast to our study, Abdullah et al. (2016) previously reported higher UA and AUA activities with IC_{50} values equal to 103.18 ± 1.70 μM and 136.92 ± 0.04 μM, respectively [67]. Such discrepancies in results clearly demand more in-depth studies.

2.4. Chemometric Analysis

Hierarchical agglomeration cluster analysis (CA) was used to investigate the similarity between the compounds tested and thus better understand and interpret their overall and multidirectional effects on the human BPH cellular model and inflammation. Results obtained for six compounds tested at two different concentrations, 10 and 20 μg/mL, respectively, were included, as only these concentrations gave results for all parameters. These parameters were as follows: antiproliferative effects after 24 h, 48 h, and 72 h of incubation; 5-α-reductase inhibitory activity; PSA secretion; effect on release of IL-6; TNF-α and NO in LPS-stimulated macrophages; anti-hyaluronidase activity; and inhibition of albumin denaturation. Exceptionally, for studies on LPS-stimulated macrophages, results obtained for UA at concentrations of 5 μg/mL and 10 μg/mL were analysed.

Cluster analysis showed specific similarity for two compounds (UO and AUA) at two concentrations (10 and 20 μg/mL) (Figure 5). The samples corresponding to these compounds at these concentrations formed two separate two-element clusters, with no other samples. A third such cluster occurred for compound AHUA but was below the Mojena's criterion. In the remaining cases, mixed clusters (containing samples of different compounds at different concentrations) were found. Overall, UO, βSIT and AHUA are distinctly distant from UA, UAL, and AUA regardless of the concentration at which they were used, indicating the differences between the effects that these compounds displayed.

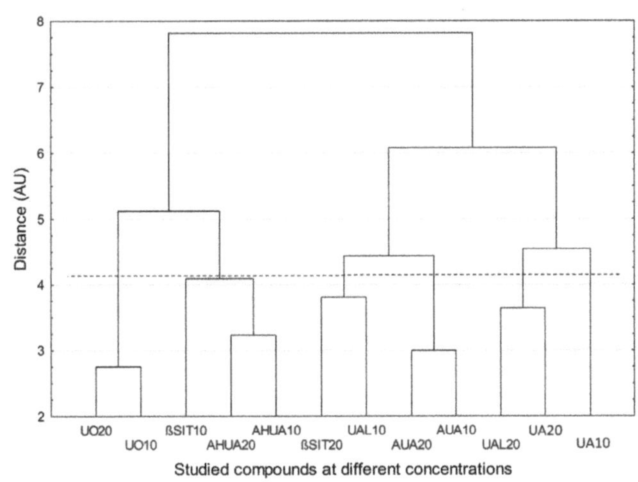

Figure 5. Dendrogram of similarity among investigated chemical compounds at different concentrations (function of the distance: Euclidean distance; method of grouping: Ward method; the added dashed horizontal line indicates that the grouping has been stopped according to the Mojena's rate).

Numbers indicate concentrations in μg/mL. Exceptionally, in case of studies on LPS-stimulated macrophages, the results obtained for UA at concentrations of 5 μg/mL and 10 μg/mL were included and assigned as UA10 and UA20, respectively. Abbreviations: UA, ursolic acid; UAL, ursolic aldehyde; AUA, 3-*O*-β-acetoxy-ursolic acid; AHUA, 3-*O*-β-acetoxy-19α-hydroxy-ursolic acid; βSIT, β-sitosterol; UO, uvaol.

3. Materials and Methods

3.1. Reagents and Instruments

Dulbecco's Modified Eagle's Medium F12 HAM (DMEM/F12), Dulbecco's Modified Eagle's Medium with 4500 mg/L glucose (DMEM high glucose), foetal bovine serum (FBS), phosphate-buffered saline (PBS), crystal violet, formaldehyde, lipopolysaccharide (LPS), DMSO, albumin from bovine serum (BSA): fraction V $\geq$ 98% (A3294), hyaluronidase from bovine testes type I-S, *Streptococcus equi* hyaluronic acid (HA), cetyltrimethylammonium bromide (CTAB), quercetin dihydrate, ursolic acid (other terpenoids were obtained by isolation), diclofenac sodium, dutasteride, and testosterone propionate were obtained from Sigma-Aldrich (Seelze, Germany). Acetate buffer pH 4.5 was purchased from J.T. Baker Chemical Co. (Phillipsburg, NJ, USA).

3.2. Cell Culture Conditions

Experiments were performed on human PNT2 prostate epithelial cells (ECACC 95012613, Merck, Darmstadt, Germany). For anti-inflammatory assay, murine RAW264.7 macrophages were used. The cells were cultured in a humidified atmosphere with 5% CO_2 at 37 °C in DMEM/F12 (PNT2) or DMEM high glucose (RAW 264.7) supplemented with 10% (FBS), 100 IU/mL penicillin, and 10 μg/mL streptomycin. The tested compounds were diluted in the culture media from freshly made stock solution (10 mg/mL in acetone) to the working concentrations.

3.3. Determination of PSA and 5-α-reductase

The experiment was performed according to Nakayama et al. (2021) [68]. Briefly, PNT2 cells were seeded onto 96 multi-well plates (1.5×10^5 cells/well) for 24 h and then treated with the tested compounds at the concentrations of 5, 10, 20, and 40 μg/mL for 72 h. Dutasteride (10 μM) was used as a reference drug. Cell culture supernatants were collected and used for quantitative analysis of PSA and 5-α-reductase level, which was performed using a Human ELISA kits, according to the manufacturer's protocol. The analyses were performed in triplicates, and the absorbance was measured using a microplate reader (SynergyTM HT—BioTek, Winooski, VT, USA). The results were determined as % of control.

3.4. Proliferation Assay

The cells were seeded onto 96-well plates (1×10^3 cells/well) and incubated for 24 h. Then, the medium was replaced with fresh medium containing 0.5 μM of testosterone propionate (T) in order to stimulate cell proliferation, as observed in prostate hyperplasia, and the tested compounds were added. Dutasteride (10 μM) was used as a reference drug. After 24, 48, and 72 h of incubation, the cell number was determined using crystal violet assay, as described previously [69]. Briefly, the cells were washed with PBS and fixed with 3.7% formaldehyde. Then, crystal violet solution was added for 10 min, followed by washing with PBS. Crystal violet was extracted from cells using 1.33% citric acid and 1.09% sodium citrate in water/methanol (1:1) solution. The absorbance was measured at 570 nm. The proliferation rate was determined as a % of control.

3.5. Inhibition of Albumin Denaturation

The protective effects of compounds against albumin heat-induced denaturation were determined as was described previously [70]. Tested compounds were dissolved in DMSO and tested at concentrations from 10 to 1000 μg/mL. Briefly, examined substances were preincubated in 25 °C for 15 min in the presence of BSA. Next, the reaction mixtures were incubated in 70 °C for 5 min for proteins denaturation. After cooling the samples, the turbidity was measured at 660 nm using microplate reader (SynergyTM HT—BioTek). Diclofenac sodium was used as reference drug. The product control solution was prepared to diminish the sample background, and the absorbance of the medium was performed as a blind control of experiment. All assays were conducted in triplicate. The percent of inhibition of protein denaturation was calculated as follows:

$$\% \text{ inhibition} = 100 - [(AS - APc)/(AC - AB)] \times 100$$

AS—absorbance of the tested substance;
APc—absorbance of the product control solution;
AC—absorbance in the absence of inhibitor;
AB—absorbance of blind control.

The half-maximal inhibitory concentration value IC_{50} was determined.

3.6. Determination of NO, IL-6 and TNF-α Release

Prior to the anti-inflammatory experiments, the toxicity of the tested compounds to RAW 264.7 macrophages was determined. The cells were seeded onto 96 multi-well plates (1.5×10^5 cells/well) and incubated with the tested compounds (0–100 μg/mL) for 24 h. Next, cell viability was tested with the MTT assay. All analyses were performed in triplicate, and the results are expressed as % of cell viability (mean ± SD). For further anti-inflammatory assay, the concentrations of 5 and 10 μg/mL for UA and 10 and 20 μg/mL for the other compounds were chosen as nontoxic.

For anti-inflammatory assays, RAW 264.7 cells were seeded onto 96 multi-well plates (1.5×10^5 cells/well) and pre-treated with the tested compounds for 1 h, followed by the addition of 10 ng/mL of LPS to induce inflammation process, as described previously [69]. Dexamethasone (0.5 μg/mL) was used as a reference drug. The incubation was continued for the next 24 h. Cell culture supernatants were used for further analysis. The nitric oxide level was determined using Griess Reagent Kit (Promega Corporation (Madison, Winooski, VT, USA), according to the manufacturer's protocol. The cytokine (TNF-α, IL-6) release level was performed using Human ELISA kits (Bioassay Technology Laboratory, Shanghai, China), according to the manufacturer's protocol. The analyses were performed in triplicates, and the absorbance was measured using a microplate reader (SynergyTM HT—BioTek). The results were determined as % of control.

3.7. Anti-Hyaluronidase Assay

Hyaluronidase inhibitory activity was evaluated on 96-well microplates using turbidimetric method as we described previously [70]. Compounds were dissolved in DMSO and tested at concentration range from 10 to1000 μg/mL. Briefly, substances were preincubated at 37 °C for 10 min with the presence of incubation buffer, enzyme, and acetate buffer (pH 4.5). Then, HA was added, and incubation continued for further 45 min. In the next step, a CTAB solution was added to precipitate undigested HA. The amount of undigested HA is proportional to the turbidity; thus, enzymatic activity was quantified spectrophotometrically at 600 nm using microplate reader (SynergyTM HT—BioTek). The absorbance in the presence of enzyme and substrate (control I) and in the absence of enzyme (control II) was measured. Product control solution, with HA instead of buffer, was prepared to deduct the sample background. The absorbance of the medium was performed as a blind control of the experiment. All experiments were conducted in triplicate. Quercetin was used as reference substance. The inhibition in percentage was calculated using the following formula:

$$\% \text{ inhibition} = \{[As - (APc - AB)] - AI\}/\{[AII - (APc - AB)] - AI\} \times 100$$

AI—absorbance of enzyme + substrate (control I);
AII—absorbance in the absence of enzyme (control II);
As—absorbance of sample solution;

APc—absorbance of the product control solution;
AB—absorbance of a blank control of experiment.

The half-maximal inhibitory concentration value IC_{50} was estimated.

3.8. Statistical Analysis

Comparison of means was carried out using IBM SPSS Statistics 29.0 for Windows. The data were analysed by one-way ANOVA followed by a T3 Dunnett post hoc test. The CA analysis was performed using Euclidean distance as a measure of distance between objects and Ward's method of grouping objects. The number of clusters was set according to the Mojena's rate. Prior to CA analysis, the data were standardized (z-transformed) to obtain zero mean and unit variance for each parameter. CA analysis was conducted by means of STATISTICA v.13.3. package (TIBCO Software Inc., Palo Alto, CA, USA). The same software was also used for the graphic representation of results.

4. Conclusions

In the current study, ursane-type triterpenoids and β-stitosterol isolated from the fruits of *Sorbus* were subjected to a series of in vitro experiments that were designed to assess their effects on various aspects of BPH pathogenesis and development. Not only did *Sorbus* compounds show antiandrogenic and antiproliferative effects on testosterone-treated PNT2 cells, reversing the stimulating effect of the hormone, but they also demonstrated an impact on inflammation, which is believed to play an important role in the development and progression of BPH and the clinical symptoms associated with BPH.

Considering the overall anti-BPH potential of *Sorbus* compounds, as shown by chemometric analysis, UA, UAL, and AUA formed a clearly distinct group from UO, βSIT, and AHUA, suggesting differences in the action of these compounds. Indeed, UA, UAL, and/or AUA tended to be among the most active compounds in the assays that were proposed and conducted in this study. Moreover, with the exception of βSIT, the examined *Sorbus* terpenoids were based on the ursane skeleton, i.e., UA and its derivatives. Thus, our study showed that chemical structure significantly influences activity. However, more targeted structure–activity studies are needed to better understand this relationship.

In conclusion, our study showed that *Sorbus* terpenoids exhibit antiandrogenic, antiproliferative, and anti-inflammatory properties and may represent an interesting target in the development of new anti-BPH therapies. Furthermore, the fruits of *S. intermedia* are edible and may be a valuable component of the daily diet.

Author Contributions: Conceptualization, A.S., A.G. and I.P.; methodology, A.G., K.G., P.P. and P.Z.; software, A.S. and P.Z.; formal analysis, A.S., K.G., P.P. and P.Z.; investigation, A.G., K.G. and P.P.; writing—original draft preparation, A.S.; writing—review and editing, A.S., A.G., K.G., P.P., P.Z. and I.P.; visualization, A.S.; supervision, I.P. All authors have read and agreed to the published version of the manuscript.

Funding: This research received no external funding.

Institutional Review Board Statement: Not applicable.

Informed Consent Statement: Not applicable.

Data Availability Statement: The data are contained within the article.

Acknowledgments: The study was created with the use of equipment (Biotek Synergy microplate reader) co-financed by the qLIFE Priority Research Area under the program "Excellence Initiative—Research University" (No. 06/IDUB/2019/94) at Jagiellonian University.

Conflicts of Interest: The authors declare no conflict of interest.

References

1. Devlin, C.M.; Simms, M.S.; Maitland, N.J. Benign prostatic hyperplasia—What do we know? *BJU Int.* **2020**, *127*, 389–399. [CrossRef]
2. Csikós, E.; Horváth, A.; Ács, K.; Papp, N.; Balázs, V.L.; Dolenc, M.S.; Kenda, M.; Glavač, N.K.; Nagy, M.; Protti, M.; et al. Treatment of Benign Prostatic Hyperplasia by Natural Drugs. *Molecules* **2021**, *26*, 7141. [CrossRef] [PubMed]
3. Allkanjari, O.; Vitalone, A. What do we know about phytotherapy of benign prostatic hyperplasia? *Life Sci.* **2015**, *126*, 42–56. [CrossRef]
4. La Vignera, S.; Condorelli, R.A.; Russo, G.I.; Morgia, G.; Calogero, A.E. Endocrine control of benign prostatic hyperplasia. *Andrology* **2016**, *4*, 404–411. [CrossRef] [PubMed]
5. Chughtai, B.; Forde, J.C.; Thomas, D.D.M.; Laor, L.; Hossack, T.; Woo, H.H.; Te, A.E.; Kaplan, S.A. Benign prostatic hyperplasia. *Nat. Rev. Dis. Prim.* **2016**, *2*, 16031. [CrossRef] [PubMed]
6. Tsunemori, H.; Sugimoto, M. Effects of inflammatory prostatitis on the development and progression of benign prostatic hyperplasia: A literature review. *Int. J. Urol.* **2021**, *28*, 1086–1092. [CrossRef]
7. Cao, D.; Sun, R.; Peng, L.; Li, J.; Huang, Y.; Chen, Z.; Chen, B.; Li, J.; Ai, J.; Yang, L.; et al. Immune Cell Proinflammatory Microenvironment and Androgen-Related Metabolic Regulation During Benign Prostatic Hyperplasia in Aging. *Front. Immunol.* **2022**, *13*, 842008. [CrossRef] [PubMed]
8. Madersbacher, S.; Alivizatos, G.; Nordling, J.; Sanz, C.R.; Emberton, M.; de la Rosette, J.J. EAU 2004 Guidelines on Assessment, Therapy and Follow-Up of Men with Lower Urinary Tract Symptoms Suggestive of Benign Prostatic Obstruction (BPH Guidelines). *Eur. Urol.* **2004**, *46*, 547–554. [CrossRef]
9. De La Rosette, J.J.; Alivizatos, G.; Madersbacher, S.; Perachino, M.; Thomas, D.; Desgrandchamps, F.; De Wildt, M. EAU Guidelines on Benign Prostatic Hyperplasia (BPH). *Eur. Urol.* **2001**, *40*, 256–263. [CrossRef] [PubMed]
10. Lee, S.W.H.; Chan, E.M.C.; Lai, Y.K. The global burden of lower urinary tract symptoms suggestive of benign prostatic hyperplasia: A systematic review and meta-analysis. *Sci. Rep.* **2017**, *7*, 7984. [CrossRef] [PubMed]
11. GBD 2017 Disease and Injury Incidence and Prevalence Collaborators. Global, regional, and national incidence, prevalence, and years lived with disability for 354 diseases and injuries for 195 countries and territories, 1990–2017: A systematic analysis for the Global Burden of Disease Study 2017. *Lancet* **2018**, *392*, 1789–1858. [CrossRef]
12. Shin, I.-S.; Lee, M.-Y.; Jung, D.-Y.; Seo, C.-S.; Ha, H.-K.; Shin, H.-K. Ursolic acid reduces prostate size and dihydrotestosterone level in a rat model of benign prostatic hyperplasia. *Food Chem. Toxicol.* **2012**, *50*, 884–888. [CrossRef] [PubMed]
13. Liu, J.; Tamura, S.; Kurashiki, K.; Shimizu, K.; Noda, K.; Konishi, F.; Kumamoto, S.; Kondo, R. Anti-Androgen Effects of Extracts and Compounds from *Ganoderma lucidum*. *Chem. Biodivers.* **2009**, *6*, 231–243. [CrossRef] [PubMed]
14. Liu, J.; Kurashiki, K.; Shimizu, K.; Kondo, R. Structure–activity relationship for inhibition of 5α-reductase by triterpenoids isolated from Ganoderma lucidum. *Bioorganic Med. Chem.* **2006**, *14*, 8654–8660. [CrossRef] [PubMed]
15. Nizomov, S.A.; Sorokina, I.V.; Zhukova, N.A.; Borisov, S.A.; Tolstikova, T.G.; Semenov, D.E.; Bakarev, M.A. Prostatotropic Action of Glycyrrhizic Acid Disodium Salt in Benign Prostatic Hyperplasia Models. *Bull. Exp. Biol. Med.* **2020**, *169*, 114–118. [CrossRef]
16. Smith, D.K.; Dds, S.L.H.; Wang, J.; Kallifatidis, G.; Morera, D.S.; Jordan, A.R.; Terris, M.K.; Klaassen, Z.; Bollag, R.; Lokeshwar, V.B.; et al. Promotion of epithelial hyperplasia by interleukin-8—CXCR axis in human prostate. *Prostate* **2020**, *80*, 938–949. [CrossRef]
17. Cheon, S.-Y.; Jin, B.-R.; Kim, H.-J.; An, H.-J. Oleanolic Acid Ameliorates Benign Prostatic Hyperplasia by Regulating PCNA-Dependent Cell Cycle Progression *In Vivo* and *In Vitro*. *J. Nat. Prod.* **2020**, *83*, 1183–1189. [CrossRef]
18. Sołtys, A.; Galanty, A.; Zagrodzki, P.; Grabowska, K.; Malarz, J.; Podolak, I. Sorbus intermedia (EHRH.) PERS. fruits as a novel source of biologically active triterpenoids—Comparative studies of ursolic acid derivatives with cytotoxic potential. *Biomed. Pharmacother.* **2022**, *154*, 113592. [CrossRef] [PubMed]
19. Sołtys, A.; Galanty, A.; Zagrodzki, P.; Podolak, I. Relationship between Maturity Stage, Triterpenoid Content and Cytotoxicity of *Sorbus intermedia* (EHRH.) PERS. Fruits—A Chemometric Approach. *Chem. Biodivers.* **2021**, *18*, e2100552. [CrossRef] [PubMed]
20. Roehrborn, C.G. Pathology of benign prostatic hyperplasia. *Int. J. Impot. Res.* **2008**, *20*, S11–S18. [CrossRef] [PubMed]
21. Love, H.D.; Booton, S.E.; Boone, B.E.; Breyer, J.P.; Koyama, T.; Revelo, M.P.; Shappell, S.B.; Smith, J.R.; Hayward, S.W. Androgen Regulated Genes in Human Prostate Xenografts in Mice: Relation to BPH and Prostate Cancer. *PLoS ONE* **2009**, *4*, e8384. [CrossRef] [PubMed]
22. Wilt, T.J.; Ishani, A.; MacDonald, R.; Stark, G.; Mulrow, C.D.; Lau, J. Beta-sitosterols for benign prostatic hyperplasia. *Cochrane Database Syst. Rev.* **1999**, *2011*, CD001043. [CrossRef] [PubMed]
23. Wilt, T.J.; Macdonald, R.; Ishani, A. β-sitosterol for the treatment of benign prostatic hyperplasia. *BJU Int.* **1999**, *83*, 976–983. [CrossRef] [PubMed]
24. Cabezal, M.; Bratoeff, E.; Heuze, I.; Ramírez, E.; Sánchez, M.; Flores, E. Effect of Beta-Sitosterol as Inhibitor of 5 Al-pha-Reductase in Hamster Prostate. *Proc. West. Pharmacol. Soc.* **2003**, *46*, 153–155.
25. Cicero, A.F.; Allkanjari, O.; Busetto, G.M.; Cai, T.; Larganà, G.; Magri, V.; Perletti, G.; Della Cuna, F.S.R.; Russo, G.I.; Stamatiou, K.; et al. Nutraceutical treatment and prevention of benign prostatic hyperplasia and prostate cancer. *Arch. Ital. Urol. Androl.* **2019**, *91*, 139–152. [CrossRef]
26. Azizi, A.; Mumin, N.H.; Shafqat, N. Phytochemicals with Anti 5-alpha-reductase Activity: A Prospective for Prostate Cancer Treatment. *F1000Research* **2021**, *10*, 221. [CrossRef]

27. Sudeep, H.V.; Thomas, J.V.; Shyamprasad, K. A double blind, placebo-controlled randomized comparative study on the efficacy of phytosterol-enriched and conventional saw palmetto oil in mitigating benign prostate hyperplasia and androgen deficiency. *BMC Urol.* **2020**, *20*, 86. [CrossRef]
28. Kassi, E.; Papoutsi, Z.; Pratsinis, H.; Aligiannis, N.; Manoussakis, M.; Moutsatsou, P. Ursolic acid, a naturally occurring triterpenoid, demonstrates anticancer activity on human prostate cancer cells. *J. Cancer Res. Clin. Oncol.* **2007**, *133*, 493–500. [CrossRef]
29. Mu, D.; Zhou, G.; Li, J.; Su, B.; Guo, H. Ursolic acid activates the apoptosis of prostate cancer via ROCK/PTEN mediated mitochondrial translocation of cofilin-1. *Oncol. Lett.* **2017**, *15*, 3202–3206. [CrossRef]
30. Choi, Y.H.; Baek, J.H.; Yoo, M.A.; Chung, H.Y.; Kim, N.D.; Kim, K.W. Induction of apoptosis by ursolic acid through activation of caspases and down-regulation of c-IAPs in human prostate epithelial cells. *Int. J. Oncol.* **2000**, *17*, 565–571. [CrossRef]
31. Shin, S.W.; Kim, S.Y.; Park, J.-W. Autophagy inhibition enhances ursolic acid-induced apoptosis in PC3 cells. *Biochim. Biophys. Acta (BBA)—Mol. Cell Res.* **2012**, *1823*, 451–457. [CrossRef]
32. Harmand, P.-O.; Duval, R.; Liagre, B.; Jayat-Vignoles, C.; Beneytout, J.-L.; Delage, C.; Simon, A. Ursolic acid induces apoptosis through caspase-3 activation and cell cycle arrest in HaCat cells. *Int. J. Oncol.* **2003**, *23*, 105–112. [CrossRef]
33. Es-Saady, D.; Simon, A.; Ollier, M.; Maurizis, J.C.; Chulia, A.J.; Delage, C. Inhibitory effect of ursolic acid on B16 proliferation through cell cycle arrest. *Cancer Lett.* **1996**, *106*, 193–197. [CrossRef]
34. Weng, H.; Tan, Z.-J.; Hu, Y.-P.; Shu, Y.-J.; Bao, R.-F.; Jiang, L.; Wu, X.-S.; Li, M.-L.; Ding, Q.; Wang, X.-A.; et al. Ursolic acid induces cell cycle arrest and apoptosis of gallbladder carcinoma cells. *Cancer Cell Int.* **2014**, *14*, 96. [CrossRef] [PubMed]
35. Zhao, H.; Tang, S.; Tao, Q.; Ming, T.; Lei, J.; Liang, Y.; Peng, Y.; Wang, M.; Liu, M.; Yang, H.; et al. Ursolic Acid Suppresses Colorectal Cancer by Down-Regulation of Wnt/β-Catenin Signaling Pathway Activity. *J. Agric. Food Chem.* **2023**, *71*, 3981–3993. [CrossRef] [PubMed]
36. Bao, X.; Zhang, Y.; Zhang, H.; Xia, L. Molecular Mechanism of β-Sitosterol and its Derivatives in Tumor Progression. *Front. Oncol.* **2022**, *12*, 926975. [CrossRef] [PubMed]
37. Awad, A.B.; Burr, A.T.; Fink, C.S. Effect of resveratrol and β-sitosterol in combination on reactive oxygen species and prostaglandin release by PC-3 cells. *Prostaglandins Leukot. Essent. Fat. Acids* **2005**, *72*, 219–226. [CrossRef] [PubMed]
38. AlQathama, A.; Shao, L.; Bader, A.; Khondkar, P.; Gibbons, S.; Prieto, J.M. Differential Anti-Proliferative and Anti-Migratory Activities of Ursolic Acid, 3-*O*-Acetylursolic Acid and Their Combination Treatments with Quercetin on Melanoma Cells. *Biomolecules* **2020**, *10*, 894. [CrossRef]
39. Bonel-Pérez, G.C.; Pérez-Jiménez, A.; Gris-Cárdenas, I.; Parra-Pérez, A.M.; Lupiáñez, J.A.; Reyes-Zurita, F.J.; Siles, E.; Csuk, R.; Peragón, J.; Rufino-Palomares, E.E. Antiproliferative and Pro-Apoptotic Effect of Uvaol in Human Hepatocarcinoma HepG2 Cells by Affecting G0/G1 Cell Cycle Arrest, ROS Production and AKT/PI3K Signaling Pathway. *Molecules* **2020**, *25*, 4254. [CrossRef]
40. Allouche, Y.; Warleta, F.; Campos, M.; Sánchez-Quesada, C.; Uceda, M.; Beltrán, G.; Gaforio, J.J. Antioxidant, Antiproliferative, and Pro-apoptotic Capacities of Pentacyclic Triterpenes Found in the Skin of Olives on MCF-7 Human Breast Cancer Cells and Their Effects on DNA Damage. *J. Agric. Food Chem.* **2011**, *59*, 121–130. [CrossRef]
41. Amico, V.; Barresi, V.; Condorelli, D.; Spatafora, C.; Tringali, C. Antiproliferative Terpenoids from Almond Hulls (*Prunus dulcis*): Identification and Structure−Activity Relationships. *J. Agric. Food Chem.* **2006**, *54*, 810–814. [CrossRef]
42. Gandaglia, G.; Briganti, A.; Gontero, P.; Mondaini, N.; Novara, G.; Salonia, A.; Sciarra, A.; Montorsi, F. The role of chronic prostatic inflammation in the pathogenesis and progression of benign prostatic hyperplasia (BPH). *BJU Int.* **2013**, *112*, 432–441. [CrossRef]
43. Robert, G.; Descazeaud, A.; Nicolaïew, N.; Terry, S.; Sirab, N.; Vacherot, F.; Maillé, P.; Allory, Y.; de la Taille, A. Inflammation in benign prostatic hyperplasia: A 282 patients' immunohistochemical analysis. *Prostate* **2009**, *69*, 1774–1780. [CrossRef]
44. Vickman, R.E.; Aaron-Brooks, L.; Zhang, R.; Lanman, N.A.; Lapin, B.; Gil, V.; Greenberg, M.; Sasaki, T.; Cresswell, G.M.; Broman, M.M.; et al. TNF is a potential therapeutic target to suppress prostatic inflammation and hyperplasia in autoimmune disease. *Nat. Commun.* **2022**, *13*, 2133. [CrossRef]
45. Vickman, R.E.; Franco, O.E.; Hayward, S.W. Could TNF-antagonists be a novel treatment strategy for BPH patients? *Cell Stress* **2022**, *6*, 65–67. [CrossRef]
46. Zhou, R.-Y.; Tong, Y.; Guo, Y.-J.; Zhang, Q.; Bi, H.-X.; Kai, K. Combined treatment with dihydrotestosterone and lipopolysaccharide modulates prostate homeostasis by upregulating TNF-α from M1 macrophages and promotes proliferation of prostate stromal cells. *Asian J. Androl.* **2022**, *24*, 513–520. [CrossRef]
47. Du, S.-Y.; Huang, H.-F.; Li, X.-Q.; Zhai, L.-X.; Zhu, Q.-C.; Zheng, K.; Song, X.; Xu, C.-S.; Li, C.-Y.; Li, Y.; et al. Anti-inflammatory properties of uvaol on DSS-induced colitis and LPS-stimulated macrophages. *Chin. Med.* **2020**, *15*, 43. [CrossRef]
48. Wang, J.; Jin, M.; Jin, C.; Ye, C.; Zhou, Y.; Wang, R.; Cui, H.; Zhou, W.; Li, G. A new pentacyclic triterpenoid from the leaves of *Rhododendron dauricum* L. with inhibition of NO production in LPS-induced RAW 264.7 cells. *Nat. Prod. Res.* **2019**, *34*, 3313–3319. [CrossRef]
49. Yang, Z.-G.; Matsuzaki, K.; Takamatsu, S.; Kitanaka, S. Inhibitory Effects of Constituents from Morus alba var. multicaulis on Differentiation of 3T3-L1 Cells and Nitric Oxide Production in RAW264.7 Cells. *Molecules* **2011**, *16*, 6010–6022. [CrossRef]
50. Jiang, X.; Shen, P.; Zhou, J.; Ge, H.; Raj, R.; Wang, W.; Yu, B.; Zhang, J. Microbial transformation and inhibitory effect assessment of uvaol derivates against LPS and HMGB1 induced NO production in RAW264.7 macrophages. *Bioorganic Med. Chem. Lett.* **2022**, *58*, 128523. [CrossRef]

51. Hu, T.; He, X.-W.; Jiang, J.-G. Functional Analyses on Antioxidant, Anti-inflammatory, and Antiproliferative Effects of Extracts and Compounds from *Ilex latifolia* Thunb., a Chinese Bitter Tea. *J. Agric. Food Chem.* **2014**, *62*, 8608–8615. [CrossRef]
52. Kim, M.-H.; Kim, J.N.; Han, S.N.; Kim, H.-K. Ursolic acid isolated from guava leaves inhibits inflammatory mediators and reactive oxygen species in LPS-stimulated macrophages. *Immunopharmacol. Immunotoxicol.* **2015**, *37*, 228–235. [CrossRef] [PubMed]
53. Zhou, J.-X.; Wink, M. Evidence for Anti-Inflammatory Activity of Isoliquiritigenin, 18β Glycyrrhetinic Acid, Ursolic Acid, and the Traditional Chinese Medicine Plants Glycyrrhiza glabra and Eriobotrya japonica, at the Molecular Level. *Medicines* **2019**, *6*, 55. [CrossRef] [PubMed]
54. Lee, J.Y.; Choi, J.K.; Jeong, N.-H.; Yoo, J.; Ha, Y.S.; Lee, B.; Choi, H.; Park, P.-H.; Shin, T.-Y.; Kwon, T.K.; et al. Anti-inflammatory effects of ursolic acid-3-acetate on human synovial fibroblasts and a murine model of rheumatoid arthritis. *Int. Immunopharmacol.* **2017**, *49*, 118–125. [CrossRef]
55. Ding, Y.; Nguyen, H.T.; Kim, S.I.; Kim, H.W.; Kim, Y.H. The regulation of inflammatory cytokine secretion in macrophage cell line by the chemical constituents of Rhus sylvestris. *Bioorganic Med. Chem. Lett.* **2009**, *19*, 3607–3610. [CrossRef]
56. Theocharis, A.D.; Skandalis, S.S.; Gialeli, C.; Karamanos, N.K. Extracellular matrix structure. *Adv. Drug Deliv. Rev.* **2016**, *97*, 4–27. [CrossRef]
57. Karamanos, N.K.; Theocharis, A.D.; Piperigkou, Z.; Manou, D.; Passi, A.; Skandalis, S.S.; Vynios, D.H.; Orian-Rousseau, V.; Ricard-Blum, S.; Schmelzer, C.E.; et al. A guide to the composition and functions of the extracellular matrix. *FEBS J.* **2021**, *288*, 6850–6912. [CrossRef]
58. Girish, K.S.; Kemparaju, K.; Nagaraju, S.; Vishwanath, B.S. Hyaluronidase Inhibitors: A Biological and Therapeutic Perspective. *Curr. Med. Chem.* **2009**, *16*, 2261–2288. [CrossRef]
59. Kovar, J.L.; Johnson, M.A.; Volcheck, W.M.; Chen, J.; Simpson, M.A. Hyaluronidase Expression Induces Prostate Tumor Metastasis in an Orthotopic Mouse Model. *Am. J. Pathol.* **2006**, *169*, 1415–1426. [CrossRef]
60. Benitez, A.; Yates, T.J.; Lopez, L.E.; Cerwinka, W.H.; Bakkar, A.; Lokeshwar, V.B. Targeting Hyaluronidase for Cancer Therapy: Antitumor Activity of Sulfated Hyaluronic Acid in Prostate Cancer Cells. *Cancer Res.* **2011**, *71*, 4085–4095. [CrossRef] [PubMed]
61. Khammee, T.; Rujitanapanich, S.; Chunhakant, S.; Jaratrungtawee, A.; Kuno, M. In Vitro and in Silico Evaluations of Chemical Constituents from the Rhizomes of Aglaonema simplex (Blume) Blume as Hyaluronidase Inhibitor. *Thai J. Sci. Technol.* **2020**, *9*, 269–277. [CrossRef]
62. Süntar, I.; Akkol, E.K.; Keles, H.; Yesilada, E.; Sarker, S.D.; Baykal, T. Comparative evaluation of traditional prescriptions from Cichorium intybus L. for wound healing: Stepwise isolation of an active component by in vivo bioassay and its mode of activity. *J. Ethnopharmacol.* **2012**, *143*, 299–309. [CrossRef] [PubMed]
63. Chaiyana, W.; Anuchapreeda, S.; Punyoyai, C.; Neimkhum, W.; Lee, K.-H.; Lin, W.-C.; Lue, S.-C.; Viernstein, H.; Mueller, M. Ocimum sanctum Linn. as a natural source of skin anti-ageing compounds. *Ind. Crops Prod.* **2018**, *127*, 217–224. [CrossRef]
64. Neimkhum, W.; Anuchapreeda, S.; Lin, W.-C.; Lue, S.-C.; Lee, K.-H.; Chaiyana, W. Effects of *Carissa carandas* Linn. Fruit, Pulp, Leaf, and Seed on Oxidation, Inflammation, Tyrosinase, Matrix Metalloproteinase, Elastase, and Hyaluronidase Inhibition. *Antioxidants* **2021**, *10*, 1345. [CrossRef]
65. Nema, N.K.; Maity, N.; Sarkar, B.K.; Mukherjee, P.K. Matrix metalloproteinase, hyaluronidase and elastase inhibitory potential of standardized extract of *Centella asiatica*. *Pharm. Biol.* **2013**, *51*, 1182–1187. [CrossRef]
66. Michel, P.; Owczarek, A.; Matczak, M.; Kosno, M.; Szymański, P.; Mikiciuk-Olasik, E.; Kilanowicz, A.; Wesołowski, W.; Olszewska, M.A. Metabolite Profiling of Eastern Teaberry (*Gaultheria procumbens* L.) Lipophilic Leaf Extracts with Hyaluronidase and Lipoxygenase Inhibitory Activity. *Molecules* **2017**, *22*, 412. [CrossRef] [PubMed]
67. Abdullah, N.H.; Thomas, N.F.; Sivasothy, Y.; Lee, V.S.; Liew, S.Y.; Noorbatcha, I.A.; Awang, K. Hyaluronidase Inhibitory Activity of Pentacylic Triterpenoids from *Prismatomeris tetrandra* (Roxb.) K. Schum: Isolation, Synthesis and QSAR Study. *Int. J. Mol. Sci.* **2016**, *17*, 143. [CrossRef]
68. Nakayama, A.; Ide, H.; Lu, Y.; Takei, A.; Fukuda, K.; Osaka, A.; Arai, G.; Horie, S.; Okada, H.; Saito, K. Effects of Curcumin Combined With the 5-alpha Reductase Inhibitor Dutasteride on LNCaP Prostate Cancer Cells. *In Vivo* **2021**, *35*, 1443–1450. [CrossRef]
69. Galanty, A.; Zagrodzki, P.; Gdula-Argasińska, J.; Grabowska, K.; Koczurkiewicz-Adamczyk, P.; Wróbel-Biedrawa, D.; Podolak, I.; Pękala, E.; Paśko, P. A Comparative Survey of Anti-Melanoma and Anti-Inflammatory Potential of Usnic Acid Enantiomers—A Comprehensive In Vitro Approach. *Pharmaceuticals* **2021**, *14*, 945. [CrossRef]
70. Grabowska, K.; Wróbel, D.; Żmudzki, P.; Podolak, I. Anti-inflammatory activity of saponins from roots of *Impatiens parviflora* DC. *Nat. Prod. Res.* **2020**, *34*, 1581–1585. [CrossRef]

pharmaceuticals

MDPI

Article

New Approaches on the Anti-Inflammatory and Cardioprotective Properties of *Taraxacum officinale* Tincture

Alexandra Epure [1,†], Alina E. Pârvu [2,†], Laurian Vlase [3,*], Daniela Benedec [1], Daniela Hanganu [1], Ovidiu Oniga [4], Ana-Maria Vlase [5], Irina Ielciu [5], Anca Toiu [1] and Ilioara Oniga [1]

1 Department of Pharmacognosy, Faculty of Pharmacy, "Iuliu Hațieganu" University of Medicine and Pharmacy, 8 V. Babeș Street, 400012 Cluj-Napoca, Romania
2 Department of Physiopathology, Faculty of Medicine, "Iuliu Hațieganu" University of Medicine and Pharmacy, 8 V. Babeș Street, 400012 Cluj-Napoca, Romania
3 Department of Pharmaceutical Technology and Biopharmacy, "Iuliu Hațieganu" University of Medicine and Pharmacy, 8 V. Babeș Street, 400012 Cluj-Napoca, Romania
4 Department of Pharmaceutical Chemistry, Faculty of Pharmacy, "Iuliu Hațieganu" University of Medicine and Pharmacy, 8 V. Babeș Street, 400012 Cluj-Napoca, Romania
5 Department of Pharmaceutical Botany, Faculty of Pharmacy, "Iuliu Hațieganu" University of Medicine and Pharmacy, 23 Gheorghe Marinescu Street, 400337 Cluj-Napoca, Romania
* Correspondence: laurian.vlase@umfcluj.ro
† These authors contributed equally to this work.

Abstract: The present research investigated the in vivo anti-inflammatory and cardioprotective activities, as well as the antioxidant potential of *Taraxacum officinale* tincture (TOT), in relation to the polyphenolic composition. Chromatographic and spectrophotometric techniques were used to determine the polyphenolic profile of TOT and the antioxidant activity was preliminarily assessed in vitro by DPPH• and FRAP spectrophotometric methods. The in vivo anti-inflammatory and cardioprotective activities were studied in rat turpentine-induced inflammation and in rat isoprenaline-induced myocardial infarction (MI) models. The main polyphenolic compound identified in TOT was cichoric acid. The oxidative stress determinations showed the capacity of the dandelion tincture not only to decrease the total oxidative stress (TOS), the oxidative stress index (OSI), and the total antioxidant capacity (TAC), but also the malondialdehide (MDA), thiols (SH), and nitrites/nitrates (NOx) levels both in inflammation and MI models. In addition, aspartate aminotransferase (AST), alanine aminotransferase (ALT), creatin kinase-MB (CK-MB), and nuclear factor kappa B (NF-κB) parameters were decreased by the administration of the tincture. The results show that *T. officinale* could be considered a valuable source of natural compounds with important benefits in pathologies linked to oxidative stress.

Keywords: *Taraxacum officinale*; cardioprotective; anti-inflammatory; antioxidant; polyphenols; cichoric acid

check for updates

Citation: Epure, A.; Pârvu, A.E.; Vlase, L.; Benedec, D.; Hanganu, D.; Oniga, O.; Vlase, A.-M.; Ielciu, I.; Toiu, A.; Oniga, I. New Approaches on the Anti-Inflammatory and Cardioprotective Properties of *Taraxacum officinale* Tincture. *Pharmaceuticals* **2023**, *16*, 358. https://doi.org/10.3390/ph16030358

Academic Editor: Diana Roxana Pelinescu

Received: 19 January 2023
Revised: 22 February 2023
Accepted: 23 February 2023
Published: 26 February 2023

1. Introduction

Cardiovascular diseases are the main cause of global disability, and health predictions show that they will also be the most prominent cause of death in 2030. CVDs include myocardial infarction (MI), congestive heart failure, coronary heart disease angina, and peripheral arterial disease [1].

MI can be mediated via several biochemical mechanisms, such as reactive oxygen species (ROS), defective antioxidant enzymes, oxidative stress, and inflammatory process. In pathologic conditions, the disproportion between ROS and antioxidants promotes myocardial cell damage, necrosis, and apoptosis [2]. Therefore, it is necessary to improve the myocardial redox status by protecting the antioxidants and stabilizing the oxidants. Several factors are crucial in the progression of MI and reperfusion injury: oxidative stress, induction of inflammation, inflammatory cell infiltration, and activation of adaptive immune response. Therefore, using agents that can reduce their pathological mechanisms during MI and reperfusion injury by regulating these pathological mediators is a priority [3].

Taraxacum officinale (L.) Weber ex F.H.Wigg. (dandelion) is a perennial herbaceous flowering species of the Asteraceae family [4]. The chemical composition consists of phenolic compounds (polyphenolic acids, flavonoids, coumarins, tannins), sesquiterpene lactones (taraxacin, lactucopicrin, and cichorin, found mostly in the roots), triterpenes (α-amyrin, β-amyrin, lupeol, taraxol, taraxasterol), sterols (stigmasterol, β-sitosterol), polysaccharides (especially inulin in roots), minerals, amino acids, and vitamins that can be found in all the organs of the species [5]. Several hydroxybenzoic acids (protocatechuic, vanillic, syringic and gallic acids), hydroxycinnamic acids (*p*-coumaric, caffeic, ferulic, and synapic acids), and derivatives (chlorogenic, caftaric, cichoric acids) were identified in *T. officinale* [6], and cichoric acid was the main compound found in the aerial parts [7]. Dandelion also contains flavonoids, such as quercetin, kaempferol, apigenin, luteolin, catechins, hyperoside, isoquercitrin, quercitrin, rutin [8]. *T. officinale* has important pharmacological effects: antioxidant (aqueous extracts of *Taraxaci herba sin flos, T. radix, T. flos*), antihyperglycemic (ethanolic extracts *T. radix*), cholagogue (methanolic extracts *T. folium*), diuretic (aqueous extracts *T. herba, T. radix*), anti-inflammatory (methanolic etracts, *T. flos*), immunomodulatory and anti-allergic (isolated compounds from *T. officinale*), anti-thrombotic (ethanolic extracts of *T. radix*) and prebiotic (aqueous extracts of *T. radix*) [4].

The in vivo hypolipidemic, anti-obesity, and hepatoprotective properties of *T. officinale* extracts are linked to the antioxidant activity, due to the rich polyphenolic composition that may correct oxidative stress, with positive outcomes in numerous diseases (chronic inflammations, neurodegenerative disorders and metabolic syndrome) [2]. Additionally, the effects of the polyphenols on the cardiovascular system are based on mechanisms such as antihypertensive, anti-atherosclerotic, and anti-inflammatory effects; improving the lipid profile; and a direct effect on endothelial cells [9].

Several studies suggested that *T. officinale* presented valuable anti-inflammatory potential [10] and various in vivo studies evaluated the hypolipidemic properties of lowering hypertension and decreasing lipid peroxidation [11]. The aim of the present study is the in vivo evaluation of anti-inflammatory and cardioprotective effects of *T. officinale* tincture (TOT) based on the antioxidant mechanism of the contained compounds. The novelty of the present study consists of the fact that, to the best of our knowledge, it represents the first report on the TOT cardioprotective activity evaluated in vivo on an isoprenaline-induced myocardial infarction model. Moreover, the present study aims to highlight the in vivo anti-inflammatory activity of TOT.

2. Results and Discussion

2.1. Total Polyphenolic Content (TPC), Total Flavonoidic Content (TFC), and Total Caffeic Acid Derivatives Content (TCADC)

The results obtained by the spectrophotometric analysis of polyphenols from TOT are presented in Table 1.

Table 1. The polyphenols content of TOT.

Extract	TPC (mg GAE/g d.w.)	TFC (mg RE/g d.w.)	TCADC (mg CAE/g d.w.)
TOT	26.75 ± 0.73	6.28 ± 0.32	16.74 ± 0.80

Note: Values are expressed as mean of 3 determinations ± SD. GAE—gallic acid equivalents, RE—rutin equivalents, CAE—caffeic acid derivates.

The TPC in TOT was comparable with some of the data reported by other authors (18.53 mg GAE/g d.w., for 20% ethanolic extract; 33.53 mg GAE/g d.w., for 40% ethanolic extract; 33.90 mg GAE/g d.w., for aqueous ethanol extract 1:1) [12]. Other researchers reported different values for TPC from *T. herba*, depending on the extraction conditions and solvents: 123.42 mg GAE/g d.w. (for 60% ethanolic extract); 70.46 mg GAE/g d.w. (for 80% ethanolic extract); 41.47–691.6 mg GAE/g d.w. for aqueous and hydroalcoholic extracts from Pakistan; 33.94 mg GAE/g d.w. (for 1:1 aqueous ethanolic extract) and 23.27 mg GAE/g d.w. (for 80% ethanolic extract, plants harvested from USA) [12].

Regarding the TFC, TOT was characterised by lower values than those reported in USA as 14.00 mg RE/g d.w. for aqueous ethanolic extracts 1:1 and 12.35 mg RE/g d.w. for 80% ethanol extracts, and also by those reported for dandelion aerial parts gathered in Malaysia with values ranging from 12.82 to 55.81 mg RE/g d.w for 20 to 80% ethanol extracts [12,13]. Differences could be determined by the particularities of the analytical method and the conditions of obtaining the raw material and extracts.

To the best of our knowledge, the TCADC content in TOT was not reported before and it was evaluated in this study for the first time.

Our experiments showed that *T. officinale* harvested from Romania is a raw material rich in polyphenols, with important content of phenol acids derivatives and flavonoids.

2.2. HPLC-UV-MS Analysis

In order to characterise the TOT, HPLC-UV-MS analyses were conducted in several stages, aiming to identify and quantify polyphenolic compounds. In the first stage, phenolic acids with highly hydrophilic character were determined, then other polyphenolic compounds, and in the end, cichoric acid was identified and quantified (Table 2).

Table 2. Phenolic compounds identified in TOT by HPLC-UV-MS.

Polyphenols	$[M-H]^-$	Retention Time (min) Rt ± SD	TOT (μg/g d.w.)
Protocatechuic acid	153	2.80 ± 0.05	9.20 ± 0.09
Vanillic acid	167	6.70 ± 0.07	2.00 ± 0.02
Syringic acid	197	8.40 ± 0.09	0.90 ± 0.01
Ferulic acid	193	12.80 ± 0.10	53.60 ± 0.37
Cichoric acid	473	1.12 ± 0.01 *	12,124.89 ± 76.38
Rutin	609	20.20 ± 0.15	14.51 ± 0.10
Quercitrin	447	23.64 ± 0.13	26.08 ± 0.22
Luteolin	285	29.10 ± 0.19	44.08 ± 0.30
Apigenin	269	33.10 ± 0.15	5.79 ± 0.05

Note: Values are the mean ± SD (n = 3). * Determined by a different method characterised by different experimental conditions.

The chromatograms are available in Supplementary Material (Figures S1–S3).

Cichoric acid (2, 3 dicaffeoyl-tartaric acid) was the main polyphenolic compound identified in TOT. These results are in accordance with current data that shows *T. officinale* as a source of cichoric acid [14].

Different amounts, between 0.9 and 9.2 μg/g d.w. were determined in TOT for syringic, vanillic, and protocatechuic acids, while ferulic acid was present in higher concentration (53.60 μg/g d.w.). Among the flavonoids, quercetin-O-glycosides, such as quercitrin (present in higher quantity) and rutin were identified and quantified, together with free aglycons, luteolin, and apigenin, in quantities between 5–44 μg/g d.w. In dandelion extracts analysed by Xue et al., cichoric acid was present at a quantity of 24,031 μg/g d.w. for 50% ethanol extract, consistent to our TOT HPLC analysis. Generally, the concentrations of some phenolic acids and flavonoids were comparable to our findings [13]. Previous analysis of our team quantified cichoric acid at a value of 7163.11 μg/g d.w. for a tincture, prepared with the vegetable material gathered from a lower altitude, a grassland region, characterised by a dry climate with higher temperatures and lower levels of precipitation [15]. The material analysed in the present study was harvested from a plateau area, a higher altitude distinguished by moderate temperatures and higher levels of precipitation. The differences could be explained by the influence of the pedo-climatic conditions on the biosynthesis of the compounds in plants.

Cichoric acid has been described to exhibit many pharmacological activities with benefits in various pathologies. Some properties such as antimicrobial, anti-inflammatory, anti-tumoral, anti-diabetic, neuroprotective, and hyaluronidase inhibiting activities, among others, are reported [14,16].

Based on the current research, the *T. officinale* aerial parts could also be considered a valuable source of polyphenols such as cichoric acid.

2.3. Antioxidant Activity

The in vitro antioxidant activity of TOT was evaluated using two assays: DPPH• and FRAP. The results obtained for the antioxidant capacity determinations are summarised in Table 3.

Table 3. Antioxidant activity of TOT.

Sample	DPPH· EC50 (μg/mL)	FRAP (μM TE/g)
TOT	165.93 ± 6.94	52.49 ± 1.57

Note: Values are expressed as mean of 3 determinations ± SD.

For the DPPH assay, the EC50 value has an opposite relation with the antioxidant capacity; a greater antioxidant capacity is achieved with the decrease of EC50 value.

The assessment of antioxidant activity by the FRAP method revealed a good antioxidant capacity of TOT, comparable with some others' reports. The data from the literature shows very different values that quantify the antioxidant activity of dandelion extracts by the FRAP method, ranging between 3.32 and 131.5 μM TE/g d.w., using different concentrations of extracts (40%, 70%, and 96% ethanolic extracts of *T. herba* from plant material harvested in Bulgaria and Poland) [17,18].

Our results for the in vitro antioxidant activity evaluation were in agreement with the TPC of the tested extracts.

The current knowledge regarding the relationship between polyphenols and antioxidant activity hints that the antioxidant activity of *T. officinale* extract is associated with the quantity of total phenolic acids, which was higher than flavonoids content in our tincture [13].

2.4. Pharmacological Studies

2.4.1. The Evaluation of In Vivo Anti-Inflammatory Effects

Three concentrations (100, 50, and 25 mg/mL) of the TOT sample (namely TOT 100, TOT 50, and TOT 25, respectively) were administrated on rats in a turpentine oil-induced acute inflammation in vivo experiment. To assess the effects, serum oxidative stress markers were analysed. The results are summarised in Table 4.

The experiment evaluated the oxidative stress through general tests such as TOS, OSI, and TAC and by specific tests as MDA, SH, and Nox. In addition, the NF-κB parameter, which has an important role in pro-inflammatory cytokines signalling, was assessed.

Turpentine oil is an inflammatory catalytic agent that activates phagocytes by boosting the Nox and ROS species production [19]. When an inflammation process is triggered, TOS, and subsequently, OSI parameters manifest increased values comparative to the healthy animals, as seen in our experiment ($p < 0.001$). Additionally, the same pattern applies to Nox and MDA levels whilst the TAC and SH levels decrease ($p < 0.001$). NF-κB controls the transcription of pro-inflammatory cytokines among other substances mediating inflammatory responses (NF-κB signalling in inflammation), thus, a current inflammation status is correlated with higher NF-κB concentrations ($p < 0.05$). All three TOT samples lowered TOS and OSI parameters ($p < 0.001$). The Nox levels were improved by TOT 25 ($p < 0.01$), and TOT 100 improved both SH (0.01) and NF-κB levels ($p < 0.05$). The TAC and MDA concentrations were not influenced by any TOT sample.

Table 4. Serum oxidative stress markers in rat turpentine-induced inflammation model.

GROUPS	TOS (μM H_2O_2 E/L)	OSI	TAC (mM TE/L)	NOx (μM/L)	MDA (nM/L)	SH (mM/L)	NF-κB (ng/mL)
CONTROL	5.13 ± 0.84	4.70 ± 0.77	1.0901 ± 0.001	32.67 ± 2.38	1.91 ± 0.19	0.52 ± 0.05	2.2 ± 0.22
INFLAMM	8.55^{b} ± 0.73	8.54^{b} ± 0.66	1.0873 ± 0.001	45.34^{b} ± 3.53	3.00^{b} ±0.21	0.25^{b} ± 0.02	4.17^{a} ± 0.99
DICLOFENAC	7.84 ± 0.35	7.84 ± 0.32	1.0870 ± 0.000	41.48 ± 2.11	2.94 ± 0.39	0.31 ± 0.04	2.41^{f} ± 0.32
TOT 100	4.92^{e} ± 0.24	4.92^{e} ± 0.22	1.0871 ± 0.001	37.59^{g} ± 5.43	3.20^{g} ± 0.66	0.20^{d} ± 0.02	2.92^{cg} ± 0.60
TOT 50	5.15^{e} ± 0.72	5.15^{e} ± 0.67	1.0878 ± 0.001	37.00^{g} ± 3.89	2.91^{g} ± 0.39	0.24 ± 0.02	3.64 ± 0.51
TOT 25	4.63^{e} ± 0.30	4.63^{e} ± 0.27	1.0886 ± 0.000	30.32^{d} ± 7.13	3.06^{g} ± 0.72	0.27^{g} ± 0.11	3.52 ± 0.37

Note: Values are expressed as mean ± SD (n = 5). ^{a} $p < 0.05$, ^{b} $p < 0.001$ versus CONTROL; ^{c} $p < 0.05$, ^{d} $p < 0.01$, ^{e} $p < 0.001$ versus INFLAMM. ^{f} $p < 0.05$ versus INFLAMM. ^{g} $p > 0.05$ versus DICLOFENAC. TOS—total oxidative status, OSI—oxidative stress index, TAC—total antioxidant capacity, Nox—total levels of nitrites and nitrates, MDA—malondialdehyde, SH—total levels of thiols, NF-kB—Nuclear factor kappa-light-chain-enhancer of activated B cells, CONTROL—healthy animals group, INFLAMM group—negative control, DICLOFENAC group—positive control.

These findings show that TOT has an antioxidant effect in rats' turpentine-induced inflammation by correcting the oxidant levels and increasing the antioxidants. Furthermore, TOT has a direct anti-inflammatory effect by reducing NF-κB levels. In this study, the non-steroidal anti-inflammatory diclofenac was used as a positive control. The oxidative parameters for the animals that received the drug were not significantly modified ($p > 0.05$), but the NF-κB parameters were lowered ($p < 0.05$). This is in accordance with the drug's pharmacological mechanism, which involves the nonselective inhibition of cyclooxygenase enzymes (COX 1 and COX 2), thus inhibiting prostaglandin synthesis—molecules that play a key role in inflammation and pain [20]. We compared the results determined by TOT for NF-κB with the diclofenac group, and there were no significant differences ($p > 0.05$) suggesting that TOT has a comparable effect with diclofenac regarding the NF-κB status.

The obtained results highlight the fact that the treatment with TOT on rats with turpentine-induced inflammation, determined anti-inflammatory effects by reducing NF-κB expression and by reducing oxidants.

With the aim to assess the relationship between the parameters, a PCA correction circle was used (Figure 1). Original Pearson Correlation information is available in Supplementary Material (Tables S1–S5).

This type of representation involves grouping the different features according to their positive and negative correlation. Features with a positive correlation will be grouped together. The uncorrelated ones are orthogonal to each other, and negative correlations will be plotted on the opposing quadrants of the plot. We obtained the coordinates of the values by calculating the correlation between each original variable and their associated components [21].

In the tested groups: CONTROL, DICLOFENAC, TOT 100, TOT 50, and TOT 25, the oxidative stress parameters and the transcription factor NF-κB were overall correlated, concluding that inflammation was associated with oxidative stress.

Several studies show that inflammation and oxidative stress are interdependent. In an outgoing inflammatory process, the phagocytic cells (macrophages and neutrophils) activate. This produces large amounts of reactive oxygen species and reactive nitrogen species, including hydrogen peroxide, hydroxyl free radical, superoxide, nitric oxide, and peroxynitrite [22,23].

Various molecular mechanisms involved in the anti-inflammatory activities of polyphenols include the inhibition of certain pro-inflammatory enzymes, such as ciclo-oxigenase 2 (COX2), lipo-oxigenase (LOX), and inducible nitric oxide synthase (iNOS). Polyphenols promote the inhibition of the transcription factor NF-kB and enhance the activation of a phase-II antioxidant enzymes, the mitogen-activated protein kinase, and protein kinase-C [9].

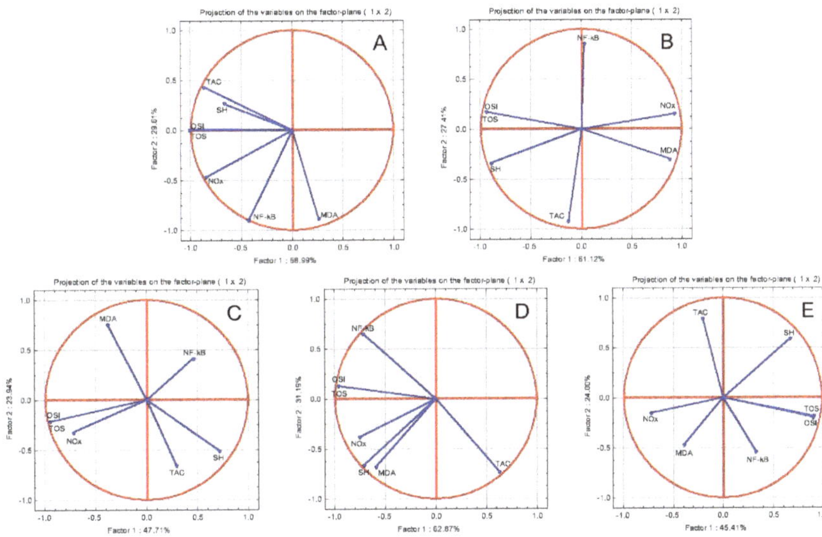

Figure 1. Anti-inflammatory oxidative stress tests PCA results: PCA correlation circles in turpentine oil-induced acute inflammation: (**A**) INFLAMMATION group: NF-kB increase was positively correlated with oxidants tests, specifically, TOS, OSI, Nox, and MDA (**B**) DICLOFENAC group: NF-kB reduction was positively correlated only with TOS, OSI, and Nox (**C**) TOT 100 group: NF-kB reduction was positively correlated only with MDA (**D**) TOT 50 group: NF-kB reduction was positively correlated only with TOS, OSI, and Nox (**E**) TOT 25 group: NF-kB reduction was positively correlated only with TOS, OSI, and MDA.

The anti-inflammatory effects of *T. officinale* extracts were evaluated in relationship with the polyphenolic composition and anti-oxidative effects. Jeon et al. describes the anti-inflammatory activity on lipopolysaccharides (LPS) which stimulated murine macrophage cells (RAW264.7) by improving NO, PGE2, and the cytokines TNF-α and IL-1 β levels, and inhibiting iNOS, COX2, and activation of MAP kinases, suggesting a direct anti-inflammatory effect in addition to the antioxidative mechanisms. Methanolic extracts of *T. officinale* also presented with NF-kB inhibition in an in vitro study conducted on LPSs in human umbilical vein endothelial cells [24].

Regarding in vivo models, aqueous extracts of *T. officinale* leaves lowered inflammation in cholecystokinin-induced acute pancreatitis in rats by blocking the production of IL-6 and TNF-α cytokines that depend upon NF-kB transcription [25]. Our TOT had an anti-inflammatory effect by inducing an important reduction of NF-kB.

In the air pouch model of carrageenan-induced inflammation, ethanolic extracts of *T. officinale* aerial parts inhibited the production of exudate and reduced the levels of leukocytes and nitric oxide within. It also exhibited a dose-dependent suppression on the vascular permeability of acetic acid abdominal induction assay in mice [10]. In turpentine oil- induced inflammation, TOT caused NOx reduction where the higher dilution had the strongest effect. Similarly, TOT lowered other oxidative stress markers, specifically, TOS, OSI, and MDA. For TOT, the anti-inflammatory activity was correlated with oxidative stress marker reduction (Figure 1). TOT had no significant activity on the antioxidant markers.

These results suggest that due to the phytochemicals present in *T. officinale*, TOT has an anti-inflammatory effect through inhibiting the activation of NF-κB and by lowering the oxidant concentrations. The phenolic compounds identified in *T. officinalis* could be involved in the anti-inflammatory effects of the extracts, as well as cichoric acid, based on its known antioxidant and anti-inflammatory properties [26]. Similar to cichoric acid, ferulic acid was found to exhibit anti-inflammatory effects, mainly due to the antioxidant properties [27]. The quantified flavonoids from our studied tincture, rutin and quercitrin, as well as luteolin and apigenin, showed anti-inflammatory effects in experimental tests [28]. Therefore, TOT anti-inflammatory activity based on the improvement of serum oxidative stress parameters could be associated with the presence of phenolic compounds with antioxidant properties, which could exhibit positive outcomes in inflammation treatment.

2.4.2. The Evaluation of In Vivo Cardioprotective Effects

TOT activity in turpentine-oil induced inflammation encouraged us to test the effect in the acute inflammation associated with the acute myocardial infarction (MI). In the present study, the TOT's effect was evaluated on acute MI induced by isoprenaline. The parameters analysed in this experiment were NF-kB as an anti-inflammatory marker, TOS, OSI, TAC, MDA, SH, and NOx as oxidative stress markers, plus serum cardiac injury marker enzymes (AST, ALT, and CK-MB). The results are summarised in Tables 5 and 6.

Table 5. Serum oxidative stress markers in rat isoprenaline (ISO)-induced MI.

GROUPS	TOS (μM H_2O_2 E/L)	OSI	TAC (mM TE/L)	NOx (μM/L)	MDA (nM/L)	SH (mM/L)	NF-κB (ng/mL)
CONTROL	5.13 ± 0.84	4.70 ± 0.77	1.0901 ± 0.001	32.67 ± 2.38	1.91 ± 0.19	0.52 ± 0.05	2.2 ± 0.22
ISO	7.43 ^{b} ± 0.11	6.83 ^{b} ± 0.10	1.0876 ± 0.00	45.51 ^{b} ± 0.37	3.41 ^{b} ± 024	0.39 ^{b} ± 0.01	3.42 ^{a} ± 0.59
TOT 100	4.50 ^{e} ± 0.12	4.13 ^{e} ± 0.11	1.0886 ± 0.00	36.49 ± 5.90	2.70 ± 0.20	0.26 ^{d} ± 0.02	2.24 ± 0.50
TOT 50	4.40 ^{e} ±0.12	4.05 ^{e} ± 0.11	1.0873 ± 0.00	34.14 ^{c} ± 3.56	2.38 ^{d} ± 0.14	0.29 ^{d} ± 0.02	1.35 ^{c} ± 0.27
TOT 25	4.19 ^{e} ± 0.13	3.85 ^{e} ± 0.12	1.0880 ± 0.00	30.55 ^{e} ± 3.28	3.26 ± 0.14	0.29 ^{d} ± 0.02	1.05 ^{c} ± 0.18

Note: Values are expressed as mean ± SD (n = 5). ^{a} $p < 0.01$, ^{b} $p < 0.001$ versus CONTROL; ^{c} $p < 0.05$, ^{d} $p < 0.01$, ^{e} $p < 0.001$ versus ISO. TOS—total oxidative status, OSI—oxidative stress index, TAC—total antioxidant capacity, NOx—total levels of nitrites and nitrates, MDA—malondialdehyde, SH—total levels of thiols, NF-kB—Nuclear factor kappa-light-chain-enhancer of activated B cells, CONTROL—healthy animals group, ISO—negative control.

Table 6. Serum cardiac injury markers in rat isoprenaline-induced MI.

GROUPS	AST (UI/L)	ALT (UI/L)	CK-MB (UI/L)
CONTROL	35.32 ± 4.89	29.10 ± 4.12	7.26 ± 1.02
ISO	30.94 ± 8.35	40.04 ^{a} ± 7.29	12.11 ^{b} ± 1.08
TOT 100	26.45 ^{c} ± 1.08	24.77 ± 0.55	8.11 ± 1.51
TOT 50	32.52 ± 2.65	26.31 ± 1.22	7.92 ^{c} ± 1.16
TOT 25	30.32 ± 3.53	26.36 ± 2.97	8.47 ± 0.95

Note: Values are expressed as mean ± SD (n = 5). ^{a} $p < 0.05$, ^{b} $p < 0.001$ versus CONTROL; ^{c} $p < 0.05$, versus ISO. AST—aspartate transaminase, ALT—Alanine transaminase, CK-MB—Creatin kinase isoenzyme MB.

The consequences of MI induced by ISO were cell injury, indicated by the serum AST, ALT, and CK-MB increase, and an inflammatory response with elevated NF-κB. At the same time, an increased oxidative stress was indicated by the increased TOS, OSI, NOx, and MDA, and TAC and SH reduction.

In MI, all three TOT concentrations reduced myocardial cells injury enzymes, had anti-inflammatory activity by reducing NF-kB, and an antioxidant effect by lowering TOS, OSI, NOx, and MDA ($p < 0.001$). TOT 50 was the most effective concentration. SH was reduced after TOT treatments. A possible explanation for the decrease of SH levels is that in MI, prophylaxis by dietary antioxidant consumption reduces the formation of ROS and RNS and does not have an impact on preventing the reduction of previously present antioxidant species [29].

Overall, TOT 50 improved foremost the parameters (although there were no significant differences between the extract groups; $p > 0.05$) which is in accordance with the general knowledge that a mix of polyphenols in small doses can potentiate the antioxidant effect rather than single entities in higher doses; also at higher concentrations, some polyphenols can exhibit pro-oxidant effects via the Fenton reaction [30].

In order to evaluate the relationship between the parameters, a PCA was performed (Figure 2). Original Pearson Correlation information is available in Supplementary Material

(Tables S6–S9). In general, in all groups there were correlations between AST, ALT, and CK-MB with the oxidative stress parameters, and also with NF-κB.

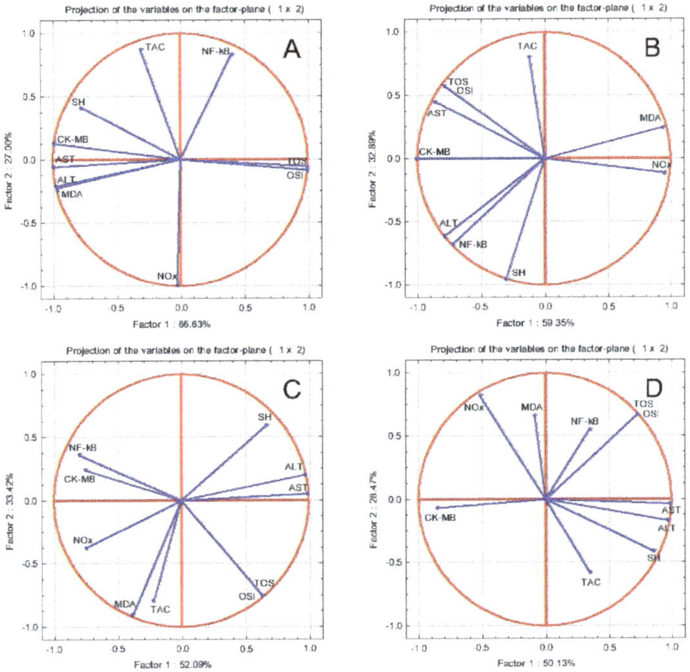

Figure 2. Anti-inflammatory, oxidative stress tests and cardiac function tests PCA correlation circles in isoprenaline-induced acute myocardial infarction: (**A**) ISO group: AST, ALT, and CK-MB correlate with MDA, and NF-κB with TOS and OSI; (**B**) TOT 100: AST, ALT, and CK-MB correlate with NF-κB, TOS, and OSI, and MDA with NOx; (**C**) TOT 50: AST and ALT correlate with TOS and OSI, CK-MB correlates with NOx and NF-κB; (**D**) TOT 25: AST and ALT correlate with TOS, OSI, and NF-κB, CK-MB correlates with NOx.

Cichoric acid, the major polyphenolic compound of *T. officinale* extract possesses cardioprotective activity, based on the improvement of cardiovascular homeostasis [31] and anti-atherosclerotic effect, based on antioxidative and anti-inflammatory mechanisms [32]. Ferulic acid also has cardioprotective effects [27].

The hydroxybenzoic acids (protocatechuic, vanillic, syringic acid) were studied in relation to cardiac and vascular applications, with positive outcomes [33,34].

Studies concerning flavonoids show positive outcomes in cardiovascular diseases, as rutin has an active role in reducing cardiac hypertrophy, and quercitrin (through its aglycon quercetin), has a role both as anti-atherosclerotic and as a cardiovascular risk improving agent [35,36]. Luteolin and apigenin exhibit cardioprotective effects with various applications in ischemia/reperfusion injury, atherosclerosis, and heart failure [36–38].

The heart is an organ that is vulnerable to oxidative stress because of the absence of antioxidant systems, therefore a systemic oxidative reduction may be favourable and diminish the myocardial injury during MI [39].

The presence of diverse polyphenolic compounds in TOT, such as phenolic acids and flavonoids with cardioprotective properties, may explain the capacity of the studied tincture to protect against MI induced by ISO, with a positive influence on serum oxidative stress parameters and serum cardiac injury markers.

Other in vivo studies regarding the cardiovascular potential of *T. officinale* discuss the positive outcomes of extract administration in relation to metabolic syndrome (hypolipidemic effect by improving cholesterol and total lipids levels, and overall reduction in aortal thickness, as well as anti-obesity effects and an overall decrease in oxidative stress) and other disorders that can lead to heart damage [11]. The present study emphasises the positive influence of the active compounds from the tincture on some parameters involved in oxidative processes during the myocardium injury in MI.

Previous studies that compare the in vitro and in vivo antioxidant activities of different plant extracts show that these are not always correlated [39]. In the present experiment, TOT had a higher in vivo antioxidant activity compared to the in vitro antioxidant capacity, which was found to be good, using the described methods.

In the present study, we pointed out that TOT has an in vivo anti-inflammatory activity on turpentine-induced inflammation and cardioprotective effects on ISO-induced MI, by reducing the oxidative stress and inflammation.

3. Materials and Methods

3.1. Chemicals and Reagents

The references used in the LC-MS analysis were purchased from Sigma-Aldrich (Schnelldorf, Germany): cichoric acid (>95%), caffeic acid (≥98%), chlorogenic acid (≥95%), ferulic acid (>95%), sinapic acid (≥98%), hyperoside (≥98%), isoquercitrin (≥90%), quercitrin (quercetin 3-rhamnoside) (≥78%), quercetin (≥95%), luteolin (≥98%), kaempferol (≥97%), apigenin (≥95%), syringic acid (≥95%), protocatechuic acid (3,4-dihydroxybenzoic acid) (≥97%), and vanillic acid (≥97%). Ammonium acetate, acetonitrile, petroleum ether, chloroform, hydrochloric acid, acetic acid, potassium hydroxide, and Folin–Ciocâlteu reagent were purchased from Merck (Darmstadt, Germany). Sodium carbonate, sodium acetate trihydrate, and anhydrous aluminium chloride were purchased from Sigma-Aldrich (Schnelldorf, Germany). 2,4,6-tri(2-pyridyl)-1,3,5-triazine (TPTZ) reagent, 2,2-diphenyl-1-picryl-hydrazyl-hydrate (DPPH) reagent, and Trolox were acquired from Sigma-Aldrich (Schnelldorf, Germany); rutin (>95 %) was from Fluka Chemie GmbH (Buchs, Switzerland). Methanol p.a., ethanol 96%, and dichloromethane were purchased from Chemical Company (Iași, Romania) and iron chloride from Merck (Darmstadt, Germany). The commercial biochemistry kits for the pharmacological investigations (kit CK-MB-LQ. Anti CK-M. Immunoinh.; kit GOT/AST-LQ. IFCC. Enzymatic–UV; GPT/ALT-LQ. IFCC. Enzymatic–UV; kit UREA-LQ. Urease-GLDH. Kinetic; kit creatinine-J. J) were purchased from S.C. DG Diagnostics S.R.L. Cluj-Napoca. The turpentine used to induce inflammation was purchased from Sigma-Aldrich (Germany). Diclofenac sodium was purchased from a local pharmacy with the trademark Refen, concentration of 75 mg/mL, 3 mL (Hemofarm Koncern A.D.) Isoprenaline used to induce myocardial ischemia was purchased from Sigma-Aldrich.

3.2. Plant Material and Extraction Procedure

T. officinale aerial parts (+*herba*) were harvested from Hunedoara County, (Lat. 45°45′18.151″ N/Long. 22°53′25.181″ E), Western Romania, during the flowering stage, from wild populations. A sample of the herbal material is available in the Pharmacognosy Department herbarium (voucher number 116). The plant material was air-dried, powdered, degreased with dichloromethane, and then was used for extraction. The tincture (1:10) was obtained from 50 g of plant material and 500 mL 70% ethanol, by maceration for 7 days at room temperature (*T. officinale* tincture = TOT) [39].

TOT was used for phytochemical analysis: total polyphenols content, total flavonoids content, and total caffeic acid derivatives content were determined. Then, HPLC-MS analysis, the antioxidant assays (DPPH• and FRAP methods), and the pharmacological experiments were performed using this tincture.

For the assessment of the anti-inflammatory and cardioprotective activities, *T. officinale* tincture (TOT 100, corresponding to 1 mg dry weight plant material/10 mL) was used, as well as two dilutions of the tincture, obtained with distilled water, TOT 50 (0.5 mg dry weight plant material/10 mL), and TOT 25 (0.25 mg dry weight plant material/10 mL).

3.3. Total Polyhenols Content Determination

The total polyphenols content (TPC) was determined using the Folin–Ciocâlteu spectrophotometric method [40]. Briefly, TOT (2 mL) was diluted in a 25 mL volumetric flask with the same solvent and then, to 2 mL of each solution, 1 mL Folin–Ciocâlteu reagent and 10 mL of distilled water were added; the mixture was diluted to 25 mL with a solution of sodium carbonate (290 g/L). After 30 min in darkness, the absorbance of the samples was measured at 760 nm using a Cary 60 UV–Vis spectrophotometer from Agilent Technologies (R^2 = 0.999). TPC is expressed as mg gallic acid equivalents (GAE)/g d.w. [15].

3.4. Total Flavonoids Content Determination

The total flavonoids content (TFC) of TOT was performed by a spectrophotometric method based on the colour reaction with $AlCl_3$ reagent. Briefly, 10 mL of the TOT was diluted to 25 mL using methanol. To 5 mL of the diluted solution, 5.0 mL sodium acetate (100 g/L) and 3.0 mL aluminium chloride (25 g/L) were added; the obtained solution was diluted with methanol up to 25 mL in a calibrated flask. The absorbance was measured at 430 nm and the results are expressed as mg rutin equivalents (RE)/g d.w. (R^2 = 0.999) [40]

3.5. Total Caffeic Acid Derivates Content Determination

The caffeic acid derivates content (TCADC) of the TOT was analysed using Arnow reagent by a spectrophotometric method. In short, in a volumetric flask, 10 mL of the TOT was mixed with methanol at 25 mL, and 5 mL of the obtained solution was diluted with ethanol 50% at 10 mL. Next, 1 mL of HCl 0.5 N, was added to 1 mL of solution, together with 1 mL of Arnow reagent and 1 mL NaOH 1 N; the mixture was increased to 10 mL with ethanol 50% [41]. The absorbance (λ = 500 nm) was used to calculate the results, expressed as mg caffeic acid equivalents (CAE)/g d.w. (R^2 = 0.994).

All phytochemical determinations were performed in triplicate.

3.6. Evaluation of the In Vitro Antioxidant Capacity

3.6.1. DPPH Radical Scavenging Activity

The antioxidant activity was assessed in vitro by different methods. The assay measures the scavenging ability of extracts towards DPPH, a stable nitrogen cantered free radical. First, 2 mL of TOT (in different concentrations) were added to 2 mL of 0.1 g/L DPPH• methanol solution. The absorbance was determined at 517 nm, after 30 min of incubation at room temperature in the dark (R^2 = 0.997). The DPPH radical scavenging activity (AA) was calculated as follows: $AA\% = (A_{control} - A_{sample}/A_{control}) \times 100$, where $A_{control}$ is the absorbance of DPPH• radical + methanol (does not contain the sample) and A_{sample} is the absorbance of DPPH• radical + sample extract. The EC50 (μg/mL) values were also calculated in order to determine the half maximal inhibitory concentration of TOT [42,43]. The experiments were performed in triplicate.

3.6.2. Ferric-Reducing Antioxidant Power Assay

This assay has been proven to measure the antioxidant capacity of plant products containing polyphenols. The absorbance was measured at 450 nm. A volume of 0.4 mL of TOT was diluted with water to 1.8 mL and mixed with 6 mL of FRAP reagent (R^2 = 0.992). The antioxidant activity was expressed as Trolox equivalents (TE) [44]. All determinations were realised in triplicate.

3.7. HPLC-UV-MS Separation

The phytochemical assay of the tincture was performed by liquid chromatography coupled with mass spectrometry (Agilent Technologies 1100 HPLC Series system coupled to an Agilent 1100 mass spectrometer (LC/MSD Ion Trap SL)). The separation was achieved using a reverse-phase analytical column (Zorbax SB-C18 100 × 3.0 mm i.d., 3.5 µm particles, t° = 48 °C). The injection volume was 5 µL and the flow rate was set to 1 mL/min. The MS system functioned using an electro spray ion source in negative mode. Chromatographic data have been interpreted using ChemStation and Data Analysis. Compound identification was performed in both UV and MS mode, by comparing their traces/spectra obtained in the experiment with spectra from the data library. The limit of detection was 0.1 µg/mL and the limit of quantification was 0.5 µg/mL. The UV trace was used for quantification of identified compounds from MS detection. Quantitative determinations were performed using an external standard method. Calibration curves in the 0.5–50 µg/mL range with good linearity ($R^2 > 0.999$) for a five-point plot were used to determine the concentration of polyphenols. The mobile phase consisted of 95/5 (*v*/*v*) ammonium acetate, 1 mM in water and acetonitrile, isocratic elution, and mobile phase flow rate of 1 mL/min. The mass spectrometer operated in negative mode and nitrogen was used as a nebulising and dry gas. The nebuliser was positioned at 65 psi with the dry gas flow at 12 L /min at 350 °C [45,46]. The determination of polyphenolic acids with higher hydrophilic character (protocatechuic, vanillic, syringic acids) was employed with the same analytical conditions, but using a different binary gradient and compound in MS mode [47]. Analysis of cichoric acid was performed in another stage of work, using the same apparatus previously described, using a different developed procedure of liquid chromatography coupled with mass spectrometry detection. The advantage was a high throughput determination, having a major edge of rapid analysis m/z 293 + m/z 311, which is specific to cichoric acid. The QuantAnalysis 1.7 software (Brucker Daltonics, Darmstadt, Germany) instrumental data system was employed for the quantification of cichoric acid using peak area and the external standard method ($R^2 > 0.999$) [48].

3.8. Pharmacological Evaluation

3.8.1. Experimental Animals

The experiments were performed on adult male Wistar albino rats, weighing 200–250 g. The animals were bred in the "Iuliu Hațieganu" University of Medicine and Pharmacy Animal Facility. Prior and during to the experiments, animals were housed in proper conditions (12 h night/day cycle, temperatures of 21–22 °C and humidity of 50–55%), with water ad libitum and free access to a standard pellet-based diet (Cantacuzino Institute, Bucharest, Romania). All the animals were sacrificed by cervical dislocation at study completion under general anaesthesia. All treatments that involved animals were in accordance with EU Directive 2010/63/EU. The experimental design was approved by the Institutional Animal Ethical Committee (IAEC) of the "Iuliu Hațieganu" University of Medicine and Pharmacy Cluj-Napoca and by the National Sanitary Veterinary and Food Safety Agency (no. 171/13.07.2019).

3.8.2. Protocols

The Evaluation of In Vivo Anti-Inflammatory Effects

For the evaluation of anti-inflammatory effects, 6 groups of animals (n = 5) were used: (1)—the negative control group (CONTROL)—on day 1 received 0.9% saline solution i.m. (6 mL/kg b.w.) and 1 mL orally by gavage, followed by daily administration of 1 mL saline solution, orally; (2)—inflammation group (TURPENTINE)—on day 1 received i.m. turpentine oil (6 mL/kg b.w.) and 0.9% saline solution by gavage, followed by daily administrations of saline solution 0.9% by gavage; (3)—positive control (DICLOFENAC)—on day 1 received i.m. turpentine oil (6 mL/kg b.w.) and sodium diclofenac 10 mg/kg b.w. orally, followed by daily administrations of diclofenac; (4)—TOT 100 group—on day 1 received i.m. turpentine oil (6 mL/kg b.w.) and 1 mL TOT 100 orally, followed by daily administration of 1 mL TOT 100 by gavage; (5)—TOT 50 group—on day 1 received i.m. turpentine oil (6 mL/kg b.w.) and 1 mL orally TOT 50, followed by daily administration of TOT 50 by gavage; (6)—TOT 25 group—on day 1 received i.m. turpentine oil (6 mL/kg b.w.)

and 1 mL orally TOT 25, followed by daily administration of 1 mL TOT 25, orally [49,50]. All treatments were performed for seven days.

The Evaluation of In Vivo Cardioprotective Effects

In order to evaluate the cardioprotective effects, pre-treatments with TOT were analysed. The animals were divided into 5 groups (n = 5): (1)—negative control (CONTROL); (2)—isoprenaline (ISO); (3)—received TOT 100; (4)—received TOT 50 and (5)—received TOT 25. For seven days the animals received by gavage (orally p.o. 1 mL/day) water, in groups CONTROL and ISO, respectively, the different concentrations of TOT in groups 3, 4, and 5 (p.o. 1 mL/day). Except the CONTROL group, on days 8 and 9 animals received isoprenaline (subcutaneously s.c. 150 mg/kg b.w.) to induce experimental MI [30,51]. On day 10, blood samples were collected by retro-orbital puncture under general anaesthesia induced by a mixture of ketamine (70 mg/kg b.w.) and xilazine (10 mg/kg b.w.) [52]. Serum was separated and stored at −80 °C until the oxidative stress and cardiac markers analysis. Cardiac markers, AST, ALT, and CK-MB, were assessed using commercial kits.

The Evaluation of Oxidative Stress Parameters

TOS was assessed using a colorimetric method based on the oxidation of a ferrous ion to a ferric ion in the presence of various oxidant species [53]. The results were expressed in µmol H_2O_2 equivalents/L. TAC was measured using a colorimetric assay described by Erel and expressed as mmol Trolox equivalents/L [54]. OSI was calculated as the ratio between TOS and TAC [55]. As a lipid peroxidation marker, MDA was determined using the thiobarbituric acid assay. The MDA serum concentration was expressed as nM/L [56]. The serum NO concentration was assessed using the Griess reaction and expressed as nitrite µM/L [57]. Serum total thiols were expressed as mM GSH/L and were determined using Ellman's reagent [58].

The NF-kB was determined using a NF-kB ELISA KIT, (ER1186, Fine Biotech, and Wuhan, China) according to the manufacturer instructions.

3.8.3. Statistical Analysis

The statistical analysis was performed using Excel and Statistica 12.0 software. The results are expressed as means ± standard deviation. The data were compared by using a one-way analysis of variance (ANOVA) test and post hoc Bonferroni–Holm test. The correlation between the parameters of the same group was assessed by Pearson's coefficient (r) in accordance to the Colton scale. The level of significance was established at $p < 0.05$. Multivariate analysis of the parameters was performed using Principal Component Analysis (PCA). On the PCA correlation circle when two vectors are close, forming a small angle, the two variables are positively correlated, if they meet each other at 90°, they are not likely to be correlated, when they diverge and form a large angle (close to 180°), they are negative correlated [21].

4. Conclusions

The present study was focused on the common dandelion, *T. officinale*, harvested from the Romanian spontaneous flora. The polyphenolic composition revealed cichoric acid identified in the largest amount, but others phenolic acids (protocatechuic, vanillic, syringic, and ferulic acids), and also flavonoids (rutin, quercitrin, luteolin, and apigenin) were present.

The tested extract (TOT) had a good influence on serum oxidative markers in rat turpentine-induced inflammation model and also in rat ISO-induced MI (TOS, OSI, TAC, MDA, SH, NOx, 513 NF-κB). The pharmacological effects are due to the active principles from the extract; there were identified polyphenols, but also other compounds (e.g., terpenes) that may be present in the tincture. These compounds determined an anti-inflammatory effect in the tested model by lowering the oxidative stress parameters involved in the pro-inflammatory cytokines signalling, based on antioxidant mechanism. The cardioprotective effect of the dandelion tincture was evaluated by the decreasing of the serum cardiac injury enzymes (AST and CK-MB), but also by the diminution of the serum oxidative stress markers levels, increased by isoprenaline in induced MI (TOS, OSI, NOx, MDA).Through our results, we aim to set a basis for further pharmacological inquiries, to extend indigenous medicinal plants uses to new therapeutic directions, combining safety and efficacy.

Supplementary Materials: The following supporting information can be downloaded at: https://www.mdpi.com/article/10.3390/ph16030358/s1, Figure S1: Chromatogram for 1. ferulic acid, 2. rutin, 3. quercitrin, 4. luteolin, 5. apigenin; Figure S2. Chromatogram for cichoric acid; Figure S3. Chromatogram for protocatechuic, vanilic, syringic acids; Table S1. Pearson Correlations Antiinflammatory activity Taraxaci herba Negative Control; Table S2. Pearson Correlations Antiinflammatory activity Taraxaci herba Positive Control; Table S3. Pearson Correlations Antiinflammatory activity Taraxaci herba TOT; Table S4. Pearson Correlations Antiinflammatory activity Taraxaci herba TOT 1:2; Table S5. Pearson Correlations Antiinflammatory activity Taraxaci herba TOT 1:3; Table S6. Pearson Correlations Cardioprotective activity Taraxaci herba Negative Control; Table S7. Pearson Correlations Cardioprotective activity Taraxaci herba TOT; Table S8. Pearson Correlations Cardioprotective activity Taraxaci herba TOT 1:2; Table S9. Pearson Correlations Cardioprotective activity Taraxaci herba TOT 1:3.

Author Contributions: A.E., A.E.P., L.V., D.B., D.H. and I.O. conceived and designed the structure of the manuscript, A.E., A.E.P., L.V., D.B., D.H., I.I., O.O. and A.-M.V. contributed to the phytochemical analysis and pharmacological protocols, A.E., A.E.P., A.T. and I.O. critically reviewed the manuscript. All authors have read and agreed to the published version of the manuscript.

Funding: This research received no external funding.

Institutional Review Board Statement: The study was conducted according to the guidelines of the Declaration of Helsinki, and approved by the Ethics Committee of "Iuliu Hațieganu" University, Cluj-Napoca (protocol code 171/13.06.2019).

Informed Consent Statement: Not applicable.

Data Availability Statement: Data is contained within the article and supplementary material.

Conflicts of Interest: The authors declare no conflict of interest.

References

1. Khan, I.A.; Hussain, M.; Hussain, N.; Alqahtani, A.M.; Alqahtani, T. Cardioprotective Effect of Rumex vesicarius Linn. Leaf Extract against Catecholamine-Induced Cardiotoxicity. *Molecules* **2022**, *27*, 3383. [CrossRef] [PubMed]
2. Majewski, M.; Lis, B.; Juśkiewicz, J.; Ognik, K.; Borkowska-Sztachańska, M.; Jedrejek, D.; Stochmal, A.; Olas, B. Phenolic Fractions from Dandelion Leaves and Petals as Modulators of the Antioxidant Status and Lipid Profile in an In Vivo Study. *Antioxidants* **2020**, *9*, 131. [CrossRef]
3. Wang, T.; Wu, S.; Ibrahim, I.A.A.; Fan, L. Cardioprotective Role of Swertiamarin, a Plant Glycoside Against Experimentally Induced Myocardial Infarction via Antioxidant and Anti-inflammatory Functions. *Appl. Biochem. Biotechnol.* **2022**, *194*, 1–15. [CrossRef] [PubMed]
4. Schütz, K.; Carle, R.; Schieber, A. Taraxacum—A review on its phytochemical and pharmacological profile. *J. Ethnopharmacol.* **2006**, *107*, 313–323. [CrossRef] [PubMed]
5. Singh, A.S.; Malhotra, S.; Subban, R. Dandelion (Taraxacum officinale)—Hepatoprotective Herb with Therapeutic Potential. *Pharmacogn. Rev.* **2008**, *2*, 163–167.
6. Liu, R.H. Potential synergy of phytochemicals in cancer prevention: Mechanism of action. *Int. J. Funct. Nutr.* **2004**, *134* (Suppl. S12), 3479–3485. [CrossRef]
7. Lis, B.; Jedrejek, D.; Moldoch, J.; Stochmal, A.; Olas, B. The anti-oxidative and hemostasis-related multifunctionality of L-chicoric acid, the main component of dandelion: An in vitro study of its cellular safety, antioxidant and anti-platelet properties, and effect on coagulation. *J. Funct. Foods* **2019**, *62*, 103524. [CrossRef]
8. Schütz, K.; Kammerer, D.R.; Carle, R.; Schieber, A. Characterization of phenolic acids and flavonoids in dandelion (Taraxacum officinale WEB. ex WIGG.) root and herb by high-performance liquid chromatography/electrospray ionization mass spectrometry. *J. Mass Spectrom.* **2005**, *19*, 179–186. [CrossRef]
9. Andriantsitohaina, R.; Auger, C.; Chataigneau, T.; Étienne-Selloum, N.; Li, H.; Martínez, M.C.; Schini-Kerth, V.B.; Laher, I. Molecular mechanisms of the cardiovascular protective effects of polyphenols. *Br. J. Nutr.* **2012**, *108*, 1532–1549. [CrossRef]

10. Jeon, H.J.; Kang, H.J.; Jung, H.J.; Kang, Y.S.; Lim, C.J.; Kim, Y.M.; Park, E.H. Anti-inflammatory activity of Taraxacum officinale. *J. Ethnopharmacol.* **2008**, *115*, 82–88. [CrossRef]
11. Olas, B. New Perspectives on the Effect of Dandelion, Its Food Products and Other Preparations on the Cardiovascular System and Its Diseases. *Nutrients* **2022**, *14*, 1350. [CrossRef]
12. Aabideen, Z.U.; Mumtaz, M.W.; Akhtar, M.T.; Mukhtar, H.; Raza, S.A.; Touqeer, T.; Saari, N. Anti-Obesity Attributes; UHPLC-QTOF-MS/MS-Based Metabolite Profiling and Molecular Docking Insights of Taraxacum officinale. *Molecules* **2020**, *25*, 4935. [CrossRef]
13. Xue, Y.; Zhang, S.; Du, M.; Zhu, M.-J. Dandelion extract suppresses reactive oxidative species and inflammasome in intestinal epithelial cells. *J. Funct. Foods* **2017**, *29*, 10–18. [CrossRef]
14. Lee, J.; Scagel, C.F. Chicoric acid: Chemistry, distribution, and production. *Front. Chem.* **2013**, *1*, 40. [CrossRef]
15. Epure, A.; Pârvu, A.; Vlase, L.; Benedec, D.; Hanganu, D.; Vlase, A.M.; Oniga, I. Polyphenolic compounds, antioxidant activity and nephroprotective properties of Romanian Taraxacum officinale. *Farmacia* **2022**, *70*, 47–53. [CrossRef]
16. Peng, Y.; Sun, Q.; Park, Y. The Bioactive Effects of Chicoric Acid as a Functional Food Ingredient. *J. Med. Food* **2019**, *22*, 645–652. [CrossRef]
17. Ivanov, I. Polyphenols Content and Antioxidant Activities of Taraxacum officinale F.H. Wigg (Dandelion) Leaves. *Int. J. Pharmacogn. Phytochem. Res.* **2014**, *6*, 889–893.
18. Nowak, A.; Duchnik, W.; Zielonka-Brzezicka, J.; Muzykiewicz, A.; Florkowska, K.; Klimowicz, A.; Kucharski, Ł.; Wysocka, D.; Dziedzic, A. The antioxidant activity of ethanolic and aqueous extracts of dandelion (Taraxacum officinale L.). *Pomeranian J. Life Sci.* **2019**, *65*, 83–88. [CrossRef]
19. Pavel, I.; Parvu, A.E.; Dehelean, C.; Vlase, L.; Csuk, R.; Muntean, D. Assessment of the antioxidant effect of a maslinic acid derivative in an experimental model of acute inflammation. *Farmacia* **2017**, *65*, 390–395.
20. Gan, T.J. Diclofenac: An update on its mechanism of action and safety profile. *Curr. Med. Res. Opin.* **2010**, *26*, 1715–1731. [CrossRef]
21. Abdi, H.; Williams, L.J. Principal Component Analysis. *Wiley Interdiscip. Rev. Comput. Stat.* **2010**, *2*, 433–459. [CrossRef]
22. Mittal, M.; Siddiqui, M.R.; Tran, K.; Reddy, S.P.; Malik, A.B. Reactive oxygen species in inflammation and tissue injury. *Antioxid. Redox Signal.* **2014**, *20*, 1126–1167. [CrossRef] [PubMed]
23. Biswas, S.K. Does the Interdependence between Oxidative Stress and Inflammation Explain the Antioxidant Paradox? *Oxidative Med. Cell. Longev.* **2016**, *2016*, 5698931. [CrossRef] [PubMed]
24. Jeon, D.; Kim, S.J.; Kim, H.S. Anti-inflammatory evaluation of the methanolic extract of Taraxacum officinale in LPS-stimulated human umbilical vein endothelial cells. *BMC Complement. Altern. Med.* **2017**, *17*, 508. [CrossRef]
25. Seo, S.W.; Koo, H.N.; An, H.J.; Kwon, K.B.; Lim, B.C.; Seo, E.A.; Ryu, D.G.; Moon, G.; Kim, H.Y.; Kim, H.M.; et al. Taraxacum officinale protects against cholecystokinin-induced acute pancreatitis in rats. *World J. Gastroenterol.* **2005**, *11*, 597–599. [CrossRef]
26. Thygesen, L.; Thulin, J.; Mortensen, A.; Skibsted, L.H.; Molgaard, P. Antioxidant activity of cichoric acid and alkamides from Echinacea purpurea, alone and in combination. *Food Chem.* **2007**, *101*, 74–81. [CrossRef]
27. Alam, M.A. Anti-hypertensive Effect of Cereal Antioxidant Ferulic Acid and Its Mechanism of Action. *Front. Nutr.* **2019**, *6*, 121. [CrossRef]
28. Guardia, T.; Rotelli, A.E.; Juarez, A.O.; Pelzer, L.E. Anti-inflammatory properties of plant flavonoids. Effects of rutin, quercetin and hesperidin on adjuvant arthritis in rat. *Farmaco* **2001**, *56*, 683–687. [CrossRef]
29. Senoner, T.; Dichtl, W. Oxidative Stress in Cardiovascular Diseases: Still a Therapeutic Target? *Nutrients* **2019**, *11*, 2090. [CrossRef]
30. Balea, Ş.S.; Pârvu, A.E.; Pop, N.; Marín, F.Z.; Pârvu, M. Polyphenolic Compounds, Antioxidant, and Cardioprotective Effects of Pomace Extracts from Fetească Neagră Cultivar. *Oxidative Med. Cell. Longev.* **2018**, *2018*, 8194721. [CrossRef]
31. Wu, H.; Luo, D.; Li, C.; Zhang, H.; Shunxian, A.; Zhang, Y.; Sun, C. Chicoric Acid Improves Heart and Blood Responses to Hypobaric Hypoxia in Tibetan Yaks. *Am. J. Chin. Med.* **2018**, *46*, 339–355. [CrossRef]
32. Tsai, K.L.; Kao, C.L.; Hung, C.H.; Cheng, Y.H.; Lin, H.C.; Chu, P.M. Chicoric acid is a potent anti-atherosclerotic ingredient by anti-oxidant action and anti-inflammation capacity. *Oncotarget* **2017**, *8*, 29600–29612. [CrossRef]
33. Baniahmad, B.; Safaeian, L.; Vaseghi, G.; Rabbani, M.; Mohammadi, B. Cardioprotective effect of vanillic acid against doxorubicin-induced cardiotoxicity in rat. *Res. Pharm. Sci.* **2020**, *15*, 87–96. [CrossRef]
34. Manjunatha, S.; Shaik, A.H.; Maruthi Prasad, E.; Al Omar, S.Y.; Mohammad, A.; Kodidhela, L.D. Combined cardio-protective ability of syringic acid and resveratrol against isoproterenol induced cardio-toxicity in rats via attenuating NF-kB and TNF-α pathways. *Sci. Rep.* **2020**, *10*, 3426. [CrossRef]
35. Siti, H.N.; Jalil, J.; Asmadi, A.Y.; Kamisah, Y. Roles of rutin in cardiac remodeling. *J. Funct. Foods* **2020**, *64*, 103606. [CrossRef]
36. Deng, Q.; Li, X.X.; Fang, Y.; Chen, X.; Xue, J. Therapeutic Potential of Quercetin as an Antiatherosclerotic Agent in Atherosclerotic Cardiovascular Disease: A Review. *J. Evid. Based Complement. Altern. Med.* **2020**, *2020*, 5926381. [CrossRef]
37. Luo, Y.; Shang, P.; Li, D. Luteolin: A Flavonoid that Has Multiple Cardio-Protective Effects and Its Molecular Mechanisms. *Front. Pharmacol.* **2017**, *8*, 692. [CrossRef]
38. Mahajan, U.B.; Chandrayan, G.; Patil, C.R.; Arya, D.S.; Suchal, K.; Agrawal, Y.O.; Ojha, S.; Goyal, S.N. The Protective Effect of Apigenin on Myocardial Injury in Diabetic Rats mediating Activation of the PPAR-γ Pathway. *Int. J. Mol. Sci.* **2017**, *18*, 756. [CrossRef]

39. Epure, A.; Pârvu, A.E.; Vlase, L.; Benedec, D.; Hanganu, D.; Gheldiu, A.-M.; Toma, V.A.; Oniga, I. Phytochemical Profile, Antioxidant, Cardioprotective and Nephroprotective Activity of Romanian Chicory Extract. *Plants* **2021**, *10*, 64. [CrossRef]
40. Toiu, A.; Mocan, A.; Vlase, L.; Pârvu, A.E.; Vodnar, D.C.; Gheldiu, A.M.; Moldovan, C.; Oniga, I. Phytochemical Composition, Antioxidant, Antimicrobial and in Vivo Anti-inflammatory Activity of Traditionally Used Romanian Ajuga laxmannii (Murray) Benth. ("Nobleman's Beard"—Barba Împăratului). *Front. Pharmacol.* **2018**, *9*, 7. [CrossRef]
41. Hanganu, D.; Niculae, M.; Ielciu, I.; Olah, N.-K.; Munteanu, M.; Burtescu, R.; Ştefan, R.; Olar, L.; Pall, E.; Andrei, S.; et al. Chemical Profile, Cytotoxic Activity and Oxidative Stress Reduction of Different *Syringa vulgaris* L. Extracts. *Molecules* **2021**, *26*, 3104. [CrossRef]
42. Oniga, I.; Puşcaş, C.; Silaghi-Dumitrescu, R.; Olah, N.-K.; Sevastre, B.; Marica, R.; Marcus, I.; Sevastre-Berghian, A.C.; Benedec, D.; Pop, C.E.; et al. Origanum vulgare ssp. vulgare: Chemical Composition and Biological Studies. *Molecules* **2018**, *23*, 2077. [CrossRef] [PubMed]
43. Savran, A.; Zengin, G.; Aktumsek, A.; Mocan, A.; Glamoćlija, J.; Ćirić, A.; Soković, M. Phenolic compounds and biological effects of edible Rumex scutatus and Pseudosempervivum sempervivum: Potential sources of natural agents with health benefits. *Food Funct.* **2016**, *7*, 3252–3262. [CrossRef] [PubMed]
44. Benedec, D.; Hanganu, D.; Lorena, F.; Oniga, I.; Brindusa, T.; Olah, N.-K.; Gheldiu, A.-M.; Raita, O.; Vlase, L. Chemical, antioxidant and antibacterial studies of Romanian Heracleum sphondylium. *Farmacia* **2017**, *65*, 252–256.
45. Pop, A.; Bogdan, C.; Fizesan, I.; Iurian, S.; Carpa, R.; Bacali, C.; Vlase, L.; Benedec, D.; Moldovan, M.L. In Vitro Evaluation of Biological Activities of Canes and Pomace Extracts from Several Varieties of *Vitis vinifera* L. for Inclusion in Freeze-Drying Mouthwashes. *Antioxidants* **2022**, *11*, 218. [CrossRef] [PubMed]
46. Vlase, L.; Mocan, A.; Hanganu, D.; Benedec, D.; Gheldiu, A.-M.; Crişan, G. Comparative study of polyphenolic content, antioxidant and antimicrobial activity of four Galium species (Rubiaceae). *Dig. J. Nanomater. Biostructures* **2014**, *9*, 1085–1094.
47. Moldovan, M.L.; Carpa, R.; Fizeşan, I.; Vlase, L.; Bogdan, C.; Iurian, S.M.; Benedec, D.; Pop, A. Phytochemical Profile and Biological Activities of Tendrils and Leaves Extracts from a Variety of *Vitis vinifera* L. *Antioxidants* **2020**, *9*, 373. [CrossRef]
48. Keul, A.; Vlase, L.; Crăciunaş, C. Clonal propagation and production of cichoric acid in three species of Echinaceae. *In Vitro Cell. Dev. Biol. Plant* **2012**, *48*, 249–258. [CrossRef]
49. Parvu, A.E.; Parvu, M.; Vlase, L.; Miclea, P.; Mot, A.C.; Silaghi-Dumitrescu, R. Anti-inflammatory effects of *Allium schoenoprasum* L. leaves. *J. Physiol. Pharmacol.* **2014**, *65*, 309–315.
50. Balea, Ş.S.; Pârvu, A.E.; Pârvu, M.; Vlase, L.; Dehelean, C.A.; Pop, T.I. Antioxidant, Anti-Inflammatory and Antiproliferative Effects of the *Vitis vinifera* L. var. Fetească Neagră and Pinot Noir Pomace Extracts. *Front. Pharmacol.* **2020**, *11*, 990. [CrossRef]
51. Balea, Ş.; Pârvu, A.E.; Pop, N.; Marín, F.Z.; Andreicuț, A.; Pârvu, M. Phytochemical profiling, antioxidant and cardioprotective properties of pinot noir cultivar pomace extracts. *Farmacia* **2018**, *66*, 432–441. [CrossRef]
52. Sarac, F.; Yeniocak, S.; Erbin, A.; Yucetas, E.; Altundal, K.; Ucpinar, B.; Saygili, A.; Koldas, M. Ischemia Modified Albumin and D-dimer in the Diagnosis of Testicular Torsion: An Experimental Model. *Urol. J.* **2019**, *16*, 567–571.
53. Erel, O. A new automated colorimetric method for measuring total oxidant status. *Clin. Biochem.* **2005**, *38*, 1103–1111. [CrossRef]
54. Erel, O. A novel automated method to measure total antioxidant response against potent free radical reactions. *Clin. Biochem.* **2004**, *37*, 112–119. [CrossRef]
55. Toiu, A.; Mocan, A.; Vlase, L.; Pârvu, A.E.; Vodnar, D.C.; Gheldiu, A.-M.; Moldovan, C.; Oniga, I. Comparative Phytochemical Profile, Antioxidant, Antimicrobial and In Vivo Anti-Inflammatory Activity of Different Extracts of Traditionally Used Romanian *Ajuga genevensis* L. and *A. reptans* L. (Lamiaceae). *Molecules* **2019**, *24*, 1597. [CrossRef]
56. Draper, H.H.; Squires, E.J.; Mahmoodi, H.; Wu, J.; Agarwal, S.; Hadley, M. A comparative evaluation of thiobarbituric acid methods for the determination of malondialdehyde in biological materials. *Free. Radic. Biol. Med.* **1993**, *15*, 353–363. [CrossRef]
57. Miranda, K.M.; Espey, M.G.; Wink, D.A. A Rapid, Simple Spectrophotometric Method for Simultaneous Detection of Nitrate and Nitrite. *Nitric Oxide* **2001**, *5*, 62–71. [CrossRef]
58. da Costa, C.M.; dos Santos, R.C.C.; Lima, S.E. A simple automated procedure for thiol measurement in human serum samples. *J. Bras. Patol. Med.* **2006**, *42*, 345–350. [CrossRef]

pharmaceuticals

Article

Margaritaria nobilis L.f. (Phyllanthaceae) Ethanolic Extract: Low Acute Oral Toxicity and Antinociceptive Activity

Fabiana Menezes S. Camara [1], Brenda Costa da Conceição [1,2], Eloise Karoline S. Cardoso [1,2], Johan Carlos C. Santiago [3], Carlos Alberto B. Albuquerque [3], Washington L. Pereira [4], Marta C. Monteiro [2], Consuelo Y. Yoshioka e Silva [2,3], Milton Nascimento da Silva [2,3], Cristiane F. Maia [1,2] and Eneas A. Fontes-Junior [1,2,*]

1 Laboratory of Inflammation and Behavioral Pharmacology (Lafico), Health Science Institute, Federal University of Pará, Belém 66075110, PA, Brazil
2 Pharmaceutical Sciences Post-Graduation Program, Health Sciences Institute, Federal University of Pará, Belém 66075110, PA, Brazil
3 Laboratory of Liquid Chromatography (Labcrol), Exact and Natural Sciences Institute, Federal University of Pará, Belém 66075110, PA, Brazil
4 Animal Pathology Laboratory, Amazon Federal Rural University, Belém 66077830, PA, Brazil
* Correspondence: efontes@ufpa.br

Abstract: *Margaritaria nobilis* L.f. (Phyllanthaceae), a native Brazilian tree occurring mainly in the Amazon, is used in folk medicine for the treatment of abscesses (bark) and cancer-like symptoms (leaves). The present study evaluates the safety of its acute oral administration and its effects on nociception and plasma leakage. The chemical constitution of the leaf's ethanolic extract is determined by ultra-performance liquid chromatography–high-resolution mass spectrometry (LC-MS. Its acute oral toxicity is evaluated in female rats at a dose of 2000 mg/kg, evaluating the occurrence of deaths and Hippocratic, behavioral, hematological, biochemical, and histopathological changes, as well as food and water consumption and weight gain. Antinociceptive activity is evaluated in male mice with acetic-acid-induced peritonitis (APT) and formalin (FT) tests. An open field (OF) test is performed to verify possible interferences in the animals' consciousness or locomotion. LC-MS analysis shows the presence of 44 compounds classified as phenolic acid derivatives, flavonoids and O-glycosylated derivatives, and hydrolyzable tannins. No deaths or significant behavioral, histological, or biochemical changes are observed in the toxicity assessment. In nociception tests, *M. nobilis* extract significantly reduces abdominal contortions in APT, demonstrating selectivity for inflammatory components (FT second phase), not interfering in neuropathic components (FT first phase) or consciousness and locomotion levels in OF. Additionally, *M. nobilis* extract inhibits plasma acetic-acid-induced leakage. These data demonstrate the low toxicity of *M. nobilis* ethanolic extract, as well as its effectiveness in modulating inflammatory nociception and plasma leakage, possibly related to the flavonoids and tannins present in its composition.

Keywords: *Margaritaria nobilis*; natural products; medicinal plants; toxicity; nociception; pain; antinociceptive

Citation: Camara, F.M.S.; da Conceição, B.C.; Cardoso, E.K.S.; Santiago, J.C.C.; Albuquerque, C.A.B.; Pereira, W.L.; Monteiro, M.C.; Yoshioka e Silva, C.Y.; da Silva, M.N.; Maia, C.F.; et al. *Margaritaria nobilis* L.f. (Phyllanthaceae) Ethanolic Extract: Low Acute Oral Toxicity and Antinociceptive Activity. *Pharmaceuticals* **2023**, *16*, 689. https://doi.org/10.3390/ph16050689

Academic Editor: Diana Roxana Pelinescu

Received: 6 March 2023
Revised: 14 April 2023
Accepted: 20 April 2023
Published: 3 May 2023

1. Introduction

Treatment of inflammation and pain management is still a major challenge for global public health. Although they are physiological processes, they are closely linked to the genesis and evolution of numerous pathologies [1]. Therefore, adequately treating inflammation and pain can positively impact patients' clinical improvement and quality of life. This reality clarifies that analgesic and anti-inflammatory drugs are among the most prescribed and used in many countries, except for those where traditional medicine and medicinal plants are widely included in national public health policies [2].

Orthodox therapy, with steroidal (SAIDs) or non-steroidal (NSAIDs) anti-inflammatory drugs, opioids, antidepressants, antiepileptics, and sedatives, although efficient in many cases, find limitations in a significant variety of diseases, mainly when the risks outweigh the observed benefits [3]. In acute conditions, the need for high doses, which affect the stomach, liver, blood, and kidneys, is the main concern, in addition to respiratory and CNS depression, with possible cardiac changes as well [2,4]. Chronic conditions, however, which require prolonged or even continuous use of these medications, are the most challenging. In addition to worsening the adverse reactions already mentioned, they are associated with the induction of metabolic and immunological disorders, tolerance, dependence, and important behavioral impairments [3,4].

Therefore, the search for innovative therapeutic strategies, ranging from herbal medicines to new, safer, and more effective drugs, becomes imperative. In this sense, the medicinal culture of traditional populations proves to be an important source of information about plant species with pharmacological potential. It also provides clues about the safety of their use, reaffirming the value of ethnopharmacological knowledge [5].

Plant species *Margaritaria nobilis* L.f. (synonyms: *Phyllanthus antillanus* (A. Juss.) Müll. Arg.; *P. nobilis* (L.f.) Müll. Arg.; *P. ibonensis* Rusby; *Bradleia sinica* Müll. Arg.; *Cicca antillana* A.Juss.; *C. pavoniana* Baill.; *C. chinensis* Baill.; *C. surinamensis* Miq.; *C. sinica* Baill.; *Diasperus antillanus* (A. Juss.) Kuntze; *Margaritaria adelioides* Rich. ex Baill.; *M. alternifolia* L.; among others), popularly known as 'botãozinho' (little button), 'figueirinha' (little fig tree), 'cabelo-de-cotia' (agouti hair), 'café-bravo' (wild coffee) and 'fruto-de-jacamin' (jacamin fruit), for a long time was considered part of the genus Phyllanthus (family: Phyllanthaceae). Recently, however, it was reclassified based on phylogenetic studies, being now placed in the genus Margaritaria (family: Phyllanthaceae). This tree is native to Brazilian territory and distributed in all of the region's states, emphasizing the Amazon [6]. In traditional medicine, its bark is cited for treating abscesses, fruits as a tonic, and leaves for treating cancer-like symptoms [7].

There are few studies regarding the chemical constitution of *M. nobilis*, such as the study by Moraes et al. [8], which addresses the phytochemical profile and leishmanicidal activity, and the study recently published with comprehensive characterization, indicating the presence of derivatives of phenolic acids, flavonoids, and hydrolyzable tannins [7]. In the genus *Margaritaria*, the presence of phenolic derivatives, glycosylated flavonoids (*M. discoidea*), and alkaloids (*M. indica*) has been described, together with their potential as antioxidant, antinociceptive, and anti-inflammatory agents [9–13]. The present study, therefore, aims to investigate the safety of acute oral administration of *M. nobilis* leaves ethanolic extract (MnE), considering behavioral, anatomical, histological, and biochemical aspects. It also seeks to elucidate its properties on nociception and inflammatory components.

2. Results

2.1. MnE Phytochemical Composition

Through the LC-MS technique, 44 chemical compounds of the sample under study were annotated. The chromatogram of total MnE ions can be seen in Figure 1; these compounds were classified into three major groups: phenolic acids, flavonoids, and hydrolyzable tannins, which can be seen in Table 1 with information on their *m/z* ratio, retention time, and their fragmentation patterns. Most of the annotated compounds are hydrolyzable tannins, a subclass of ellagitannins.

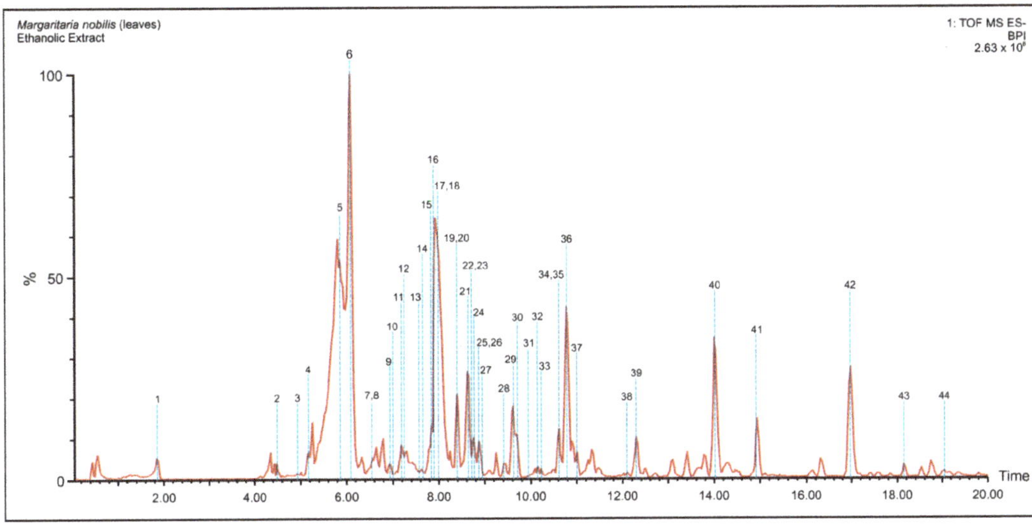

Figure 1. Total ion current chromatograms of *Margaritaria nobilis* ethanolic extract (MnE). The blue dashed lines indicate the retention time (RT) in which the compounds were recorded, being numbered in correspondence with Table 1.

Table 1. Compounds identified by LC-HRMS-ToF in *Margaritaria nobilis* ethanolic extract (MnE).

Peak	RT (min)	Compound	Formula	[M−H]−Exp.	Error (ppm)
1	1.84	Gallic acid	$C_7H_6O_5$	169.0138	0.6
2	4.51	*O*-Coumaroylgalactaric acid	$C_{15}H_{16}O_{10}$	355.0661	1.1
3	4.91	*O*-Feruloylgalactaric acid	$C_{16}H_{18}O_{11}$	385.0766	1.3
4	5.17	Methyl gallate	$C_8H_8O_5$	183.0285	4.4
5	5.92	Galloyl-DHHDP-HHDP-glucose	$C_{41}H_{28}O_{27}$	951.0703	3.9
6	6.09	Galloyl-HHDP-glucose	$C_{27}H_{22}O_{18}$	633.0710	2.8
7	6.56	Galloyl-Che-HHDP-glucose Isomer I	$C_{41}H_{30}O_{27}$	953.0888	0.8
8	6.56	Trigalloyl-glucose	$C_{27}H_{24}O_{18}$	635.0866	2.8
9	6.92	*p*-Coumaric acid	$C_9H_8O_3$	163.0389	3.7
10	6.98	Quercetin 3-*O*-glucosyl-glucoside	$C_{27}H_{30}O_{17}$	625.1368	5.9
11	7.18	Ethyl gallate	$C_9H_{10}O_5$	197.0445	2.5
12	7.24	Phyllanthusiin C Isomer	$C_{40}H_{30}O_{26}$	925.0983	3.6
13	7.55	Ellagic acid *O*-xyloside	$C_{19}H_{14}O_{12}$	433.0410	0.7
14	7.67	Quercetin 3-*O*-xylosyl-glucoside	$C_{26}H_{28}O_{16}$	595.1321	3.7
15	7.84	Quercetin 3-*O*-rhamnosyl-glucoside	$C_{27}H_{30}O_{16}$	609.1427	4.8
16	7.87	Ellagic acid *O*-rhamnoside	$C_{20}H_{16}O_{12}$	447.0585	4.7
17	7.96	Galloyl-Che-HHDP-glucose Isomer II	$C_{41}H_{30}O_{27}$	953.0904	0.8
18	8.01	Ellagic acid	$C_{14}H_6O_8$	300.9972	4.0
19	8.39	Digalloyl-HHDP-glucose	$C_{34}H_{26}O_{22}$	785.0847	1.3
20	8.39	Methyl neochebulagate Isomer	$C_{42}H_{34}O_{28}$	985.1155	0.3
21	8.62	Quercetin 3-*O*-glucoside Isomer I	$C_{21}H_{20}O_{12}$	463.0890	2.8
22	8.73	Excoecariphenol C Isomer	$C_{37}H_{30}O_{24}$	857.1077	3.3
23	8.73	Tetragalloyl-glucose	$C_{34}H_{28}O_{22}$	787.0977	2.2
24	8.76	Kaempferol 3-*O*-rhamnosyl-glucoside	$C_{27}H_{30}O_{15}$	593.1528	3.7
25	8.87	Kaempferol 3-*O*-xylosyl-glucoside	$C_{26}H_{28}O_{15}$	579.1376	4.0
26	8.87	Quercetin 3-*O*-glucoside Isomer II	$C_{21}H_{20}O_{12}$	463.0898	4.5
27	8.93	Di-*O*-Methyl ellagic acid *O*-glucoside	$C_{22}H_{20}O_{13}$	491.0852	5.3
28	9.41	Quercetin 3-*O*-rhamnosyl-xyloside	$C_{26}H_{28}O_{15}$	579.1350	0.0
29	9.61	Quercetin 3-*O*-xyloside	$C_{20}H_{18}O_{11}$	433.0765	1.4
30	9.70	Kaempferol 3-*O*-glucoside Isomer I	$C_{21}H_{20}O_{11}$	447.0935	1.8

Table 1. *Cont.*

Peak	RT (min)	Compound	Formula	[M−H]−Exp.	Error (ppm)
31	9.95	Galloyl-HHDP-di-deoxyglucose	$C_{27}H_{24}O_{16}$	603.0945	6.8
32	10.15	Trigalloyl-dideoxyglucose	$C_{27}H_{24}O_{16}$	603.1013	4.5
33	10.24	Kaempferol 3-*O*-glucoside Isomer II	$C_{21}H_{20}O_{11}$	447.0914	2.9
34	10.61	Kaempferol 3-*O*-rhamnosyl-xyloside	$C_{26}H_{27}O_{14}$	563.1431	5.3
35	10.69	Kaempferol 3-*O*-xyloside	$C_{20}H_{18}O_{10}$	417.0836	3.4
36	10.78	Methylellagic acid *O*-rhamnoside	$C_{21}H_{18}O_{12}$	461.0736	3.5
37	11.01	Phyllanthusiin A Isomer	$C_{41}H_{28}O_{27}$	951.0743	0.3
38	12.10	Trigalloyl-HHDP-glucose	$C_{41}H_{30}O_{26}$	937.0962	1.6
39	12.29	Phyllanthusiin U Isomer	$C_{40}H_{28}O_{26}$	923.0801	1.1
40	14.00	Quercetin	$C_{15}H_{10}O_7$	301.0334	4.7
41	14.91	Methylquercetin 3-*O*-glucoside	$C_{22}H_{22}O_{12}$	477.1018	3.1
42	16.91	Kaempferol	$C_{15}H_{10}O_6$	285.0399	0.0
43	18.14	Galloyl-Cinnamoyl-HHDP-glucose	$C_{36}H_{28}O_{19}$	763.1154	0.9
44	19.04	Tri-*O*-methyl ellagic acid	$C_{17}H_{12}O_8$	343.0450	1.2

2.2. *MnE Acute Oral Toxicity*

2.2.1. MnE Does Not Change Hippocratic Signs, Motor Behavior, or Emotionality, or Cause Deaths

No significant changes in Hippocratic markers were observed in the four hours (15, 30, 60, and 240 min) following oral administration of MnE (2000 mg/kg), as well as in the 14 subsequent days. Similarly, oral administration of the extract did not alter motor parameters, such as total ambulation (6.72 ± 1.48 m; p = 0.960; Figure 2A) and locomotion speed (0.023 ± 0.005 m/s; p = 0.932; Figure 2B), or emotionality, such as freezing time (220.78 ± 14.93 s; p = 0.619; Figure 2C) and locomotion in the central area (0.29 ± 0.10 m; p = 0.928; Figure 2D) in the open field test, when compared to control animals (total locomotion = 6.84 ± 1.50 m; speed = 0.023 ± 0.005 m/s; freezing = 232.35 ± 16.39 s; central locomotion = 0.27 ± 0.11 m). No deaths were observed during the entire toxicity test period.

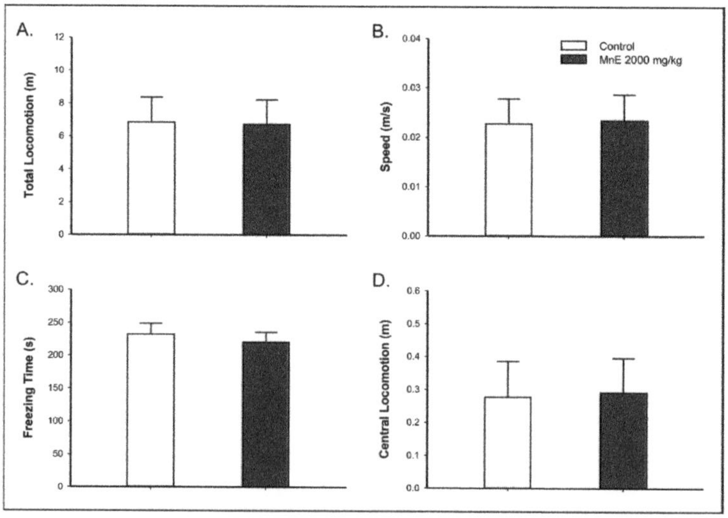

Figure 2. Effects of *Margaritaria nobilis* leaf's ethanolic extract (MnE) acute oral administration (2000 mg/kg) on rats' behavior in open field test, considering (**A**) total locomotion, (**B**) average locomotion speed, (**C**) freezing time, and (**D**) central locomotion. Data are presented as mean ± SEM (n = 5/group). Student's t-test.

2.2.2. MnE Does Not Alter Water or Feed Consumption, Weight Gain, or Organ Histology

Animals treated with MnE (2000 mg/kg; v.o.) showed feed and water consumption patterns equivalent to those observed in the control group (Figure 3A,B), as well as weight gain. No changes in morphology, color, or size were observed during macroscopic analysis of the heart, liver, stomach, and kidneys.

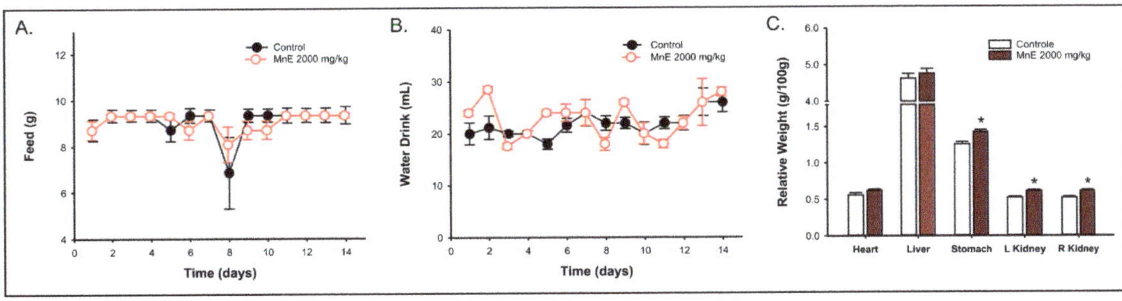

Figure 3. Effects of *Margaritaria nobilis* leaf's ethanolic extract (MnE) on daily (**A**) feed and (**B**) water intake in the 14 days following the acute limit dose (2000 mg/kg) administration and (**C**) on the relative organ-body weights on the 14th day. Data expressed as mean ± SEM (n = 5/group). (**A**) Friedman test; (**B**) one-way RM-ANOVA; (**C**) Student t-test (* $p < 0.05$ vs. control group).

The stomach (1.427 ± 0.024 g/100 g b.w.; p = 0,00316) and kidney (L—0.610 ± 0.015 g/100 g b.w.; p = 0,00362; R—0.614 ± 0.015 g/100 g b.w.; p = 0,00303) relative weights of the animals treated with MnE were higher than the control group (stomach—1.256 ± 0.033 g/100 g b.w.; L kidney—0.521 ± 0.016 g/100 g b.w.; R kidney—0.519 ± 0.017 g/100 g b.w.). The heart and liver have relative weights equivalent to controls (Figure 3C). The histological evaluation did not reveal alterations indicative of toxicity, with histological normality being identified in all kidney and heart samples. Some animals (two controls and three treated) had mild chronic superficial gastritis not associated with acute treatment. The glandular and non-glandular regions of the stomach showed histological normality. Liver tissue, in general, was preserved, with histological normality. In some animals treated with MnE, the discreet presence of cytoplasmic vacuoles was observed but without repercussions for cell function (Figure S1).

2.2.3. MnE Does Not Interfere with the WBC or Liver and Kidney Function Markers

MnE limit dose (2000 mg/kg; v.o.) administration did not change the leukogram of the animals (Table 2). Both groups had normocytic and normochromic red blood cells and no changes were observed in platelets. In addition, no changes were observed in the liver (AST and ALT; Figure 4A) and renal (urea and creatinine; Figure 4B) function markers, which showed plasma concentration equivalent to the control group.

Table 2. Effects of *Margaritaria nobilis* leaf's ethanolic extract (MnE) acute oral administration (2000 mg/kg) on differential leukocyte count. Data presented as mean ± SEM (n = 4–5/group). Student's t-test.

Leukocytes	Control (%)	MnE 2000 mg/kg (%)	p
Segmented neutrophils	37.50 ± 0.64	33.60 ± 2.09	0.153
Banded neutrophils	0.50 ± 0.29	0.20 ± 0.20	0.407
Lymphocytes	59.50 ± 1.19	64.40 ± 2.48	0.147
Monocytes	2.50 ± 0.65	1.80 ± 0.37	0.356
Basophils	0.00 ± 0.00	0.00 ± 0.00	1.000
Eosinophils	0.00 ± 0.00	0.00 ± 0.00	1.000

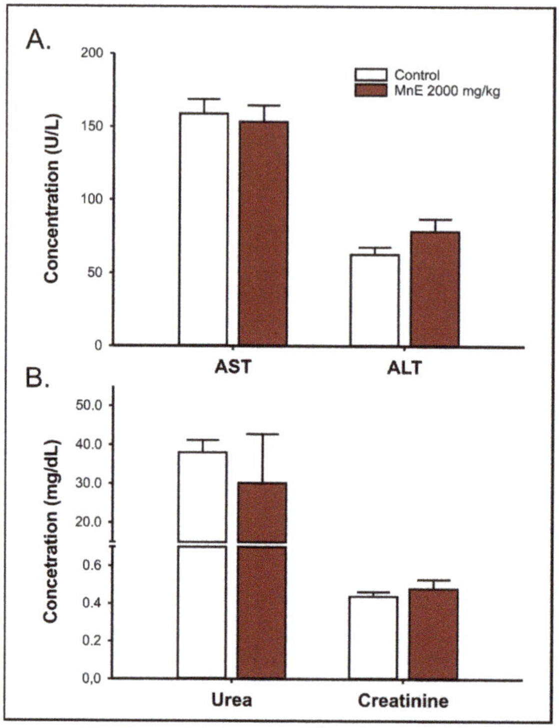

Figure 4. Effects of acute oral administration of *Margaritaria nobilis* ethanolic extract (MnE) (2000 mg/kg) on (**A**) alanine aminotransferase (ALT) and aspartate aminotransferase (AST) activity; and (**B**) urea and creatinine serum concentrations. Data expressed as mean ± SEM (n = 5/group). Student's t-test.

2.3. MnE Antinociceptive and Antiinflamatory Activity

2.3.1. MnE Reduces ACA-Induced Abdominal Writhing and Plasma Leakage

Intraperitoneal administration of ACA (0.6%) produced 67.57 ± 3.33 writhes in the observation interval. The lowest MnE dose (200 mg/kg, v.o.), despite reducing the average of writhing (48.33 ± 10.54), did not present a statistically significant effect ($p = 0.219$). Doses of 400 and 800 mg/kg, on the other hand, reduced acetic-acid-induced writhing by 64.73% (23.83 ± 8.14; $p = 0.001$) and 72.87% (18.33 ± 8.76; $p < 0.001$), respectively, showing an effect like the indomethacin (10 mg/kg; 19.14 ± 2.56) (Figure 5A) and with an estimated median effective dose (ED_{50}) of 338.298 mg/kg ($r^2 = 0.882$) (Figure 5B).

Regarding proteins concentration in peritoneal fluid, the ACA (0.6%; ip.) administration increased it by around 144% (0.996 ± 0.065 mg/mL; $p = 0.01$) compared to the white group (0.408 ± 0.02 mg/mL), which did not receive ACA. Treatment with MnE at 200 mg/kg reduced the mean protein concentration (0.540 ± 0.105 mg/mL), but its effect was not statistically significant ($p = 0.087$) when compared to the ACA group. The dose of 400 mg/kg reversed the ACA-induced damage by 59.43% (0.404 ± 0.044 mg/mL; $p = 0.003$), showing an effect like indomethacin (10 mg/kg; 0.624 ± 0.044 mg/mL). At 800 mg/kg, MnE reduced the elevation in peritoneal fluid protein levels by 63.33% (0.365 ± 0.019 mg/mL; $p < 0.001$), showing an effect superior ($p = 0.022$) to that produced by the standard drug (Figure 5C).

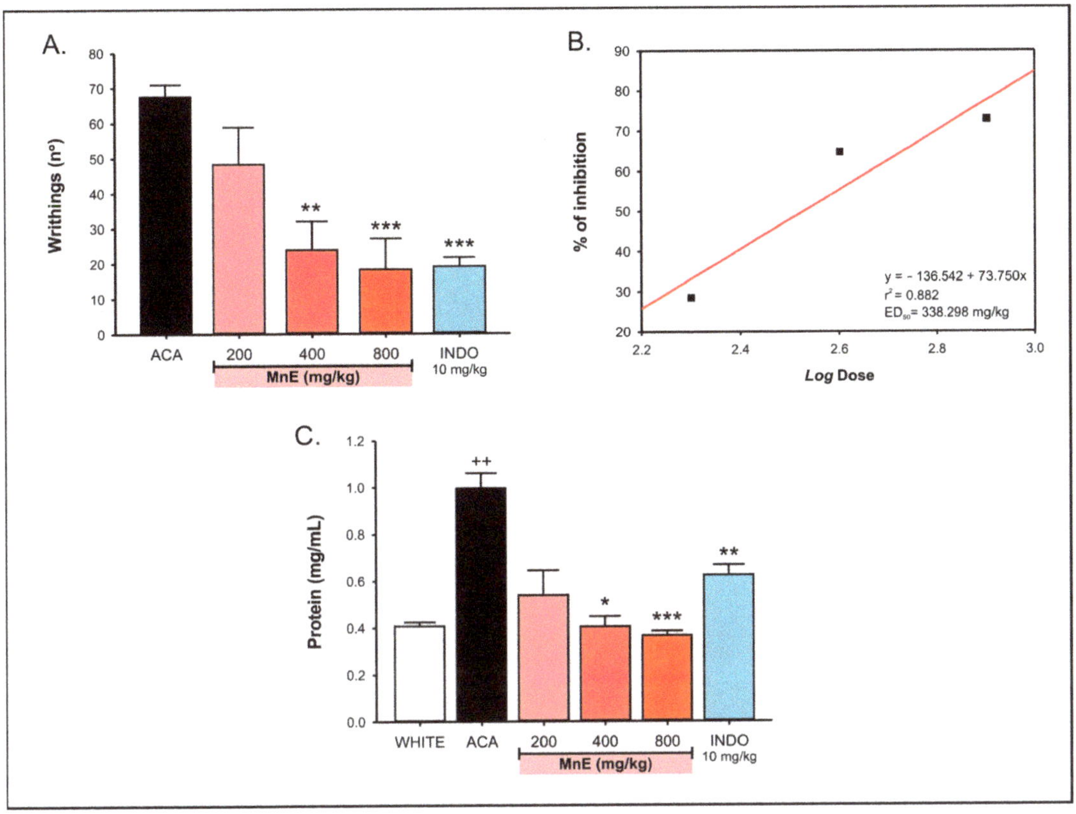

Figure 5. Effect of *Margaritaria nobilis* ethanolic extract (MnE) (200, 400, and 800 mg/kg; v.o.) on (**A**) acetic acid (0.6%)-induced abdominal writhing, with (**B**) identification of its median effective dose (ED_{50}) by linear regression and (**C**) plasma leakage. Data expressed as mean ± SEM (A: n = 6–7/group; B: 5–6/group). $^{++}$ $p < 0.01$ vs. WHITE; * $p < 0.05$, ** $p < 0.01$, *** $p < 0.001$ vs. ACA; (ANOVA, Dunnett's test).

2.3.2. MnE Reduces the Nociception in the Inflammatory Phase of the Formalin Test

Treatment with MnE (400 mg/kg; v.o.; 82.33 ± 14.96 s) did not interfere with formalin-induced licking time (83.33 ± 14.87 s; p = 0.998) during the neuropathic phase. In the inflammatory phase, on the other hand, MnE (400 mg/kg; v.o.; 129.17 ± 12.29 s) significantly reduced the licking time of the animals compared to the control group (171.67 ± 14.85 s; p = 0.031) (Figure 6A).

Additionally, the treatment with MnE (400 mg/kg; 19.06 ± 3.15 m) did not interfere with the locomotor capacity or exploration of the animals, when compared to the control group (17.69 ± 2.27 m; p = 0.809), excluding possible interferences in antinociceptive tests, such as sedation or motor impairment (Figure 6B).

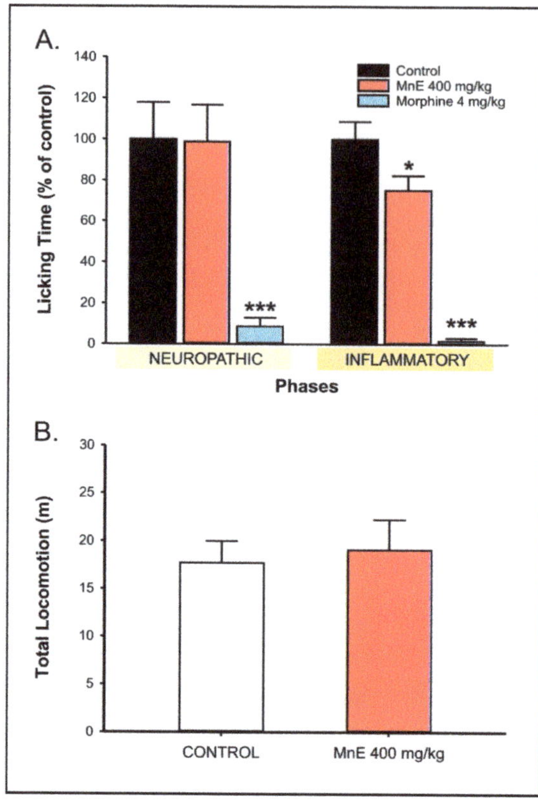

Figure 6. Effect of *Margaritaria nobilis* ethanolic extract (MnE) (200, 400, and 800 mg/kg; v.o.) on (**A**) neuropathic and inflammatory nociception in formalin test (**B**) and total locomotion in the open field. Data expressed as mean ± SEM (A: $n = 6$/group; B: $n = 7$/group). * $p < 0.05$, *** $p < 0.001$ vs. ACA; ((**A**) ANOVA, Dunnett's test; (**B**) Student t-test).

3. Discussion

The present study demonstrates for the first time that MnE (ethanolic extract of *M. nobilis* leaves) has low toxicity when orally administered in an acute pattern, as well as its ability to modulate nociception, possibly by peripheral mechanisms related to inflammation, and to inhibit plasma leakage. Such effects are probably related to its secondary metabolites identified by our group, such as flavonoids, hydrolyzed tannins, and phenolic acids. We described the annotation of 44 phytoconstituents (Table 1) present in the MnE [7]. On this aspect, the metabolic annotation has expedited the clarification of the chemical-pharmacological signature of complex matrices, without ignoring the unequivocal identification that may allow bioactivity of greater magnitudes [14]. Thus, because some of these compounds are structural scaffolds for promising pharmacological segments, we contend that they may explain the observed activities interactively, such as gallic acid derivatives (1), quercetin O-glycosylated derivatives (40), and similar compounds (10, 14, 15, 21, 26, 28, 29, and 41), kaempferol (42) and its glycosylated derivatives, ellagitannins (analogs of 5–7, 17, 31, 38, and 48), and gallotannins (analogs of 8, 23, and 32).

However, research on the efficacy of natural products with therapeutic potential must be preceded by an investigation of the safety of their administration [15]. Secondary metabolites present in numerous genera are extremely important for their biological benefits, but their toxic capacity becomes a limiting factor for the use because of their risk [16].

In this sense, the present study is based on OECD guideline 425 and resolution n° 90 of the Brazilian health surveillance agency (ANVISA), which deals with the tests necessary for the registration of herbal medicines [17,18]. A limit dose (2000 mg/kg) of MnE was then administered orally, with subsequent evaluation of behavioral, physiological, anatomical, histological, and biochemical markers of toxicity.

Considering the lethality parameter, the limit dose administration did not cause immediate deaths over the 14 days of observation, suggesting the relative safety of the natural product. This observation is reinforced by the lack of physiological or behavioral changes that would indicate toxicity. Animals orally treated with MnE did not show changes in nutritional behavior or Hippocratic signs, nor in locomotion and emotionality parameters, evaluated in the open field. Such observations indicate that the extract, even in high doses, does not interfere with vital physiological functions, such as feeding, hydration, and weight gain, or alter the animals' consciousness, mobility, and emotionality [19,20].

Even at therapeutic levels, several analgesic or anti-inflammatory drugs promote significant changes in the heart, stomach, liver, and/or kidneys, a feature that often limits the dose or duration of treatment [21,22]. The macroscopic and histological evaluation of these organs showed the absence of lesions when using the limit dose. These findings were reinforced by the absence of changes in biochemical markers, such as ALT, the most specific indicator of hepatocellular injury, AST, which can reveal lesions in the liver but also in other tissues, and urea and creatinine, markers of renal injury [15]. Similarly, the administration of the MnE limit dose did not promote hematological alterations. Given these findings and following OECD guidelines [17], MnE can be classified as a low-toxicity xenobiotic, with an estimated LD_{50} greater than 2000 mg/kg. Based on this classification, the evaluation of bioactive properties started with a dose of 200 mg/kg, corresponding to 10% of the dose tested in the toxicity assay, being adjusted based on the response, also testing doses of 400 and 800 mg/kg.

Initially, the MnE antinociceptive property was evaluated in the acetic-acid-induced peritonitis model, a classic test with high sensitivity, useful for screening drugs with antinociceptive potential. Intraperitoneal administration of acetic acid triggers local non-selective activation of cation channels, peritoneal membrane irritation, mast cells, and macrophages activation, the elevation of proinflammatory enzyme (nitric oxide synthase, cyclooxygenase, etc.) and mediator (cytokines, prostanoids, leukotrienes, bradykinin, serotonin, nitric oxide, etc.) levels [23]. The installed inflammatory process activates visceral nociceptors, causing primary hyperalgesia and intense nociception, manifested in the form of abdominal writhing [24,25]. Due to its characteristics, acetic-acid-induced peritonitis is sensitive to drugs capable to modulate inflammatory mechanisms of nociception, such as NSAIDs, although its nociception is also inhibited by other types of agents, such as opioids and antispasmodics [23,26]. Our trials demonstrated that MnE, at doses of 400 and 800 mg/kg, reduced by ~65% and ~73%, respectively, the acetic-acid-induced inflammatory nociception, with an estimated ED_{50} of 338.298 mg/kg, suggesting a potential for inflammatory pathways modulation, especially prostaglandins and nitric oxide.

A second consequence of acetic-acid-induced acute peritonitis is vasodilatation and increased vascular permeability, related to mediators such as histamine, prostaglandins, and nitric oxide. This condition favors the migration of leukocytes and plasma proteins to the peritoneal cavity [27]. In fact, in our tests, a great increase (~144%) in protein concentration was observed in the peritoneal fluid of animals administered with acetic acid. Treatment with MnE, mainly at doses of 400 and 800 mg/kg, completely inhibited plasma extravasation, promoting peritoneal protein concentrations equivalent to those of animals that were not injected with acetic acid (white group), reinforcing its ability to modulate inflammatory mechanisms such as vascular relaxation, diapedesis, and plasma leakage, which is associated with the elevation of interleukins, prostaglandins, and nitric oxide, potential targets of MnE action [28].

Next, the formalin test was carried out to evaluate MnE influences on peripheral and central components of nociception, since the intraplantar injection of formalin (2.5%)

induces a characteristic biphasic nociceptive process. In the first 5 min, the nociception observed is due to direct stimulation of sensory terminals, corresponding to the neuropathic component of the test. In this phase, the modulation of nociception has been linked mainly to drugs capable of interfering with neural mechanisms that conduct the noxious stimulus, such as opioids, anticonvulsants, and antidepressants, among others [23,27,29]. MnE did not promote alterations in the neuropathic component of formalin-induced nociception, indicating that its constituent probably does not act on nervous components. A 10-minute hiatus is observed in the formalin test, which is followed by a new period (15–30 min) of nociceptive manifestation, now caused by an inflammatory process triggered by formalin aggression, which involves prostaglandins, histamine, serotonin, and NO and plasma leakage [29,30]. MnE (400 mg/kg) significantly reduced formalin-induced inflammatory nociception.

Some of the secondary metabolites annotated by our group in the sample under study, such as ellagic acid, *p*-coumaric acid, and quercetin 3-O-xyloside, have been attributed to antioxidant activity, with the ability to scavenge reactive oxygen species. Ellagic acid is also capable of positively modulating Nrf2, a key mediator in the regulation of oxidative balance [30–33]. Oxidative stress and inflammation are closely related, as the overproduction of reactive oxygen and nitrogen species intrinsically participates in the cellular and humoral response, also influencing the inflammatory mechanisms of pain [30]. Furthermore, ellagic acid, gallic acid, methyl gallate, *p*-coumaric acid, quercetin 3-O-xyloside, quercetin, and kaempferol have also demonstrated important anti-inflammatory activity, related to the suppression of synthesis and release of pro-inflammatory cytokines (e.g., IL-1β, IL-6, IL-17, TNFα, INFγ) and elevation of anti-inflammatory cytokines (e.g., IL-4, IL-10) and to the reduction in expression and activity of pro-inflammatory enzymes (e.g., COX-2, iNOS) and mediators (e.g., PGE2, NO), in addition to the ability to modulate nuclear factor kB (NF-kB), a key element in the inflammatory response regulation, attributed to *p*-coumaric acid [30,32–40].

Such observations, in association with our findings, reinforce the thesis that the reduction in acetic-acid- and formalin-induced nociception and acetic-acid-induced leakage promised by MnE is due to the modulation of pain inflammatory components, possibly a product of the synergistic action of its metabolites. We cannot, however, rule out other accessory mechanisms, such as antihistamine, antispasmodic or adrenergic activity, which should be explored in future essays. The other compounds annotated by our group, which lack studies on their biological properties, should also be explored. The present study, by exposing the phytochemical profile of MnE, annotating 44 constituent substances, and demonstrating its low acute toxicity and antinociceptive potential, paves the way for new studies on the medicinal use of *M. nobilis*, expanding knowledge about the safety of its use, considering repeated administrations in a subchronic or chronic pattern, its antioxidant and anti-inflammatory properties, and underlying mechanisms—both of its extracts and the molecules identified.

4. Materials and Methods

4.1. Chemicals and Reagents

In the present study, all reagents were of analytical grade. As an extraction solvent, 99% absolute ethanol was used (Exodus Scientific, São Paulo, SP, Brazil). Ultrapure water, acetonitrile, and concentrated formic acid solution, obtained from Merck (Darmstadt, Germany), were used to prepare the mobile phase and solubilize the sample for LC-MS analyses. The dimethyl sulfoxide (DMSO, Sigma-Aldrich, St. Louis, MO, USA), ALT, AST, urea, and creatinine assay kits was from VIDA (Belo Horizonte, MG, Brazil). The acetic acid was from (ACA) and formaldehyde from (Vetec Química Fina, Rio de Janeiro, RJ, Brazil), the indomethacin from (Sigma-Aldrich, St. Louis, MO, USA) and morphine sulfate from (Cristália, Rio de Janeiro, RJ, Brazil).

4.2. Plant Collection and Extract Preparation Protocol

Leaf samples of *M. nobilis* (~1 kg) were collected at 1°02′08″ S and 46°49′41″ W (forest region of Bragança-PA, Brazil), identified by botanist Nascimento, E.A.P. (Brazilian Company of Agricultural Research—EMBRAPA), and a voucher specimen was deposited in its IAN herbarium (registration n° 191496).

Botanical material was washed, dried in a circulating oven (45 °C) until constant weight, and then pulverized in a ball mill (Fritsch, Idar-Oberstein, Germany). The semi-fine powder (60–100 μm) obtained was subjected to two sequential 24 h cycles of extraction with ethanol (99%; 4 L for each 1.0 kg of powder). The extract was concentrated in a rotary evaporator (Büchi, Flawil, Germany) and dried in an oven (40 °C) until constant weight.

4.3. Phytochemical Analysis

Metabolic characterization of MnE was obtained using a liquid chromatography system coupled to a time-of-flight mass spectrometer and triple quadrupole analyzer (model UPLC-QTf Xevo G2-S™, Waters, Milford, MA, USA). The chromatographic separation method was a 30 min gradient elution (0–2 min, 10–20% B; 2–30 min, 20–50% B). A 2 μL aliquot of the extract (3 $mg.mL^{-1}$) was eluted on a BEH C18 column (Waters, Wexford, Ireland; 50 × 2.1 mm i.d., particle size 1.7 μm) at 40 °C and under a flow of 0.3 mL/mL of ultrapure water (solvent A) and acetonitrile (solvent B) + 0.1% (*v*/*v*) formic acid. The analyzed ionization mode was negative with a mass range of 100 to 1200 *m/z* and a scanning interval of 0.1 s. The source was maintained at 120 °C and the gas flow was adjusted to 50 $L.h^{-1}$. The desolvation gas was set to 800 $L.h^{-1}$ and 450 °C. The capillary voltage was set at 2.0 kV with cone voltage at 80 V. Data acquisition and processing was performed by MassLynx software licensed from Waters®.

4.4. Animals

Female Wistar rats (150–200 g) and male mice (25–30 g) were provided by the Federal University of Pará (UFPA) vivarium. They were kept in a standardized environment, with exhaustion, acclimatization (22 ± 1 °C) and light cycle (light 6 a.m. to 6 p.m.), in polypropylene cages (39 × 32 × 16 cm; up to five/box). Water and food were available ad libitum [39]. The experimental protocols were approved by the UFPA ethics committee (protocol n° 9568260617) and performed in a sound-attenuated laboratory under low-intensity light (12 lux), between 12:00 a.m. and 5:00 p.m.

4.5. Drug Solutions and Administration

MnE was solubilized in saline solution added with 4% DMSO. Standard drugs were dissolved in 0.9% saline. ACA and formaldehyde were dissolved in distilled water. Solutions were prepared with an administration pattern according to body weight (mice: 0.1 mL/10 g; rats: 0.1 mL/100 g). The oral (v.o.; gavage—MnE and indomethacin), subcutaneous (sc.—morphine), intraperitoneal (ip.—ACA), and intraplantar (formaldehyde) routes were adopted.

4.6. Acute Oral Toxicity

4.6.1. Treatment and Hippocratic Screening

The toxicity study was conducted by Guidelines for Testing Chemicals n° 425 of the Organization for Economic Cooperation and Development (OECD) and resolution 90 of the Brazilian health surveillance agency [18,41]. Therefore, female rats (five/group) were treated with 0.9% saline added with 4% DMSO (control) or a limit dose (2000 mg/kg) of MnE.

Manifestation of signs of toxicity and death was then evaluated, as described by Malone (1977) [42], at intervals of 0, 15, 30, 60, and 240 min after administration, and daily thereafter for 14 days. Animals' weights, as well as the consumption of water and feed, were recorded daily too. The rats were euthanized by cervical dislocation under anesthesia at the end of the evaluation period. Blood samples were then collected by a ventricular puncture for hematological and biochemical assays. Heart, liver, stomach, and kidneys were collected for relative weight and macroscopic and histological characteristics evaluations.

4.6.2. Open Field Test (OF)

The open field test was performed 4 h after MnE limit dose (2000 mg/kg) administration to evaluate possible harmful effects on mobility and emotionality. For that, animals were positioned individually in the center of a square arena (width 100 × depth 100 × height 40 cm), the spontaneous locomotion being evaluated for 5 min [19]. The experiment was recorded by a cam positioned above the arena. Total locomotion, locomotion speed, freezing time, and locomotion in the center of the arena were evaluated using ANY-maze® (Stoelting Co., Wood Dale, IL, USA) software.

4.6.3. Blood Count

Differential leukocyte counts and evaluation of possible abnormalities in red blood cells, leukocytes, and platelets were performed using the blood smear technique. Therefore, smears were prepared immediately after blood collection by the wedge method [43]. The evaluation was performed using an optical microscope (Nikon Eclipse E200, Melville, NY, USA).

4.6.4. Biochemical Assays

Sample

Blood samples were subjected to centrifugation (10 min at 1400× *g*) for plasma separation, which was used for alanine aminotransferase (ALT) and aspartate aminotransferase (AST) activity determination and urea and creatinine concentration.

Hepatic Function Assays

Alanine Aminotransferase (ALT) Activity

Determined through the reaction between pyruvate, formed by the transfer of amino groups from alanine to ketoglutarate, catalyzed by ALT, and NADH, forming L-lactate and NAD+, catalyzed by lactate dehydrogenase (LDH). ALT catalytic concentration (U/L) was determined from the rate of consumption of NADH, measured by spectrophotometry (λ = 340 nm) [44,45].

Aspartate Aminotransferase (AST) Activity

Determined through the reaction between oxaloacetate, formed by the transfer of amino groups from aspartate to ketoglutarate, catalyzed by AST and NADH, forming L-malate and NAD+, catalyzed by malate dehydrogenase (MDH). AST catalytic concentration (U/L) was determined from the rate of NADH consumption, measured by spectrophotometry (λ = 340 nm) [45,46].

Kidney Function Assays

Creatinine

Serum creatinine was determined by its reaction with picric acid (Jaffe reaction), which forms a yellowish-red chromogen whose intensity, measured by spectrophotometry (λ = 510 nm), is proportional to its concentration (mg/dL) [47].

Urea

Urea concentration was measured in serum, based on the reaction between ammonia, formed by hydrolysis of urea in an aqueous medium, with alpha-ketoglutarate and NADH, catalyzed by glutamate dehydrogenase (GLDH), which forms glutamate and NAD+. NADH consumption, measured by spectrophotometry (λ = 340 nm), is proportional to urea concentration [48].

4.6.5. Histopathological Analysis

The heart, liver, stomach, and kidneys were weighed, fixed in buffered formalin (10%), and embedded in paraffin. Sections of 5 µm (thick) were then obtained, which were subjected to an alcohol-xylene series and mounted on slides. Staining was performed with hematoxylin and eosin (H&E), and the examination was made under an optical microscope (Nikon Eclipse E200).

4.7. Antinociceptive Activity

4.7.1. Acetic-Acid-Induced Peritonitis

Based on the Koster et al. (1959) [49] model, cavity inflammation was induced by ip injection of ACA (0.6% *v*/*v*) in mice (n = 7/group) pretreated (60 min) with saline added 4% DMSO (Control); MnE increasing doses (200, 400, and 800 mg/kg) or indomethacin (INDO, 10 mg/kg). Nociception was then measured by the number of writhes manifested between 10 to 30 min after ACA administration.

After euthanasia by cervical dislocation under anesthesia, 3 mL of saline was injected into the peritoneal cavity of the animals. Peritoneal wash was collected, and protein concentration (mg/mL) was measured by Lory's (1951) method [50], as an indirect marker of plasma leakage. A group that was not given ACA (white) was added to determine baseline protein concentrations.

4.7.2. Formalin Test (FT)

To assess the participation of neural and/or inflammatory processes, biphasic nociception was induced by plantar injection (sc; right hind paw) of 20 µL of formalin (0.92%) [51] of mice pretreated (60 min) with saline added 4% DMSO (Control), MnE (400 mg/kg), or morphine (4 mg/kg). Morphine was administrated 30 min before the noxious stimulus.

Nociception was measured through the time expended licking the formalin-injected paw, being evaluated in two phases: phase I (0–5 min), neuropathic nociception, which is triggered by direct stimulation of sensory terminals by formalin; and phase II (15–30 min) is inflammatory nociception, generated by the consequent inflammatory process [22].

4.7.3. Open Field (OF) Test

To verify possible effects on consciousness or mobility that could compromise the animal's performance submitted to formalin test, control and MnE groups were evaluated in the open field, as described above, 5 min before exposure to the noxious stimulus.

4.8. Statistical Analysis

Data are presented by the mean ± standard error of the mean (SEM). The Shapiro–Wilk method was applied to evaluate the data distribution. Difference between groups with Gaussian distribution was evaluated by Student's *t*-test, one-way ANOVA, followed by Dunnett's test, or one-way RM-ANOVA, followed by Holm Sidak's test. Groups with a non-Gaussian distribution were evaluated using Mann–Whitney's test; Kruskal–Wallis's test followed by Dunn's test or Friedman's test. We considered statistically significant the differences with $p < 0.05$.

5. Conclusions

In summary, our results present the chromatographic profile of MnE (ethanolic extract of *M. nobilis*), identifying 44 secondary metabolites, classified as phenolic acids, flavonoids, and hydrolyzed tannins. We demonstrate that its LD_{50} for acute oral administration is above 2000 mg/kg, being classified as a low-toxicity xenobiotic. It also shows its activity in reducing nociception, as well as inhibiting plasma extravasation, both associated with inflammatory processes, which probably involves the ability of its secondary metabolites, such as ellagic acid, gallic acid, methyl gallate, *p*-coumaric acid, quercetin 3-O-xyloside, quercetin, and kaempferol, to reduce the synthesis and/or release of inflammatory enzymes and mediators, such as COX-2, iNOS, cytokines, prostanoids, and NO, in addition to increasing the production of anti-inflammatory cytokines and presenting antioxidant activity. We hypothesize, therefore, that the MnE antinociceptive and plasma leakage inhibition properties are the product of the synergistic action of its constituents, still considering that several of the substances annotated in the MnE still need to have their properties elucidated, as well as other mechanisms to be explored, which should compose future investigations. The present study, therefore, highlights the potential of this Amazonian species for the development of therapeutic agents devoted to the treatment of painful and inflammatory conditions.

Supplementary Materials: The following supporting information can be downloaded at: https://www.mdpi.com/article/10.3390/ph16050689/s1, Figure S1: Histological evaluation of the stomach, kidney, liver, and heart of rats acutely treated with MnE limit dose.

Author Contributions: Conceptualization, E.A.F.-J. and C.Y.Y.e.S.; formal analysis, E.A.F.-J., F.M.S.C., J.C.C.S., C.A.B.A., W.L.P., M.C.M., C.Y.Y.e.S., M.N.d.S. and C.F.M.; investigation, E.A.F.-J., F.M.S.C., J.C.C.S., C.A.B.A., W.L.P., M.C.M., C.Y.Y.e.S., M.N.d.S. and C.F.M.; methodology, E.A.F.-J., F.M.S.C., B.C.d.C., E.K.S.C. and J.C.C.S.; supervision, E.A.F.-J., and C.Y.Y.e.S.; writing—original draft, E.A.F.-J., F.M.S.C., B.C.d.C., E.K.S.C., and J.C.C.S.; writing—review and editing, J.C.C.S., C.A.B.A., C.Y.Y.e.S.; C.F.M. and E.A.F.-J. All authors have read and agreed to the published version of the manuscript.

Funding: The Research Pro-Rectory of the Federal University of Pará (PROPESP, UFPA, Brazil) provided a research grant and the article publication fee; the Fundação Amazônia de Amparo a Estudos e Pesquisas provided a research grant (EDITAL 01/2023–PROPESP and PRO463-2017).

Institutional Review Board Statement: The study was approved by the Federal University of Pará Ethics Committee (code: 9568260617), being carried out following the guidelines of the Care and Use of Laboratory Animals Guide (2011).

Informed Consent Statement: Not applicable.

Data Availability Statement: Data is contained within the article and supplementary material.

Conflicts of Interest: The authors declare no conflict of interest.

References

1. Karshikoff, B.; Tadros, M.A.; Mackey, S.; Zouikr, I. Neuroimmune Modulation of Pain across the Developmental Spectrum. *Curr. Opin. Behav. Sci.* **2019**, *28*, 85–92. [CrossRef] [PubMed]
2. Zhao, Q.; Zhu, L.; Wang, S.; Gao, Y.; Jin, F. Molecular Mechanism of the Anti-Inflammatory Effects of Plant Essential Oils: A Systematic Review. *J. Ethnopharmacol.* **2023**, *301*, 115829. [CrossRef] [PubMed]
3. Zobdeh, F.; Eremenko, I.I.; Akan, M.A.; Tarasov, V.V.; Chubarev, V.N.; Schiöth, H.B.; Mwinyi, J. Pharmacogenetics, and Pain Treatment with a Focus on Non-Steroidal Anti-Inflammatory Drugs (NSAIDs) and Antidepressants: A Systematic Review. *Pharmaceutics* **2022**, *14*, 1190. [CrossRef]
4. Kim, K.-H.; Seo, H.-J.; Abdi, S.; Huh, B. All about Pain Pharmacology: What Pain Physicians Should Know. *Korean J. Pain* **2020**, *33*, 108–120. [CrossRef] [PubMed]
5. Süntar, I. Importance of Ethnopharmacological Studies in Drug Discovery: Role of Medicinal Plants. *Phytochem. Rev.* **2020**, *19*, 1199–1209. [CrossRef]
6. WFO. *Margaritaria nobilis* L.f. Published on the Internet. 2022. Available online: http://www.worldfloraonline.org/taxon/wfo-0000236325 (accessed on 8 December 2022).
7. Santiago, J.C.C.; Albuquerque, C.A.B.; Muribeca, A.D.J.B.; Sá, P.R.C.; Pamplona, S.D.G.S.R.; Silva, C.Y.Y.E.; Ribera, P.C.; Fontes-Júnior, E.D.A.; Da Silva, M.N. *Margaritaria nobilis* L.F. (Phyllanthaceae): Ethnopharmacology and Application of Computational Tools in the Annotation of Bioactive Molecules. *Metabolites* **2022**, *12*, 681. [CrossRef]
8. Moraes, L.S.; Donza, M.R.H.; Rodrigues, A.P.D.; Silva, B.J.M.; Brasil, D.S.B.; Zoghbi, M.D.G.B.; Andrade, E.H.A.; Guilhon, G.M.S.P.; Silva, E.O. Leishmanicidal Activity of (+)-Phyllanthidine and the Phytochemical Profile of *Margaritaria nobilis* (Phyllanthaceae). *Molecules* **2015**, *20*, 22157–22169. [CrossRef]
9. Arbain, D.; Byrne, L.; Cannon, J.; Engelhardt, L.; White, A. The Alkaloids of *Margaritaria indica* (Euphorbiaceae). The Crystal Structure and Absolute Configuration of the Hydrobromide of (+)-15α-Methoxy-14,15-Dihydrophyllochrysine. *Aust. J. Chem.* **1990**, *43*, 439. [CrossRef]
10. Adedapo, A.A.; Sofidiya, M.O.; Afolayan, A.J. Anti-Inflammatory and Analgesic Activities of the Aqueous Extracts of *Margaritaria discoidea* (Euphorbiaceae) Stem Bark in Experimental Animal Models. *Regist. Behav. Tech.* **2008**, *57*, 4. [CrossRef]
11. Dickson, R.; Fleischer, T.; Ekuadzi, E.; Mensah, A.; Annan, K.; Woode, E. Antibacterial, Antioxidant and Anti-Inflammatory Properties of *Margaritaria discoidea*, a Wound Healing Remedy from Ghana. *Pharmacogn. J.* **2010**, *2*, 32–39. [CrossRef]
12. Ekuadzi, E.; Dickson, R.; Fleischer, T.; Annan, K.; Pistorius, D.; Oberer, L.; Gibbons, S. Flavonoid Glycosides from the Stem Bark of *Margaritaria discoidea* Demonstrate Antibacterial and Free Radical Scavenging Activities: Antibacterial and Antioxidant Flavonoid Glycosides. *Phytother. Res.* **2014**, *28*, 784–787. [CrossRef] [PubMed]

13. Johnson-Ajinwo, O.R.; Richardson, A.; Li, W.-W. Cytotoxic Effects of Stem Bark Extracts and Pure Compounds from *Margaritaria discoidea* on Human Ovarian Cancer Cell Lines. *Phytomedicine* **2015**, *22*, 1–4. [CrossRef] [PubMed]
14. Hell, T.; Rutz, A.; Dürr, L.; Dobrzyński, M.; Reinhardt, J.K.; Lehner, T.; Keller, M.; John, A.; Gupta, M.; Pertz, O.; et al. Combining Activity Profiling with Advanced Annotation to Accelerate the Discovery of Natural Products Targeting Oncogenic Signaling in Melanoma. *Natl. Prod. Acad. Sci. USA* **2022**, *85*, 1540–1554. [CrossRef] [PubMed]
15. Benrahou, K.; Mrabti, H.N.; Assaggaf, H.M.; Mortada, S.; Salhi, N.; Rouas, L.; El Bacha, R.; Dami, A.; Masrar, A.; Alshahrani, M.M.; et al. Acute and Subacute Toxicity Studies of *Erodium guttatum* Extracts by Oral Administration in Rodents. *Toxins* **2022**, *14*, 735. [CrossRef] [PubMed]
16. Kpemissi, M.; Metowogo, K.; Melila, M.; Veerapur, V.P.; Negru, M.; Taulescu, M.; Potârniche, A.-V.; Suhas, D.S.; Puneeth, T.A.; Vijayakumar, S.; et al. Acute and Subchronic Oral Toxicity Assessments of *Combretum micranthum* (Combretaceae) in Wistar Rats. *Toxicol. Rep.* **2020**, *7*, 162–168. [CrossRef]
17. OECD. *Test No. 420: Acute Oral Toxicity—Fixed Dose Procedure. OECD Guidelines for the Testing of Chemicals, Section 4*; OECD: Pairs, France, 2002. [CrossRef]
18. Ministério da Saúde. Resolução-Re N° 90. 16 March 2004. Available online: https://bvsms.saude.gov.br/bvs/saudelegis/anvisa/2004/rdc0090_16_03_2004.html (accessed on 5 March 2023).
19. Jothy, S.L.; Zakaria, Z.; Chen, Y.; Lau, Y.L.; Latha, L.Y.; Sasidharan, S. Acute Oral Toxicity of Methanolic Seed Extract of Cassia Fistula in Mice. *Molecules* **2011**, *16*, 5268–5282. [CrossRef]
20. Shahed-Al-Mahmud, M.; Lina, S.M.M. Evaluation of Sedative and Anxiolytic Activities of Methanol Extract of Leaves of *Persicaria hydropiper* in Mice. *Clin. Phytosci.* **2017**, *3*, 20. [CrossRef]
21. Baldo, B.A. Toxicities of Opioid Analgesics: Respiratory Depression, Histamine Release, Hemodynamic Changes, Hypersensitivity, Serotonin Toxicity. *Arch. Toxicol.* **2021**, *95*, 2627–2642. [CrossRef]
22. Bindu, S.; Mazumder, S.; Bandyopadhyay, U. Non-Steroidal Anti-Inflammatory Drugs (NSAIDs) and Organ Damage: A Current Perspective. *Biochem. Pharmacol.* **2020**, *180*, 114147. [CrossRef]
23. Lopes, K.; Oliveira, J.; Sousa-Junior, F.J.C.; Santos, T.D.F.; Andrade, D.; Andrade, S.L.; Pereira, W.L.; Gomes, P.W.P.; Monteiro, M.C.; E Silva, C.Y.Y.; et al. Chemical Composition, Toxicity, Antinociceptive, and Anti-Inflammatory Activity of Dry Aqueous Extract of *Varronia multispicata* (Cham.) Borhidi (*Cordiaceae*) Leaves. *Front. Pharmacol.* **2019**, *10*, 1376. [CrossRef]
24. Falcão, T.R.; Araújo, A.A.D.; Soares, L.A.L.; Farias, I.B.D.; Silva, W.A.V.D.; Ferreira, M.R.A.; Araújo, R.F.D., Jr.; Medeiros, J.S.D.; Lopes, M.L.D.D.S.; Guerra, G.C.B. *Libidibia ferrea* Fruit Crude Extract and Fractions Show Anti-Inflammatory, Antioxidant, and Antinociceptive Effect In Vivo and Increase Cell Viability In Vitro. *Evid.-Based Complement. Altern. Med.* **2019**, *2019*, 6064805. [CrossRef] [PubMed]
25. Dantas, L.L.S.F.R.; Fonseca, A.G.; Pereira, J.R.; Furtado, A.A.; Gomes, P.A.T.M.; Fernandes-Pedrosa, M.F.; Leite, A.C.L.; Rêgo, M.J.B.M.; Pitta, M.G.R.; Lemos, T.M.A.M. Anti-Inflammatory and Antinociceptive Effects of the Isatin Derivative (Z)-2-(5-Chloro-2-Oxoindolin-3-Ylidene)-N-Phenyl-Hydrazinecarbothioamide in Mice. *Braz. J. Med. Biol. Res.* **2020**, *53*, e10204. [CrossRef] [PubMed]
26. Islam, S.; Shajib, S.; Rashid, R.B.; Khan, M.F.; Al-Mansur, A.; Datta, B.K.; Rashid, M.A. Antinociceptive Activities of *Artocarpus lacucha* Buch-Ham (Moraceae) and Its Isolated Phenolic Compound, Catechin, in Mice. *BMC Complement. Altern. Med.* **2019**, *19*, 214. [CrossRef] [PubMed]
27. Chen, L.; Deng, H.; Cui, H.; Fang, J.; Zuo, Z.; Deng, J.; Li, Y.; Wang, X.; Zhao, L. Inflammatory Responses and Inflammation-Associated Diseases in Organs. *Oncotarget* **2018**, *9*, 7204–7218. [CrossRef]
28. Valle-Dorado, M.G.; Hernández-León, A.; Nani-Vázquez, A.; Ángeles-López, G.E.; González-Trujano, M.E.; Ventura-Martínez, R. Antinociceptive Effect of *Mansoa alliacea* Polar Extracts Involves Opioid Receptors and Nitric Oxide in Experimental Nociception in Mice. *Biomed. Pharmacother.* **2022**, *152*, 113253. [CrossRef]
29. De Lima, M.N.N.; Guimarães, B.A.; De Castro, A.L.S.; Ribeiro, K.B.; Miller, D.C.; Da Silva, P.I.C.; Freitas, J.J.S.; De Lima, A.B.; Setzer, W.N.; Da Silva, J.K.R.; et al. Chemical Composition and Antinociceptive and Anti-Inflammatory Activity of the Essential Oil of *Hyptis crenata* Pohl Ex Benth. from the Brazilian Amazon. *J. Ethnopharmacol.* **2023**, *300*, 115720. [CrossRef]
30. Da Cruz De Moraes, S.Z.; Shan, A.Y.K.V.; Oliveira Melo, M.A.; Pereira Da Silva, J.; Rocha Santos Passos, F.; De Souza Graça, A.; Araújo, B.S.D.; Quintans, J.D.S.S.; Quintans Júnior, L.J.; Oliveira Barreto, E.D.; et al. Antinociceptive and Anti-Inflammatory Effect of *Poincianella pyramidalis* (Tul.) L.P. Queiroz. *J. Ethnopharmacol.* **2020**, *254*, 112563. [CrossRef]
31. Ríos, J.-L.; Giner, R.M.; Marín, M.; Recio, M.C. A Pharmacological Update of Ellagic Acid. *Planta Med.* **2018**, *84*, 1068–1093. [CrossRef]
32. Lee, J.; Choi, J.W.; Sohng, J.K.; Pandey, R.P.; Park, Y.I. The Immunostimulating Activity of Quercetin 3-O-Xyloside in Murine Macrophages via Activation of the ASK1/MAPK/NF-$_{K}$B Signaling Pathway. *Int. Immunopharmacol.* **2016**, *31*, 88–97. [CrossRef]
33. Seo, J.Y.; Pandey, R.P.; Lee, J.; Sohng, J.K.; Namkung, W.; Park, Y.I. Quercetin 3-*O*-Xyloside Ameliorates Acute Pancreatitis in Vitro via the Reduction of ER Stress and Enhancement of Apoptosis. *Phytomedicine* **2019**, *55*, 40–49. [CrossRef]
34. Ojo, O.A.; Rotimi, D.E.; Ojo, A.B.; Ogunlakin, A.D.; Ajiboye, B.O. Gallic Acid Abates Cadmium Chloride Toxicity via Alteration of Neurotransmitters and Modulation of Inflammatory Markers in Wistar Rats. *Sci. Rep.* **2023**, *13*, 1577. [CrossRef] [PubMed]
35. Corbett, S.; Daniel, J.; Drayton, R.; Field, M.; Steinhardt, R.; Garrett, N. Evaluation of the Anti-Inflammatory Effects of Ellagic Acid. *J. PeriAnesth. Nurs.* **2010**, *25*, 214–220. [CrossRef] [PubMed]

36. Rajendran, P.; Rengarajan, T.; Nandakumar, N.; Palaniswami, R.; Nishigaki, Y.; Nishigaki, I. Kaempferol, a Potential Cytostatic and Cure for Inflammatory Disorders. *Eur. J. Med. Chem.* **2014**, *86*, 103–112. [CrossRef] [PubMed]
37. Pragasam, S.J.; Rasool, M. Dietary component *p-coumaric* acid suppresses monosodium urate crystal-induced inflammation in rats. *Inflamm. Res.* **2013**, *62*, 489–498. [CrossRef]
38. Li, Y.; Yao, J.; Han, C.; Yang, J.; Chaudhry, M.; Wang, S.; Liu, H.; Yin, Y. Quercetin, Inflammation and Immunity. *Nutrients* **2016**, *8*, 167. [CrossRef]
39. Yu, T.-Y.; Feng, Y.-M.; Kong, W.-S.; Li, S.-N.; Sun, X.-J.; Zhou, G.; Xie, R.-F.; Zhou, X. Gallic Acid Ameliorates Dextran Sulfate Sodium-Induced Ulcerative Colitis in Mice via Inhibiting NLRP3 Inflammasome. *Front. Pharmacol.* **2023**, *14*, 1095721. [CrossRef]
40. Zhou, P.; Lai, J.; Li, Y.; Deng, J.; Zhao, C.; Huang, Q.; Yang, F.; Yang, S.; Wu, Y.; Tang, X.; et al. Methyl Gallate Alleviates Acute Ulcerative Colitis by Modulating Gut Microbiota and Inhibiting TLR4/NF-κB Pathway. *Int. J. Mol. Sci.* **2022**, *23*, 14024. [CrossRef]
41. OECD. *Test No. 425: Acute Oral Toxicity: Up-and-Down Procedure*; Organisation for Economic Co-operation and Development: Paris, France, 2022. [CrossRef]
42. Malone, M.H. Pharmacological Approaches to Natural Product Screening and Evaluation. In *New Natural Products and Plant Drugs with Pharmacological, Biological or Therapeutical Activity*; Wagner, H., Wolff, P., Eds.; Springer: Berlin/Heidelberg, Germany, 1977; pp. 23–53. [CrossRef]
43. Thrall, M.A.; Weiser, G.; Allison, R.; Campbell, T.W. *Veterinary Hematology and Clinical Chemistry*, 2nd ed.; Wiley: Somerset, UK, 2012.
44. Hamada, H.; Ohkura, Y. A New Photometric Method for the Determination of Serum Glutamate Pyruvate Transaminase Activity Using Pyruvate and Glutamate as Substrates. *Chem. Pharm. Bull.* **1976**, *24*, 1865–1869. [CrossRef]
45. Arnold, P.M.; Parslow, G.R. Designing a Coupled Assay System for Aspartate Aminotransferase. *Biochem. Educ.* **1995**, *23*, 40–41. [CrossRef]
46. Yagi, T.; Kagamiyama, H.; Ohtawara, S.; Soda, K.; Nozaki, M. A New Assay for L-Aspartate: 2-Oxoglutarate Aminotransferase. *Anal. Biochem.* **1979**, *100*, 20–24. [CrossRef]
47. Bartels, H.; Böhmer, M. Micro-determination of creatinine. *Clin. Chim. Acta* **1971**, *32*, 81–85. [PubMed]
48. Hallett, C.J.; Cook, J.G.H. Reduced Nicotinamide Adenine Dinucleotide-Coupled Reaction for Emergency Blood Urea Estimation. *Clin. Chim. Acta* **1971**, *35*, 33–37. [CrossRef] [PubMed]
49. Koster, R.; Anderson, M.; De Beer, E.J. Acetic acid for analgesic screening. *Fed. Proc.* **1959**, *18*, 412–417.
50. Lowry Oliver, H.; Rosebrough, N.; Farr, A.L.; Randall, R.J. Protein measurement with the Folin phenol reagent. *J. Biol. Chem.* **1951**, *193*, 265–275. [CrossRef]
51. Hunskaar, S.; Fasmer, O.B.; Hole, K. Formalin Test in Mice, a Useful Technique for Evaluating Mild Analgesics. *J. Neurosci. Methods* **1985**, *14*, 69–76. [CrossRef] [PubMed]

pharmaceuticals

Article

Therapeutic Effect of Costunolide in Autoimmune Hepatitis: Network Pharmacology and Experimental Validation

Zheng Huang †, Shangshu Nie †, Shuhui Wang, Han Wang, Jin Gong, Wei Yan, Dean Tian and Mei Liu *

Department of Gastroenterology, Tongji Hospital of Tongji Medical College, Huazhong University of Science and Technology, Wuhan 430000, China
* Correspondence: meiliutjh@126.com
† These authors contributed equally to this work.

Abstract: Novel treatments for autoimmune hepatitis (AIH) are highly demanded due to the limitations of existing therapeutic agents. Costunolide is a promising candidate due to its anti-inflammatory and hepatoprotective function, but its effect in AIH remains obscure. In this study, we integrated network pharmacology and experimental validation to reveal the effect and mechanism of costunolide in AIH. A total of 73 common targets of costunolide and AIH were obtained from databases. Pathway enrichment analysis indicated that PI3K-AKT pathway was the core pathway of costunolide in AIH. Protein–protein interaction network analysis and molecular docking revealed that SRC and IGF1R might play critical roles. In two murine AIH models, costunolide significantly attenuated liver injury, inflammation, and fibrosis reflected by the liver gross appearance, serum transaminases, necrosis area, spleen index, immune cell infiltration, and collagen deposition. Western blot and immunohistochemistry confirmed that phosphorylated AKT, SRC, and IGF1R were upregulated in AIH models, and costunolide administration could inhibit the phosphorylation of these proteins. In summary, costunolide significantly ameliorates murine AIH. The therapeutic effect might work by suppressing the activation of PI3K-AKT pathway and inhibiting the phosphorylation of SRC and IGF1R. Our research reveals the potent therapeutic effect of costunolide in AIH and the potential role of SRC and IGF1R in AIH for the first time, which may further contribute to the novel drug development for AIH and other autoimmune diseases.

Keywords: costunolide; autoimmune hepatitis; network pharmacology; immune-mediated liver injury; PI3K-AKT; SRC; IGF1R

Citation: Huang, Z.; Nie, S.; Wang, S.; Wang, H.; Gong, J.; Yan, W.; Tian, D.; Liu, M. Therapeutic Effect of Costunolide in Autoimmune Hepatitis: Network Pharmacology and Experimental Validation. *Pharmaceuticals* **2023**, *16*, 316. https://doi.org/10.3390/ph16020316

Academic Editor: Diana Roxana Pelinescu

Received: 17 January 2023
Revised: 10 February 2023
Accepted: 15 February 2023
Published: 17 February 2023

1. Introduction

Autoimmune hepatitis (AIH) is a chronic inflammatory liver disease due to immune-mediated destruction of hepatocytes. The etiology of AIH involves the interaction of both genetic background and environmental triggers, which is not entirely understood. Its pathophysiological processes are believed to involve specific genetic traits, molecular mimicry, impaired immunoregulatory mechanisms, etc. [1]. AIH is characterized biochemically by the elevation of serum transaminase, serologically by the presence of autoantibodies and elevated immunoglobulin G (IgG) levels, and histologically by interface hepatitis [2]. It occurs worldwide in all ethnicities, affecting all ages with distinct female preponderance [2]. Although most patients respond well to the standard immunosuppressive therapy of steroids and azathioprine, insufficient response and intolerable side effects occur in 10–20% of patients [3]. Long-term or uncontrolled AIH may lead to progressive fibrosis, cirrhosis, liver failure, and even hepatocellular carcinoma [4]. Therefore, identification of novel therapeutic drugs and clarifying the underlying mechanism is of great significance for the treatment of AIH.

Traditional Chinese medicine (TCM) and natural products may provide some new options for the drug discovery of AIH [5]. Costunolide (COS) is a well-known sesquiterpene lactone in the germacranolides series isolated from Aucklandiae Radix (Mu Xiang in Chinese). Numerous preclinical studies have indicated that costunolide possesses antioxidative, anti-inflammatory, antiallergic, anticancer, and antidiabetic properties [6]. Remarkably, costunolide exerts a powerful anti-inflammation effect in multiple immune-related diseases such as carrageenan-induced paw edema and lung inflammation, ethanol-induced gastric ulcer, lipoteichoic acid-induced acute lung injury, and dextran sulfate sodium (DSS)-induced murine colitis [7–12]. Meanwhile, costunolide exhibits a hepatoprotective effect against lipopolysaccharide and D-galactosamine-induced acute liver injury and alcoholic liver injury [13,14]. It also ameliorates liver fibrosis in vitro and in vivo [15]. However, the effect of costunolide in AIH remains obscure.

Network pharmacology was proposed by integrating network biology and polypharmacology [16]. It is capable of describing complexities among biological systems, drugs, and diseases from a network perspective, which has been applied in many studies to explore the molecular mechanism of drugs [17–21]. In this study, we investigated the effect of costunolide in two murine models of AIH: Concanavalin A (ConA)-induced acute immune-mediated hepatitis and human Cytochrome P4502D6 (CYP2D6) plasmid injection-induced chronic autoimmune hepatitis [22]. In addition, we integrated network pharmacology with experimental validation to clarify underlying mechanisms. Our research indicates that costunolide might be a potential option for the therapy of AIH. The flow chart of this research is shown in Figure 1.

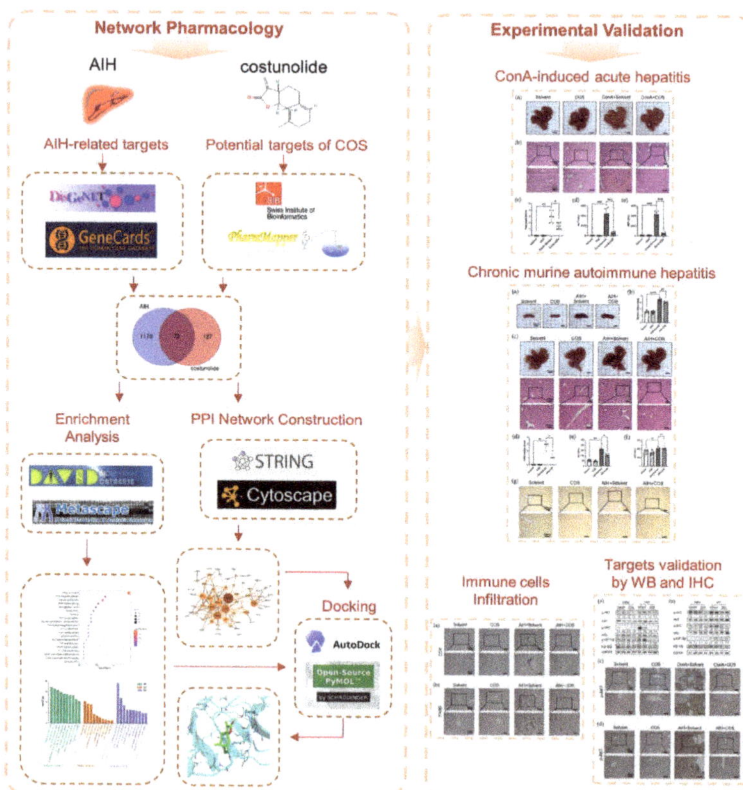

Figure 1. Flow chart of research. Detailed procedures of analysis are described in Materials and Methods. Briefly, AIH-related targets were collected from DisGeNET and GeneCards. The potential targets of costunolide were obtained from SwissTargetPrediction and PharmMapper. The potential targets of costunolide in AIH were generated by taking intersection of targets above. KEGG and GO enrichment analysis were performed using DAVID and Metascape. STRING database and Cytoscape were used for the construction of PPI network analysis. The docking between costunolide and key targets were conducted by PyMol and AutoDock. In vivo experiments were performed to verify the effect of costunolide on AIH and its underlying mechanism.

2. Results

2.1. Identification of the Potential Targets of Costunolide in AIH

Costunolide is a lactone compound isolated from Aucklandiae Radix. The two-dimensional (2D) structure of costunolide is shown in Figure 2a. The pharmacological and molecular properties of costunolide are shown in Table 1. Costunolide shows good oral bioavailability (OB) and high drug likeness (DL) in both the Traditional Chinese Medicine Systems Pharmacology Database and Analysis Platform (TCMSP) and SwissADME.

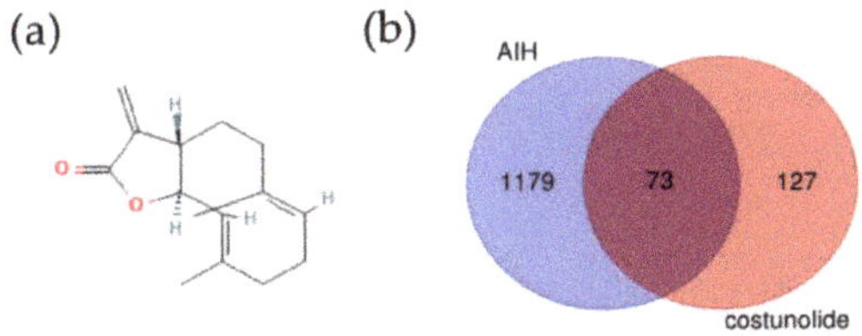

Figure 2. Potential targets of costunolide in AIH: (**a**) 2D structure of costunolide (PubChem Identifier: CID 5281437 https://pubchem.ncbi.nlm.nih.gov/compound/5281437#section=2D-Structure (accessed on 26 February 2022)), (**b**) Venn diagram of intersecting targets of costunolide and AIH.

Table 1. Pharmacological and molecular properties of costunolide.

MW	AlogP	Hdon	Hacc	OB (%)	Caco-2 (nm/s)	BBB	DL	FASA-	TPSA	RBN
232.35	4.02	0	2	29.07	1.28	1.42	0.11	0.36	26.30	0

Abbreviations: MW, molecular weight; AlogP, low lipid/water partition coefficient; Hdon, hydrogen bond donors; Hacc, hydrogen bond acceptors; OB, oral bioavailability; Caco-2, Caco-2 permeability; BBB, blood brain barrier; DL, drug likeness; FASA-, fractional water accessible surface area of all atoms with negative partial charge; TPSA, topological polar surface area; RBN, rotatable bonds number. The molecular and pharmacological properties data of costunolide were obtained from TCMSP [23] and SwissADME [24].

From SwissTargetPrediction, we got 33 potential targets, and we retrieved 174 targets from PharmMapper. After merging data, removing duplicates, and confirming in UniProt, we got 200 potential targets of costunolide. Using the search word "autoimmune hepatitis", we obtained 190 related genes in DisGeNET and 1183 genes which score >5 in GeneCards. After merging and removing duplicates, we obtained 1252 genes related to AIH. Via intersecting targets of costunolide and AIH-related genes, we obtained 73 potential targets of costunolide in AIH (Figure 2b).

2.2. Recognition of Enriched Pathway of the Potential Targets of Costunolide in AIH

The Kyoto Encyclopedia of Genes and Genomes (KEGG) enrichment analysis was performed using The Database for Annotation, Visualization and Integrated Discovery (DAVID). The top 20 clusters were selected based on the *p*-value and were presented in the bubble chart (Figure 3). The potential targets of costunolide in AIH were involved in pathways in cancer (hsa05200), PI3K-AKT signaling pathway (hsa04151), lipid and atherosclerosis (hsa05417), MAPK signaling pathway (hsa04010), proteoglycans in cancer (hsa05205), etc.

Meanwhile, based on the *p*-value of Gene Ontology (GO) enrichment analysis, the top 10 significantly enriched terms in Biological Process (BP), Cellular Component (CC), and Molecular Function (MF) categories were selected (Figure 4). Potential targets of costunolide in AIH were mainly involved in biological processes such as cellular response to lipid (GO:0071396), response to hormone (GO:0009725), and response to lipopolysaccharide (GO:0032496). The main cellular component terms were membrane raft (GO:0045121), vesicle lumen (GO:0031983), and receptor complex (GO:0043235). The main molecular function terms were nuclear receptor activity (GO:0004879), protein tyrosine kinase activity (GO:0004713), and protein domain specific binding (GO:0019904).

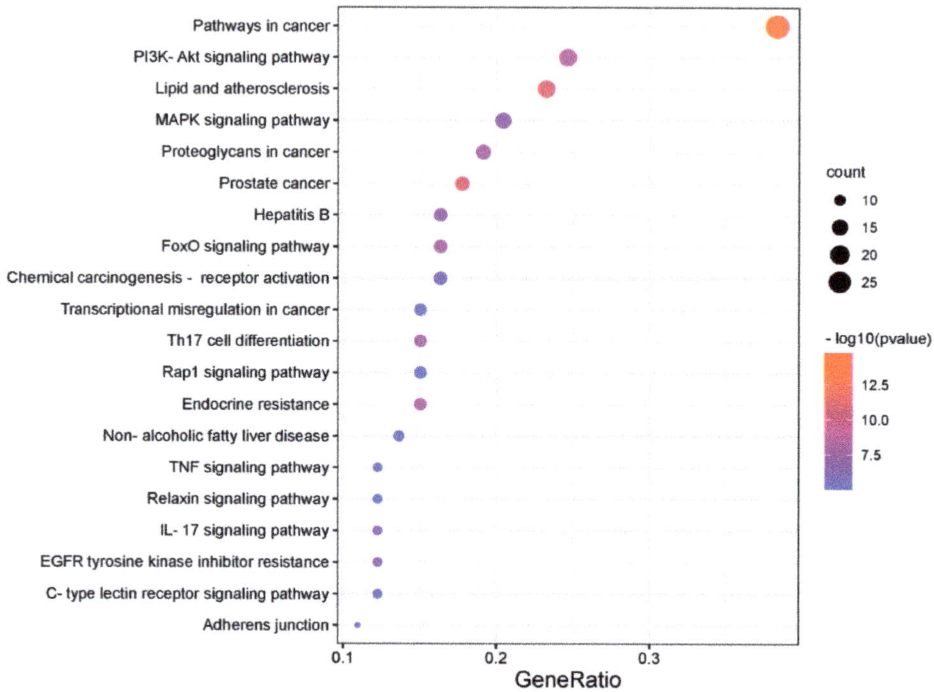

Figure 3. KEGG pathway enrichment analysis of intersecting targets of costunolide and AIH.

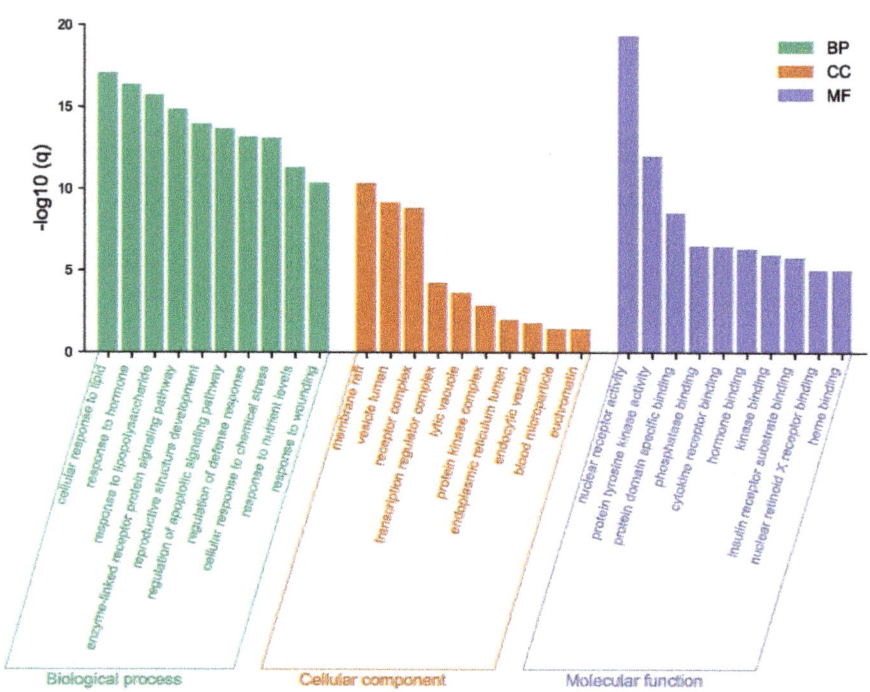

Figure 4. Gene Ontology (GO) enrichment analysis of intersecting targets of costunolide and AIH.

2.3. Topological Network Analysis of the Potential Targets of Costunolide in AIH

Seventy-three potential targets of costunolide in AIH were imported into The Search Tool of Retrieval of Interacting Genes (STRING) to construct the protein–protein interaction (PPI) network. After removing disconnected nodes, the network contained 61 nodes and 156 edges (Figure 5a). The average number of neighbors is 5.115 and the clustering coefficient is 0.317. Using plug-in CytoHubba, we found top 10 genes that might play critical roles in the therapeutic effect of costunolide in AIH, namely MAPK1, SRC, GRB2, EGFR, LCK, ESR1, JAK2, IGF1, HSP90AA1, and IGF1R, ranked by Maximal Clique Centrality (MCC) (Figure 5b).

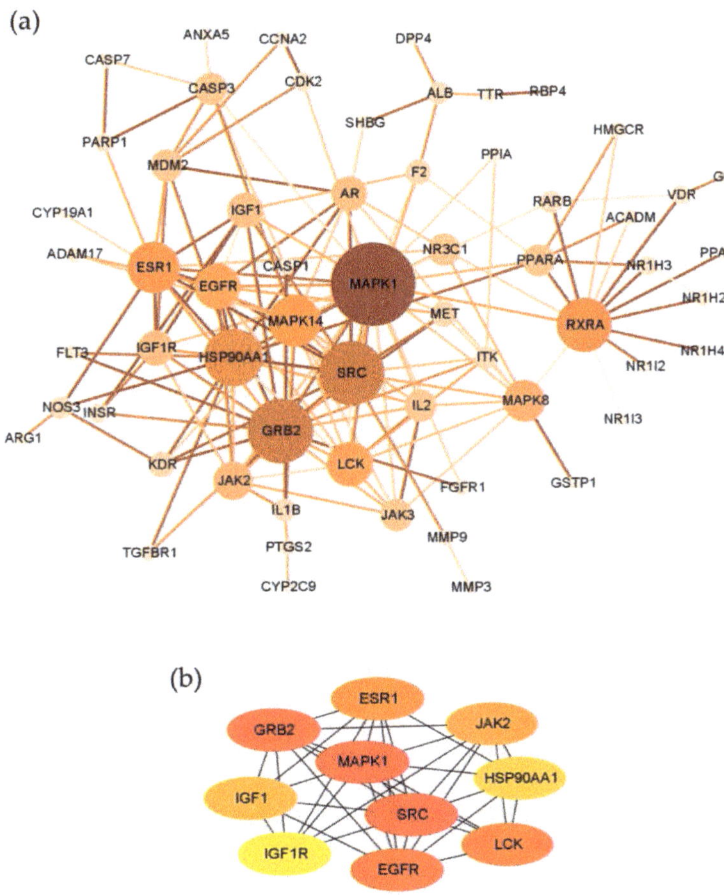

Figure 5. Protein–protein interaction network analysis. (**a**) Protein–protein interaction network of intersecting targets of costunolide and AIH visualized by Cytoscape. (**b**) The core subnetwork constructed by CytoHubba.

2.4. Molecular Docking

Based on the pathway enrichment analysis and literature review, SRC and IGF1R ranked high in the PPI network analysis and were reported to regulate the PI3K-AKT pathway in inflammation-related diseases. However, their roles in AIH were not elucidated yet. Thus, we chose SRC and IGF1R as our interested targets of costunolide in AIH. Using PyMol and Autodock, molecular docking visually showed the interaction between costunolide and SRC, IGF1R. The diagrams of drug-target binding mode were shown left and the details right (Figure 6). The dotted yellow line represents the hydrogen bond. The lower binding energy indicates higher stability. The binding energies between costunolide and SRC and between costunolide and IGF1R were −7.7 and −5.9 kcal/mol, respectively (Table 2).

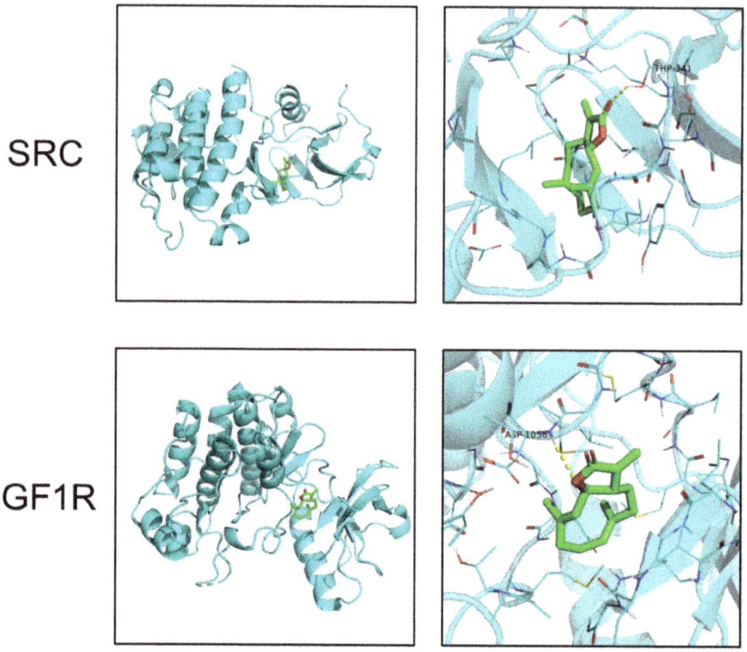

Figure 6. Molecular docking of costunolide with SRC and IGF1R.

Table 2. Binding energies of costunolide with SRC and IGF1R.

Target	PDB ID	Binding Energy (kcal/mol)
SRC	6E6E	−7.7
IGF1R	3NW7	−5.9

2.5. Costunolide Attenuated ConA-Induced Acute Hepatitis

Firstly, we administered costunolide in the ConA-induced acute hepatitis model. The liver gross appearance of ConA+Solvent mice showed hepatic congestion and suppuration on the surface, while ConA+COS liver showed a milder change (Figure 7a). In hematoxylin-eosin (H&E) staining, we observed that ConA induced necrosis of liver parenchyma. Costunolide administration significantly restrained the hepatic necrosis, while the necrosis score of ConA+Solvent and ConA+COS showed a significant difference (Figure 7b,c). Similar to the histological results, level of the liver enzyme alanine aminotransferase (ALT) and aspartate aminotransferase (AST) in serum surged in the ConA+Solvent group compared with Solvent and COS group, while costunolide significantly suppressed it (Figure 7d,e). Notably, there was no obvious difference of gross appearance, H&E staining, and liver enzyme level between the Solvent group and the COS group.

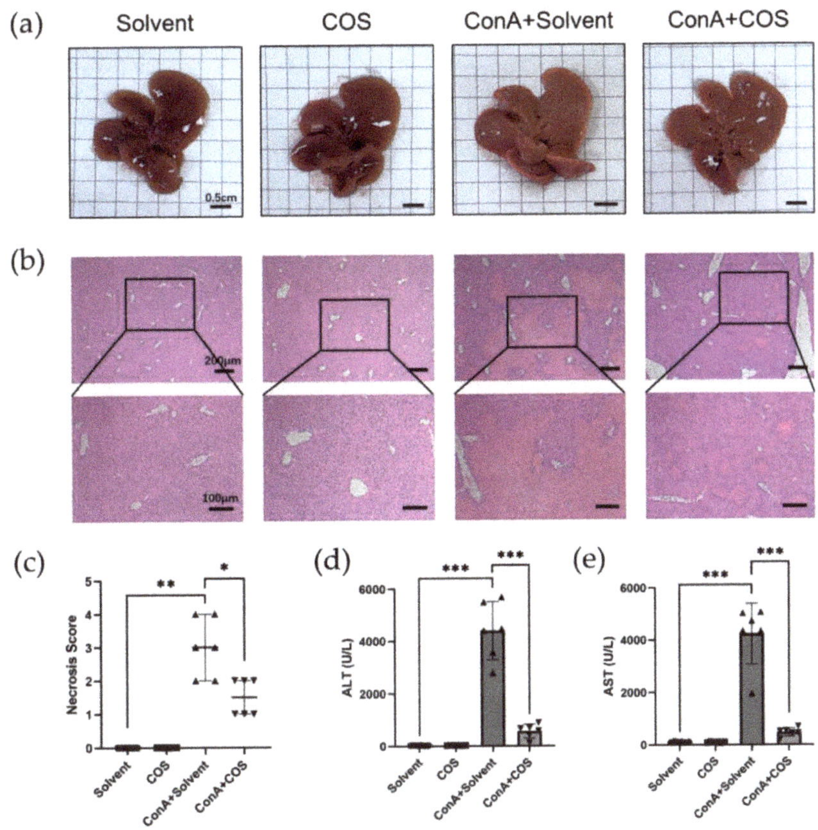

Figure 7. Costunolide attenuated ConA-induced acute hepatitis. (**a**) Liver gross appearance. (**b**) Representative H&E staining of liver sections. (**c**) Necrosis score of H&E staining of liver sections. (**d**) Effect of costunolide on serum ALT levels. (**e**) Effect of costunolide on serum AST levels. Data shown in c–e are from 6 individual mice per group. * $p < 0.05$, ** $p < 0.01$, *** $p < 0.001$.

2.6. Costunolide Ameliorated Chronic Murine Autoimmune Hepatitis

In our chronic murine AIH model, the spleens of AIH mice were enlarged significantly, while costunolide suppressed it (Figure 8a). There was a significant difference of spleen indexes between AIH+Solvent and AIH+COS mice (Figure 8b). Unlike ConA-induced acute hepatitis, there was no obvious change of the liver gross appearance in the chronic murine AIH model (Figure 8c). H&E staining of liver sections showed that costunolide inhibited immune cell infiltration in AIH mice, which is the typical characteristic of AIH (Figure 8c). There was a significant difference of inflammation score between AIH+Solvent and AIH+COS group (Figure 8d). The ALT level showed a similar trend as histological results (Figure 8e). There was no significant change of AST level in our chronic model (Figure 8f). Further, Sirius red staining of liver sections indicated that costunolide reduced liver fibrosis in the chronic AIH model (Figure 8g).

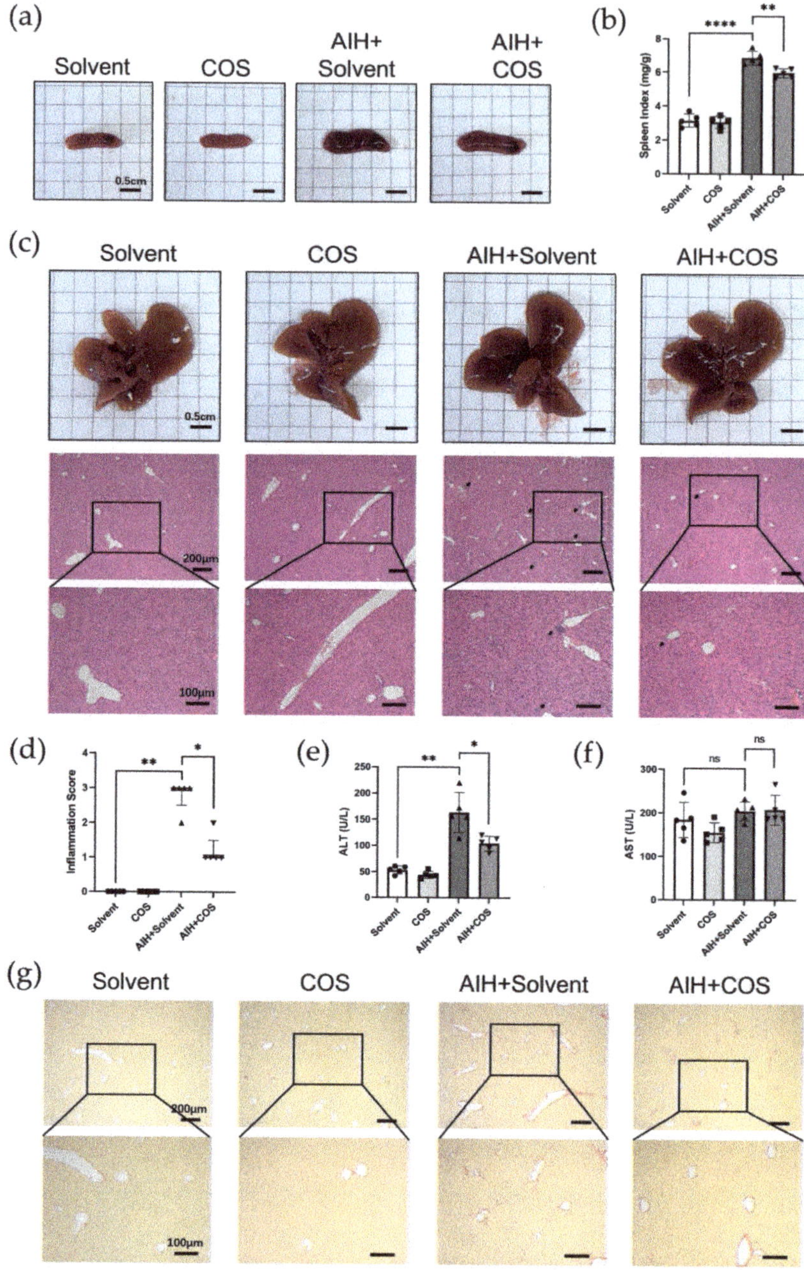

Figure 8. Costunolide ameliorated chronic murine autoimmune hepatitis. (**a**) Spleen gross appearance. (**b**) Spleen index (mg/g). (**c**) Representative liver gross appearance and H&E staining of liver sections (black arrows indicate immune cell infiltration). (**d**) Inflammation score of H&E staining of liver sections. (**e**) Effect of costunolide on serum ALT levels. (**f**) Effect of costunolide on serum ALT levels. (**g**) Representative Sirius-red staining of liver sections (Red color indicates collagen deposition). Data shown in (**b**,**d**,**e**) are from 5 individual mice per group. * $p < 0.05$, ** $p < 0.01$, **** $p < 0.0001$.

To further elucidate the landscape of immune cell infiltration in different groups, we conducted immunohistochemistry (IHC) of CD4 and F4/80 in liver sections. The IHC result revealed the increased infiltration of CD4$^+$ T cells and macrophages in AIH+Solvent mice, while costunolide suppressed this trend substantially (Figure 9a,b).

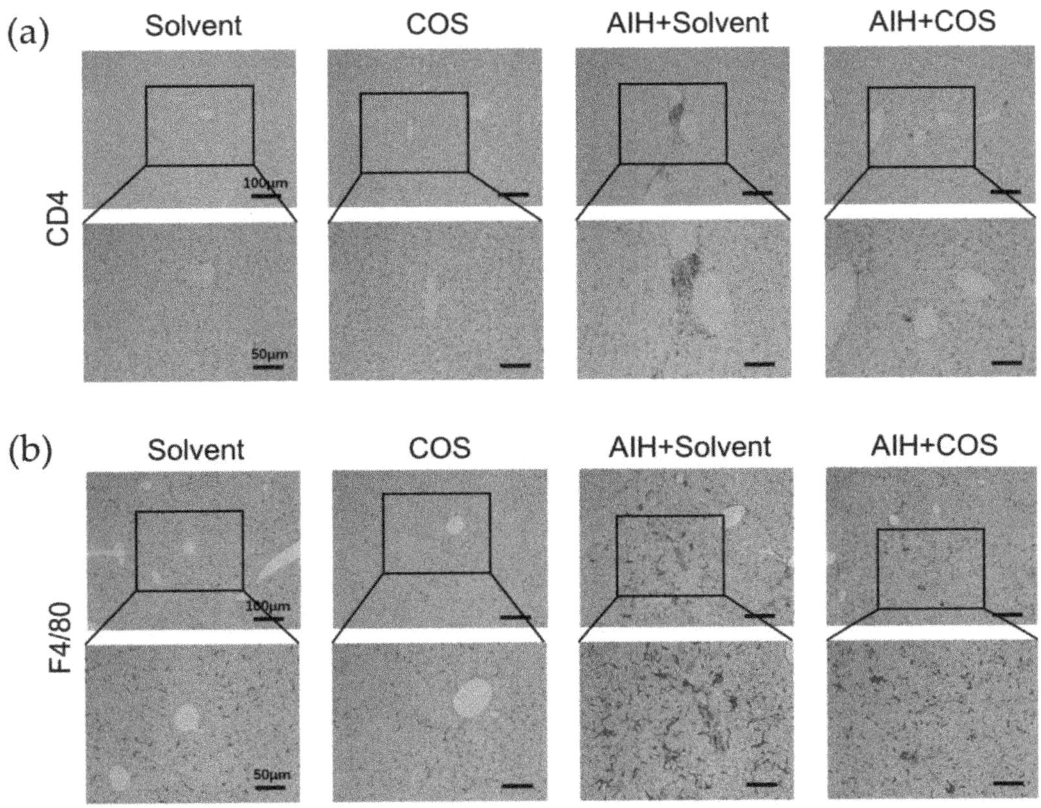

Figure 9. Costunolide inhibited immune cell infiltration in chronic murine autoimmune hepatitis. (**a**) Immunohistochemistry of CD4$^+$ T cells. (**b**) Immunohistochemistry of F4/80$^+$ macrophages.

2.7. Costunolide Inhibited the Activation of PI3K-AKT Pathway and Suppressed the Phosphorylation of SRC and IGF1R in AIH

Based on the pathway enrichment analysis, PPI network analysis and literature review, we investigated the protein level change of phosphorylated AKT, SRC, and IGF1R in liver tissue of two AIH models. The phosphorylation of AKT, SRC, and IGF1R were significantly increased in the ConA model, whilst costunolide dramatically inhibited it (Figure 10a, Supplementary Figure S1a). Similar results were found in the chronic AIH model (Figure 10b, Supplementary Figure S1b).

To further confirm this trend, we conducted IHC of phospho-AKT in liver sections. The phosphorylated AKT was mainly expressed in nuclei of hepatocytes in Solvent and COS mice, and its expression level was higher in two murine AIH models, mostly in hepatocytes nuclei and cytoplasm as well as infiltrating immune cells. The phospho-AKT expression level in ConA/AIH+COS groups were lower than that in the ConA/AIH+Solvent groups (Figure 10c,d).

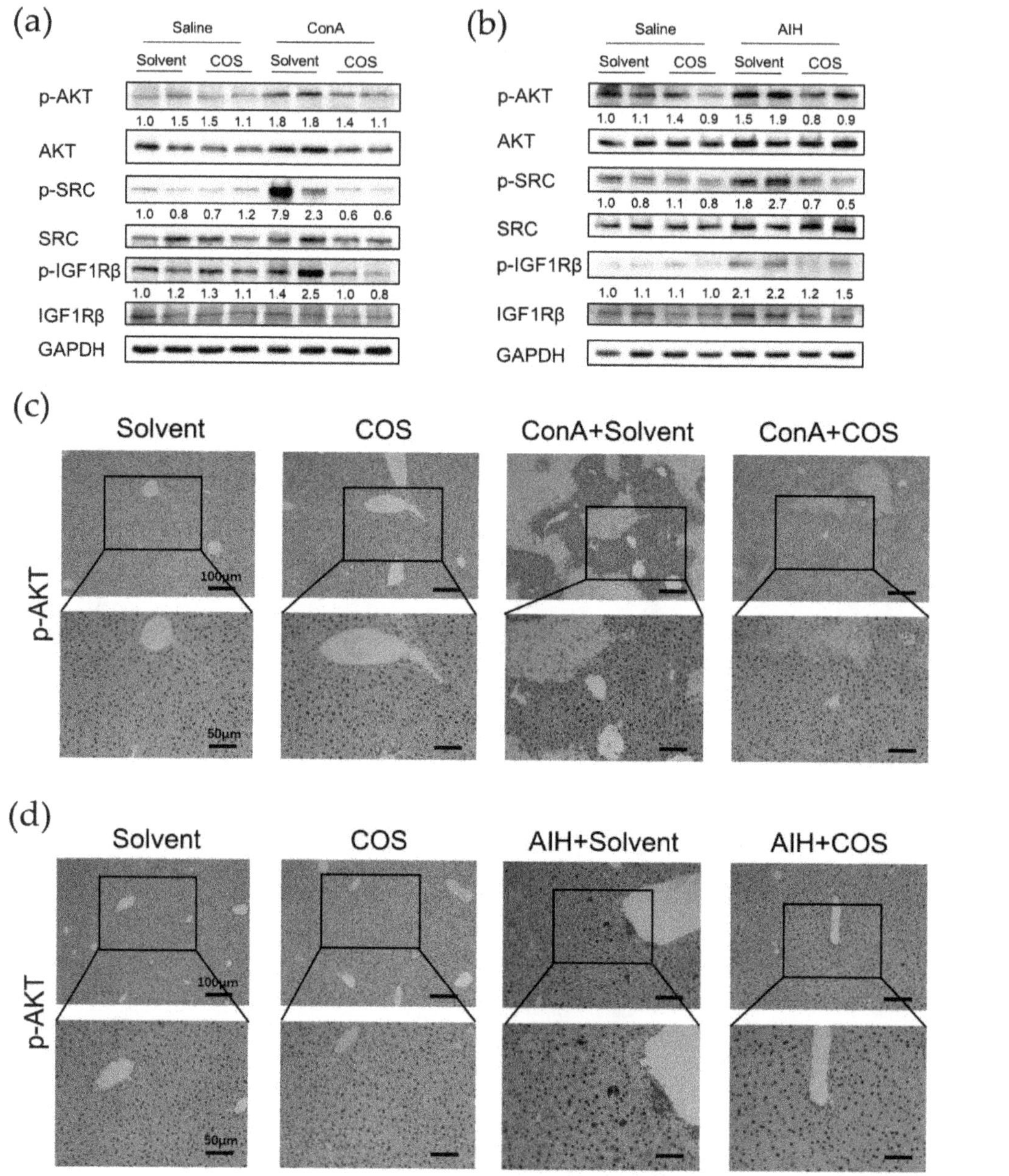

Figure 10. Costunolide inhibited the phosphorylation of AKT, SRC, and IGF1R. (**a**) Protein expression in ConA-induced acute hepatitis. Two samples from two different mice in each group are presented. The numbers indicate the ratio of pAKT/AKT, pSRC/SRC, and pIGF1Rβ/IGF1Rβ, respectively. (**b**) Protein expression in chronic murine autoimmune hepatitis. (**c**) Immunohistochemistry of phospho-AKT in ConA-induced acute hepatitis. (**d**) Immunohistochemistry of phospho-AKT in chronic murine autoimmune hepatitis.

3. Discussion

Autoimmune hepatitis (AIH) is a chronic liver disease related to immunological tolerance disorders targeting hepatocytes. Although AIH has long been considered a rare disease, several studies have indicated that the incidence and prevalence of AIH were increasing worldwide [25]. Although most patients respond well to traditional immunosuppressive therapy, many patients were troubled by the intolerance, insufficient response to medication, and recurrence after withdrawal of treatment [2]. The discovery of novel therapeutic strategies is necessary for the management of the increasing population of AIH. Costunolide is a promising candidate due to its anti-inflammatory and hepatoprotective function in other disease models, whereas there is no study about its effect in AIH yet [6,12,14]. In this study, we integrated network pharmacology and experimental validation to elucidate the therapeutic effect and underlying mechanism of costunolide against AIH.

In the current study, we identified 73 targets that might participate in the effect of costunolide on AIH. Based on the KEGG enrichment pathway analysis and literature review, we speculated that costunolide might mainly restrain the disease progress of AIH through PI3K-AKT pathway. Generally, the PI3K-AKT pathway is an important intracellular signaling pathway for cell cycle progression, cell survival, and metabolism [26]. It is reported that IHC of liver sections from AIH patients showed upregulation of phosphorylated AKT [27]. In ConA-induced hepatitis, PI3K-AKT pathway is activated in hepatocytes and dendritic cells, promoting inflammation via inducing the transcription of IL-6 and activation of $CD8^{+}$ T cell responses, respectively [28,29]. Similarly, in our experiments, we observed a significant elevated protein level of phosphorylated AKT in the liver of ConA-induced hepatitis and chronic AIH model in WB, and costunolide administration induced a decrease of it. This finding was further confirmed by IHC showing a similar trend as in WB, while the upregulated phosphorylated AKT in AIH models was mainly expressed in hepatocytes and infiltrating immune cells. In hepatocytes, the activation of AKT generally plays a protective effect [30]. However, it was recently reported that AKT activation in hepatocytes was also associated with increased inflammatory cytokine expression [28]. Activation of the PI3K-AKT-mTOR pathway and PI3K-AKT-FOXO1 signaling axis regulates the activation, differentiation, and function of T cells in different conditions [31–35]. Attenuation of PI3K signaling could lead to defects in T cell activation and neutrophil migration, restraining the inflammation in systemic lupus erythematosus (SLE) and rheumatoid arthritis (RA) models [36,37]. Several studies have reported the suppressing effect of costunolide on PI3K-AKT pathway in cancer cells and in DSS-induced colitis [11,26,38]. These findings suggest that the therapeutic effect of costunolide against AIH might be related to the inhibition of PI3K-AKT pathway, whereas the specific cell types should be studied further. Contrary to our results, it was reported that phospho-AKT was decreased in liver of ConA-induced hepatitis and in ConA-treated L02 cells compared with the saline control. This difference might be due to the different dosages of ConA and the time between ConA injection and sacrifice. Moreover, directly administrating ConA on L02 cells might not reflect the disease feature of immune-mediated hepatocyte destruction in vivo [39]. Interestingly, the top three signaling pathways of KEGG analysis included lipid and atherosclerosis pathway. It is reported that pre-existing high-fat diet-induced non-alcoholic fatty liver disease (NAFLD) in mice potentiates the severity of AIH [40]. Whether the lipid metabolism is involved in the pathophysiology of AIH and if costunolide could interfere with it awaits further research.

Based on the PPI network analysis using STRING and CytoHubba, we further identified 10 important targets in the therapeutic effect of costunolide in AIH, namely MAPK1, SRC, GRB2, EGFR, LCK, ESR1, JAK2, IGF1, HSP90AA1, and IGF1R, ranked by MCC. After comprehensive consideration of KEGG pathway analysis, GO enrichment analysis, molecular docking, and literature review, we selected SRC and IGF1R as our candidate targets for further research. SRC belongs to the non-receptor protein-tyrosine kinases family, playing key roles in cell growth, division, migration, and survival signaling pathways [41]. Multiple studies indicated that SRC regulated the PI3K-AKT pathway in inflammation-related diseases such as multiple sclerosis (MS), Grave's Disease (GD), and in the neutrophil extracellular traps (NET) formation [42–44]. In our experiments, we found that the phosphorylation level of SRC was significantly upregulated in hepatic tissue in two AIH models and was drastically decreased in the costunolide group. Similar to the present literature, our results provided clues that SRC phosphorylation rose in AIH, and it might exacerbate AIH via the PI3K-AKT pathway, which demands further rigorous research to confirm it. Moreover, another protein in SRC family, LCK, also ranked high in our network analysis and it is an important regulator in T cell proliferation and function [45]. Whether costunolide also interacts with LCK in its effect against AIH is an interesting focus for further studies. Insulin-like growth factor-1 receptor (IGF1R), the primary signaling receptor of insulin-like growth factors (IGF) system, signals through the PI3K-AKT-mTOR and RAS-RAF-MEK-ERK pathways [46]. Dysregulation of the IGF system has been directly related to altered $CD4^+$ T cell function in RA and GD [47,48]. Although there is no report about the direct contributory role of IGF1R in AIH yet, recently, its specific modulatory role in autoimmunity was confirmed by researchers in EAE mice. It augmented AKT-mTOR and STAT3 signaling, favoring Th17 cell differentiation over that of Treg cells [46]. Our results showed that phosphorylated IGF1R level was increased in the liver of AIH models, and its level decreased in the costunolide group, indicating that IGF1R phosphorylation level was altered in AIH, and costunolide might attenuate AIH via inhibiting the phosphorylation of IGF1R. Since the appropriate balance of Th17 and Treg cells maintains immune tolerance and impairment of it permits progress of AIH, the variation of IGF1R might be related to AIH progression by modulating Th17/Treg balance by regulating the PI3K-AKT pathway, which should be studied further.

In this study, we found costunolide possessed a strong therapeutic effect in two AIH models. Transaminase is the sensitive indicator of the liver injury. In our research, we observed that the administration of costunolide could lower the elevated transaminase level in two murine AIH models. The change of AST level was not statistically significant in our chronic AIH model, which might be due to the relatively overall lower degree of liver inflammation. As is shown in H&E staining, the necrosis area was also diminished by costunolide in the ConA model. The spleen index (spleen weight/ body weight) could roughly reflect the degree of chronic inflammation [49]. We found that the spleen index in AIH group was significantly higher than the Solvent group and COS group, while the spleen index of COS+AIH group was significantly lower, indicating that costunolide could suppress the chronic inflammation. This finding was further supported by the suppressed immune cell infiltration in liver observed in H&E and the reduced $CD4^+$ T cells and macrophage infiltration in liver as is shown in IHC. Based on the present literature, costunolide inhibits the production of pro-inflammatory factors in macrophages and influences the differentiation of $CD4^+$ T cells in vitro [50,51]. Since T cells and macrophages both play pivotal roles in AIH, the specific mechanism of costunolide attenuating AIH and the exact role of PI3K-AKT pathway inside await further investigation such as flow cytometry. We also observed that costunolide suppressed liver fibrosis in chronic AIH model. That antifibrotic function might mainly be due to its inhibition of inflammation, which removed the etiological factor causing liver injury [52]. It also could be related to its direct antifibrotic effect via inhibiting hepatic stellate cell (HSC) activation as reported before [15]. It is worth noting that methacrylic acid copolymer (MAC)-coated PH-responsive mesoporous silica nanoparticles (MSNs) carrying costunolide significantly repressed liver fibrogenesis at a reduced dose in vitro and in vivo [53]. Adopting similar technology might also reduce the dosage needed for the therapeutic effect of costunolide against AIH.

This study has some limitations. The binding affinity of costunolide with potential targets awaits further verification using surface plasmon resonance (SPR) or bio-layer interferometry (BLI) assays. The specific cell types in which those phosphorylated protein level increased are not identified. Since in different cell types those signaling pathways might exert distinct roles, we plan to perform immunofluorescence to reveal it. The specific molecular mechanism is not clear due to no rescue experiment being conducted. In addition, the CYP2D6 model reflects type 2 AIH. The conclusion of our study remains to be confirmed in more AIH models.

4. Materials and Methods

4.1. Chemical Information Collection and Targets Retrieval of Costunolide

The molecular and pharmacological properties data of costunolide were obtained from TCMSP [23] and SwissADME [24]. The canonical SMILES (Simplified Molecular-Input Line-Entry System), image, and SDF format file of 2D structure of costunolide were obtained from the PubChem database (PubChem Identifier: CID 5281437 https://pubchem.ncbi.nlm.nih.gov/compound/5281437#section=2D-Structure (accessed on 26 February 2022)) [54]. SwissTargetPrediction [55] and PharmMapper [56] were utilized to predict the potential targets of costunolide. For outcome of PharmMapper, targets which norm fit score > 0.25 were included in subsequent analysis. After integrating data and removing duplicates, all proteins were verified in the UniProt database [57].

4.2. Prediction of Potential Targets of Costunolide in AIH and Enrichment Analysis

AIH-related targets were collected from DisGeNET [58] and GeneCards (https://www.genecards.org/ (accessed on 21 June 2022)) [59]. In GeneCards, genes that Score > 5 were included in subsequent analysis. The potential targets of costunolide and AIH-related targets obtained in above steps were imported into an online Venn diagram drawing tool (https://bioinformatics.psb.ugent.be/webtools/Venn/ (accessed on 13 September 2022)).

KEGG enrichment analysis was conducted using DAVID [60,61]. Species was set as *Homo sapiens*. Threshold and EASE were set as 2 and 0.1, respectively. GO enrichment analysis was performed using Metascape [62]. Species was set as *Homo sapiens* and all genes in the genome were used as the enrichment background. Min overlap, p value cutoff, and min enrichment were set as 3, 0.01, and 1.5, respectively. Bubble chart and bar graph were plotted by http://www.bioinformatics.com.cn (accessed on 14 September 2022), an online platform for data analysis and visualization.

4.3. Protein–Protein Interaction Target Network Construction and Visualization

STRING database (https://cn.string-db.org/ (accessed on 14 September 2022)) was utilized to construct a PPI network [63]. The organism was set as *Homo sapiens*. The minimum required interaction score was set as 0.900 (highest confidence), and disconnected nodes in the network were hidden.

Network visualization and analysis were performed using Cytoscape version 3.9.1 [64]. The TSV-format file was downloaded from the STRING database and imported into Cytoscape. The size and color of nodes were defined continuously according to the degree (number of edges), and the color of stroke was set according to the combined score calculated by network analyzer of Cytoscape. The plug-in CytoHubba (https://apps.cytoscape.org/apps/cytohubba (accessed on 14 September 2022)) was used to find the core subnetwork. The node's scores were calculated and top 10 hub genes ranked by MCC were selected, then the network graph was constructed.

4.4. Molecular Docking

The crystal structures of candidate targets were obtained from the PDB database (https://www.rcsb.org/ (accessed on 20 September 2022)). Open-Source PyMol was used to pre-process the structure, including removing the ligand and water molecules, adding hydrogen [65]. AutoDock 4.2 was used to conduct docking between costunolide and key targets [66].

4.5. Establishment of Two Experimental AIH Models and Administration of Costunolide

C57BL/6 mice were supplied by the Laboratory Animal Centre of Tongji Hospital. All animals were housed in the SPF environment and a 12 h light/12 h dark cycle at room temperature of 22 ± 2 °C with 55 ± 2% humidity. All animals had free access to food and water. The animal study protocol was approved by the Laboratory Animal Welfare and Ethics Committee of Tongji Hospital of Tongji Medical College, Huazhong University of Science and Technology (IACUC Issue No.: TJH-202207036; date of approval: 15 July 2022). The specific animal model establishment procedure and treatment scheme was depicted in supplementary materials (Supplementary Figure S2).

Concanavalin A (ConA) was used to establish the acute immune-mediated liver injury model. ConA (C2010, CAS: 11028-71-0, LOT: 091610120V) was purchased from Sigma-Aldrich. Costunolide (CAS: #553-21-9, LOT: J22GB152180) was purchased from Shanghai YuanYe Bio-Technology Co., Ltd. Twenty-four male mice (6–8 weeks old, 20–25 g) were randomly divided into solvent group (Solvent, n = 6), costunolide group (COS, n = 6), ConA group (ConA+Solvent, n = 6), and ConA+costunolide group (ConA+COS, n = 6). A single dose of ConA (15 mg/kg) dissolved in normal saline was injected via tail vein of mice, and the same volume of normal saline injection was used as control. Based on the existing literature, costunolide (10 mg/kg) was injected intraperitoneally once daily for three days, the first dose was given two days before giving ConA [11,26]. Costunolide was dissolved in DMSO, then mixed with 40% PEG300 (CAS: 25322-68-3, MCE), 5% Tween 80 (CAS: 9005-65-6, MCE), and normal saline. Mice were sacrificed 24 h after ConA injection. After anesthesia, blood samples were collected by removing the eyeball from the socket using tissue forceps [67]. Mice were then sacrificed and livers were harvested. Part of the liver was fixed in 4% paraformaldehyde for 48 h, the rest was collected in EP tubes, cooled in liquid nitrogen, and preserved in a −80 °C refrigerator.

As described in our previous publication, a single dose of empty adenovirus and multiple high-pressure tail vein injections of human CYP2D6 plasmid were administered to establish the chronic AIH model [22]. Empty adenovirus was purchased from Shanghai DesignGene Company. Human CYP2D6 plasmid was obtained from our lab. Twenty male mice (6–8 weeks old, 20–25 g) were randomly divided into four groups: solvent group (Solvent, n = 5), costunolide group (COS, n = 5), AIH group (AIH+Solvent, n = 5), and AIH+costunolide group (AIH+COS, n = 5). In the COS group and AIH+COS group, costunolide (10 mg/kg) was injected intraperitoneally once daily from day 14 after the adenovirus injection continuously for 20 days. Mice were sacrificed at day 34 after adenovirus injection. The sacrifice and specimen taking procedure was similar to the ConA model described above. In addition, spleens were harvested to measure the spleen weight, and spleen index was calculated using spleen weight (mg) divided by mice weight (g).

4.6. Histopathology, Immunohistochemistry, and Biochemical Analysis

The fixed liver tissue was embedded in paraffine and sliced at a thickness of 4 μm. H&E staining was performed to evaluate liver injury and inflammation. Afterward, all sections were graded of whole sections blindly under optical microscope (Olympus IX71) and representative pictures were imaged. Necrosis scores and inflammation scores were calculated as below: 0, no area of necrosis/immune cells infiltration; 1, very mild, few interspersed necrosis/ infiltration; 2, mild, necrosis/infiltration area ≤30%; 3, moderate, 30% < necrosis/infiltration area ≤60%; 4, severe, necrosis/infiltration area >60%. Plus, Sirius-red staining was performed to evaluate the degree of liver fibrosis in chronic AIH model. Immunohistochemistry (IHC) was performed to detect the expression of CD4, F4/80, and phospho-AKT expression in liver tissue according to protocol using SP9000 IHC kit (ZSGB-Bio, Beijing, China). Primary antibodies: anti-CD4 (1:2000; ab183685, Abcam (Cambridge, UK)), anti-F4/80 (1:200; #70076S, CST (Danvers, MA, USA)), and anti-phospho-AKT (Ser473) (1:250; AF0016, Affinity (Cincinnati, OH, USA)) were used.

Blood samples collected were centrifuged at 3000 rpm for 15 min to obtain the serum. The liver enzyme ALT and AST in serum were measured using the Rayto automatic biochemistry analyzer Chemray 420 by Wuhan ServiceBio Company (Wuhan, China).

4.7. Western Blot

Liver tissue specimens were preserved in a −80 °C refrigerator before use. After being cut to an appropriate size, liver tissue was lysed in RIPA lysis buffer (harsh) containing protease inhibitor cocktail and phosphatase inhibitor cocktail (Wuhan Servicebio Technology Co., Ltd. (Wuhan, China)). Homogenizer was used to grind tissue into homogenate and sonication was performed using ultrasonic processor. After 40 min lysis, all samples were centrifuged at 12,000 rpm for 10 min. Protein concentration was measured using BCA kit (Wuhan Servicebio Technology Co., Ltd. (Wuhan, China)). Tissue protein (40 μg) was separated in an SDS-PAGE and was transferred to PVDF membranes. The blotted PVDF membranes were blocked with TBST containing 5% BSA at room temperature for 1 h. Then the primary antibodies were incubated overnight at 4 °C including: anti-phospho-AKT (Ser473) (1:1000; #4060T, CST), anti-AKT(1:1000; #9272S, CST), anti-phospho-SRC (Tyr416) (1:1000; #6943S, CST), anti-SRC (1:1000; #A19119, Abclonal (Wuhan, China)), anti-phospho-IGF1Rβ (Tyr1135/1136) (1:1000; #3024S, CST), anti-IGF1Rβ (1:800; #9750S, CST), anti-GAPDH (1:2000; 30202ES60, Yeason (Shanghai, China)). On the next day, secondary antibodies were incubated for 1 h at room temperature. After being washed 3 times in TBST for 10 min, membranes were exposed to hypersensitive electrochemiluminescence (ECL) kit (NCM Biotech (Suzhou, China)). Protein bands were visualized on the Tanon 5200 Multi System (Tanon Science and Technology Co., Ltd. (Shanghai, China)).

4.8. Statistical Analysis

GraphPad Prism software (version 9.0.0) was utilized to analyze data. The continuous data was presented as mean ± standard deviation (SD), and the Student's *t*-test was performed to analyze the comparison between two groups. Welch's correction was used if variances were not equal based on the F test. The discontinuous data were analyzed using the Mann–Whitney test. $p < 0.05$ was considered statistically significant.

5. Conclusions

In our present study, network pharmacology and experimental validation were integrated to investigate the effect and mechanism of costunolide in AIH. According to the results, costunolide could relieve the inflammation and fibrosis in two murine AIH models, and the therapeutic effect might work by suppressing the activation of PI3K-AKT pathway and inhibiting the phosphorylation of SRC and IGF1R. Our research reveals the effect and possible mechanism of costunolide in AIH and may further contribute to novel drug development for AIH and other autoimmune diseases.

Supplementary Materials: The following supporting information can be downloaded at: https://www.mdpi.com/article/10.3390/ph16020316/s1, Figure S1: Costunolide inhibited the phosphorylation of AKT, SRC, and IGF1R. Figure S2: Animal model and experimental design.

Author Contributions: Conceptualization, M.L. and Z.H.; methodology, Z.H. and S.W.; validation, Z.H. and S.N.; formal analysis, S.N.; writing—original draft preparation, Z.H.; writing—review and editing, J.G. and H.W.; visualization, Z.H. and S.N.; supervision, W.Y. and D.T.; funding acquisition, D.T., M.L., Z.H., and S.N. contributed equally to this work. All authors have read and agreed to the published version of the manuscript.

Funding: This research was funded by the National Natural Science Foundation of China, grant number No.81900504, No.81974071, and No.82270558.

Institutional Review Board Statement: The animal study protocol was approved by the Laboratory Animal Welfare & Ethics Committee of Tongji Hospital of Tongji Medical College, Huazhong University of Science and Technology (IACUC Issue No.: TJH-202207036; date of approval: 15 July 2022).

Informed Consent Statement: Not applicable.

Data Availability Statement: All data generated or analyzed during this study are included in this article and the supplementary material.

Acknowledgments: We are thankful for the kind help from Junyi Guo and Dewei Peng. The liver picture in the flow chart was partly generated using Servier Medical Art, provided by Servier, licensed under a Creative Commons Attribution 3.0 unported license.

Conflicts of Interest: The authors declare no conflict of interest.

References

1. Sirbe, C.; Simu, G.; Szabo, I.; Grama, A.; Pop, T.L. Pathogenesis of Autoimmune Hepatitis-Cellular and Molecular Mechanisms. *Int. J. Mol. Sci.* **2021**, *22*, 13578. [CrossRef] [PubMed]
2. Terziroli Beretta-Piccoli, B.; Mieli-Vergani, G.; Vergani, D. Autoimmmune hepatitis. *Cell. Mol. Immunol.* **2022**, *19*, 158–176. [CrossRef] [PubMed]
3. Lohse, A.W.; Sebode, M.; Jorgensen, M.H.; Ytting, H.; Karlsen, T.H.; Kelly, D.; Manns, M.P.; Vesterhus, M.; European Reference Network on Hepatological Diseases; The International Autoimmune Hepatitis Group (IAIHG). Second-line and third-line therapy for autoimmune hepatitis: A position statement from the European Reference Network on Hepatological Diseases and the International Autoimmune Hepatitis Group. *J. Hepatol.* **2020**, *73*, 1496–1506. [CrossRef] [PubMed]
4. Webb, G.J.; Hirschfield, G.M.; Krawitt, E.L.; Gershwin, M.E. Cellular and Molecular Mechanisms of Autoimmune Hepatitis. *Annu. Rev. Pathol.* **2018**, *13*, 247–292. [CrossRef]
5. Liu, J.; Ma, Z.; Li, H.; Li, X. Chinese medicine in the treatment of autoimmune hepatitis: Progress and future opportunities. *Anim. Model. Exp. Med.* **2022**, *5*, 95–107. [CrossRef]
6. Kim, D.Y.; Choi, B.Y. Costunolide-A Bioactive Sesquiterpene Lactone with Diverse Therapeutic Potential. *Int. J. Mol. Sci.* **2019**, *20*, 2926. [CrossRef]
7. Kassuya, C.A.; Cremoneze, A.; Barros, L.F.; Simas, A.S.; Lapa Fda, R.; Mello-Silva, R.; Stefanello, M.E.; Zampronio, A.R. Antipyretic and anti-inflammatory properties of the ethanolic extract, dichloromethane fraction and costunolide from Magnolia ovata (Magnoliaceae). *J. Ethnopharmacol.* **2009**, *124*, 369–376. [CrossRef]
8. Butturini, E.; Di Paola, R.; Suzuki, H.; Paterniti, I.; Ahmad, A.; Mariotto, S.; Cuzzocrea, S. Costunolide and Dehydrocostuslactone, two natural sesquiterpene lactones, ameliorate the inflammatory process associated to experimental pleurisy in mice. *Eur. J. Pharmacol.* **2014**, *730*, 107–115. [CrossRef]
9. Zheng, H.; Chen, Y.; Zhang, J.; Wang, L.; Jin, Z.; Huang, H.; Man, S.; Gao, W. Evaluation of protective effects of costunolide and dehydrocostuslactone on ethanol-induced gastric ulcer in mice based on multi-pathway regulation. *Chem. Biol. Interact.* **2016**, *250*, 68–77. [CrossRef]
10. Chen, Z.; Zhang, D.; Li, M.; Wang, B. Costunolide ameliorates lipoteichoic acid-induced acute lung injury via attenuating MAPK signaling pathway. *Int. Immunopharmacol.* **2018**, *61*, 283–289. [CrossRef]
11. Xie, F.; Zhang, H.; Zheng, C.; Shen, X.F. Costunolide improved dextran sulfate sodium-induced acute ulcerative colitis in mice through NF-kappaB, STAT1/3, and Akt signaling pathways. *Int. Immunopharmacol.* **2020**, *84*, 106567. [CrossRef] [PubMed]
12. Lv, Q.; Xing, Y.; Dong, D.; Hu, Y.; Chen, Q.; Zhai, L.; Hu, L.; Zhang, Y. Costunolide ameliorates colitis via specific inhibition of HIF1alpha/glycolysis-mediated Th17 differentiation. *Int. Immunopharmacol.* **2021**, *97*, 107688. [CrossRef] [PubMed]
13. Mao, J.; Yi, M.; Wang, R.; Huang, Y.; Chen, M. Protective Effects of Costunolide Against D-Galactosamine and Lipopolysaccharide-Induced Acute Liver Injury in Mice. *Front. Pharmacol.* **2018**, *9*, 1469. [CrossRef] [PubMed]
14. Mao, J.; Zhan, H.; Meng, F.; Wang, G.; Huang, D.; Liao, Z.; Chen, M. Costunolide protects against alcohol-induced liver injury by regulating gut microbiota, oxidative stress and attenuating inflammation in vivo and in vitro. *Phytother. Res.* **2022**, *36*, 1268–1283. [CrossRef] [PubMed]
15. Ge, M.X.; Liu, H.T.; Zhang, N.; Niu, W.X.; Lu, Z.N.; Bao, Y.Y.; Huang, R.; Yu, D.K.; Shao, R.G.; He, H.W. Costunolide represses hepatic fibrosis through WW domain-containing protein 2-mediated Notch3 degradation. *Br. J. Pharmacol.* **2020**, *177*, 372–387. [CrossRef] [PubMed]
16. Hopkins, A.L. Network pharmacology: The next paradigm in drug discovery. *Nat. Chem. Biol.* **2008**, *4*, 682–690. [CrossRef]
17. Zhang, R.; Zhu, X.; Bai, H.; Ning, K. Network Pharmacology Databases for Traditional Chinese Medicine: Review and Assessment. *Front. Pharmacol.* **2019**, *10*, 123. [CrossRef] [PubMed]
18. To, K.I.; Zhu, Z.X.; Wang, Y.N.; Li, G.A.; Sun, Y.M.; Li, Y.; Jin, Y.H. Integrative network pharmacology and experimental verification to reveal the anti-inflammatory mechanism of ginsenoside Rh4. *Front. Pharmacol.* **2022**, *13*, 953871. [CrossRef]
19. Wang, C.R.; Chen, H.W.; Li, Y.; Zhou, M.Y.; Wong, V.K.; Jiang, Z.H.; Zhang, W. Network Pharmacology Exploration Reveals Anti-Apoptosis as a Common Therapeutic Mechanism for Non-Alcoholic Fatty Liver Disease Treated with Blueberry Leaf Polyphenols. *Nutrients* **2021**, *13*, 4060. [CrossRef]
20. Li, X.; Wen, H.; Zhang, Y.; Liu, A.; Zhang, X.; Fu, M.; Pan, Y.; Xu, J.; Zhang, J. DPHB, a diarylheptane from Alpinia officinarum Hance, ameliorates insulin resistance: A network pharmacology and in vitro study. *Front. Pharmacol.* **2022**, *13*, 956812. [CrossRef]
21. Xu, Z.; Lin, S.; Gong, J.; Feng, P.; Cao, Y.; Li, Q.; Jiang, Y.; You, Y.; Tong, Y.; Wang, P. Exploring the Protective Effects and Mechanism of Crocetin From Saffron Against NAFLD by Network Pharmacology and Experimental Validation. *Front. Med.* **2021**, *8*, 681391. [CrossRef] [PubMed]
22. Wang, H.; Yan, W.; Feng, Z.; Gao, Y.; Zhang, L.; Feng, X.; Tian, D. Plasma proteomic analysis of autoimmune hepatitis in an improved AIH mouse model. *J. Transl. Med.* **2020**, *18*, 3. [CrossRef] [PubMed]

23. Ru, J.; Li, P.; Wang, J.; Zhou, W.; Li, B.; Huang, C.; Li, P.; Guo, Z.; Tao, W.; Yang, Y.; et al. TCMSP: A database of systems pharmacology for drug discovery from herbal medicines. *J. Cheminform.* **2014**, *6*, 13. [CrossRef]
24. Daina, A.; Michielin, O.; Zoete, V. SwissADME: A free web tool to evaluate pharmacokinetics, drug-likeness and medicinal chemistry friendliness of small molecules. *Sci. Rep.* **2017**, *7*, 42717. [CrossRef]
25. Lv, T.; Li, M.; Zeng, N.; Zhang, J.; Li, S.; Chen, S.; Zhang, C.; Shan, S.; Duan, W.; Wang, Q.; et al. Systematic review and meta-analysis on the incidence and prevalence of autoimmune hepatitis in Asian, European, and American population. *J. Gastroenterol. Hepatol.* **2019**, *34*, 1676–1684. [CrossRef]
26. Huang, H.; Park, S.; Zhang, H.; Park, S.; Kwon, W.; Kim, E.; Zhang, X.; Jang, S.; Yoon, D.; Choi, S.K.; et al. Targeting AKT with costunolide suppresses the growth of colorectal cancer cells and induces apoptosis in vitro and in vivo. *J. Exp. Clin. Cancer Res.* **2021**, *40*, 114. [CrossRef] [PubMed]
27. He, F.; Antonucci, L.; Yamachika, S.; Zhang, Z.; Taniguchi, K.; Umemura, A.; Hatzivassiliou, G.; Roose-Girma, M.; Reina-Campos, M.; Duran, A.; et al. NRF2 activates growth factor genes and downstream AKT signaling to induce mouse and human hepatomegaly. *J. Hepatol.* **2020**, *72*, 1182–1195. [CrossRef] [PubMed]
28. Jiang, R.; Tang, J.; Zhang, X.; He, Y.; Yu, Z.; Chen, S.; Xia, J.; Lin, J.; Ou, Q. CCN1 Promotes Inflammation by Inducing IL-6 Production via alpha6beta1/PI3K/Akt/NF-kappaB Pathway in Autoimmune Hepatitis. *Front. Immunol.* **2022**, *13*, 810671. [CrossRef]
29. Xiang, M.; Liu, T.; Tan, W.; Ren, H.; Li, H.; Liu, J.; Cao, H.; Cheng, Q.; Liu, X.; Zhu, H.; et al. Effects of kinsenoside, a potential immunosuppressive drug for autoimmune hepatitis, on dendritic cells/CD8(+) T cells communication in mice. *Hepatology* **2016**, *64*, 2135–2150. [CrossRef]
30. Morales-Ruiz, M.; Santel, A.; Ribera, J.; Jimenez, W. The Role of Akt in Chronic Liver Disease and Liver Regeneration. *Semin. Liver Dis.* **2017**, *37*, 11–16. [CrossRef]
31. Haxhinasto, S.; Mathis, D.; Benoist, C. The AKT-mTOR axis regulates de novo differentiation of $CD4^+Foxp3^+$ cells. *J. Exp. Med.* **2008**, *205*, 565–574. [CrossRef] [PubMed]
32. Delgoffe, G.M.; Pollizzi, K.N.; Waickman, A.T.; Heikamp, E.; Meyers, D.J.; Horton, M.R.; Xiao, B.; Worley, P.F.; Powell, J.D. The kinase mTOR regulates the differentiation of helper T cells through the selective activation of signaling by mTORC1 and mTORC2. *Nat. Immunol.* **2011**, *12*, 295–303. [CrossRef] [PubMed]
33. Xu, K.; Yin, N.; Peng, M.; Stamatiades, E.G.; Chhangawala, S.; Shyu, A.; Li, P.; Zhang, X.; Do, M.H.; Capistrano, K.J.; et al. Glycolytic ATP fuels phosphoinositide 3-kinase signaling to support effector T helper 17 cell responses. *Immunity* **2021**, *54*, 976–987.e7. [CrossRef] [PubMed]
34. Xu, K.; Yin, N.; Peng, M.; Stamatiades, E.G.; Shyu, A.; Li, P.; Zhang, X.; Do, M.H.; Wang, Z.; Capistrano, K.J.; et al. Glycolysis fuels phosphoinositide 3-kinase signaling to bolster T cell immunity. *Science* **2021**, *371*, 405–410. [CrossRef]
35. Herrero-Sanchez, M.C.; Rodriguez-Serrano, C.; Almeida, J.; San Segundo, L.; Inoges, S.; Santos-Briz, A.; Garcia-Brinon, J.; Corchete, L.A.; San Miguel, J.F.; Del Canizo, C.; et al. Targeting of PI3K/AKT/mTOR pathway to inhibit T cell activation and prevent graft-versus-host disease development. *J. Hematol. Oncol.* **2016**, *9*, 113. [CrossRef]
36. Maxwell, M.J.; Tsantikos, E.; Kong, A.M.; Vanhaesebroeck, B.; Tarlinton, D.M.; Hibbs, M.L. Attenuation of phosphoinositide 3-kinase delta signaling restrains autoimmune disease. *J. Autoimmun.* **2012**, *38*, 381–391. [CrossRef]
37. Camps, M.; Ruckle, T.; Ji, H.; Ardissone, V.; Rintelen, F.; Shaw, J.; Ferrandi, C.; Chabert, C.; Gillieron, C.; Francon, B.; et al. Blockade of PI3Kgamma suppresses joint inflammation and damage in mouse models of rheumatoid arthritis. *Nat. Med.* **2005**, *11*, 936–943. [CrossRef] [PubMed]
38. Tian, X.; Wang, R.; Gu, T.; Ma, F.; Laster, K.V.; Li, X.; Liu, K.; Lee, M.H.; Dong, Z. Costunolide is a dual inhibitor of MEK1 and AKT1/2 that overcomes osimertinib resistance in lung cancer. *Mol. Cancer* **2022**, *21*, 193. [CrossRef] [PubMed]
39. Fan, X.; Men, R.; Wang, H.; Shen, M.; Wang, T.; Ye, T.; Luo, X.; Yang, L. Methylprednisolone Decreases Mitochondria-Mediated Apoptosis and Autophagy Dysfunction in Hepatocytes of Experimental Autoimmune Hepatitis Model via the Akt/mTOR Signaling. *Front. Pharmacol.* **2019**, *10*, 1189. [CrossRef] [PubMed]
40. Muller, P.; Messmer, M.; Bayer, M.; Pfeilschifter, J.M.; Hintermann, E.; Christen, U. Non-alcoholic fatty liver disease (NAFLD) potentiates autoimmune hepatitis in the CYP2D6 mouse model. *J. Autoimmun.* **2016**, *69*, 51–58. [CrossRef]
41. Roskoski, R., Jr. Src protein-tyrosine kinase structure, mechanism, and small molecule inhibitors. *Pharmacol. Res.* **2015**, *94*, 9–25. [CrossRef] [PubMed]
42. Yi, H.; Bai, Y.; Zhu, X.; Lin, L.; Zhao, L.; Wu, X.; Buch, S.; Wang, L.; Chao, J.; Yao, H. IL-17A induces MIP-1alpha expression in primary astrocytes via Src/MAPK/PI3K/NF-kB pathways: Implications for multiple sclerosis. *J. Neuroimmune Pharmacol.* **2014**, *9*, 629–641. [CrossRef]
43. Hao, M.; Sun, J.; Zhang, Y.; Zhang, D.; Han, J.; Zhang, J.; Qiao, H. Exploring the Role of SRC in Extraocular Muscle Fibrosis of the Graves' Ophthalmopathy. *Front. Bioeng. Biotechnol.* **2020**, *8*, 392. [CrossRef]
44. Behnen, M.; Leschczyk, C.; Moller, S.; Batel, T.; Klinger, M.; Solbach, W.; Laskay, T. Immobilized immune complexes induce neutrophil extracellular trap release by human neutrophil granulocytes via FcgammaRIIIB and Mac-1. *J. Immunol.* **2014**, *193*, 1954–1965. [CrossRef] [PubMed]
45. Gaud, G.; Lesourne, R.; Love, P.E. Regulatory mechanisms in T cell receptor signalling. *Nat. Rev. Immunol.* **2018**, *18*, 485–497. [CrossRef] [PubMed]

46. DiToro, D.; Harbour, S.N.; Bando, J.K.; Benavides, G.; Witte, S.; Laufer, V.A.; Moseley, C.; Singer, J.R.; Frey, B.; Turner, H.; et al. Insulin-Like Growth Factors Are Key Regulators of T Helper 17 Regulatory T Cell Balance in Autoimmunity. *Immunity* **2020**, *52*, 650–667.e10. [CrossRef]
47. Pritchard, J.; Tsui, S.; Horst, N.; Cruikshank, W.W.; Smith, T.J. Synovial fibroblasts from patients with rheumatoid arthritis, like fibroblasts from Graves' disease, express high levels of IL-16 when treated with Igs against insulin-like growth factor-1 receptor. *J. Immunol.* **2004**, *173*, 3564–3569. [CrossRef]
48. Douglas, R.S.; Gianoukakis, A.G.; Kamat, S.; Smith, T.J. Aberrant expression of the insulin-like growth factor-1 receptor by T cells from patients with Graves' disease may carry functional consequences for disease pathogenesis. *J. Immunol.* **2007**, *178*, 3281–3287. [CrossRef]
49. Jia, H.Y.; Qiu, H.Y.; Zhang, M.D.; Hou, J.J.; Zhou, M.L.; Wu, Y. Lenalidomide attenuates IMQ-induced inflammation in a mouse model of psoriasis. *Biomed. Pharmacother.* **2022**, *156*, 113883. [CrossRef]
50. Pae, H.O.; Jeong, G.S.; Kim, H.S.; Woo, W.H.; Rhew, H.Y.; Kim, H.S.; Sohn, D.H.; Kim, Y.C.; Chung, H.T. Costunolide inhibits production of tumor necrosis factor-alpha and interleukin-6 by inducing heme oxygenase-1 in RAW264.7 macrophages. *Inflamm. Res.* **2007**, *56*, 520–526. [CrossRef]
51. Park, E.; Song, J.H.; Kim, M.S.; Park, S.H.; Kim, T.S. Costunolide, a sesquiterpene lactone, inhibits the differentiation of pro-inflammatory CD4(+) T cells through the modulation of mitogen-activated protein kinases. *Int. Immunopharmacol.* **2016**, *40*, 508–516. [CrossRef] [PubMed]
52. European Association for the Study of the Liver. EASL Clinical Practice Guidelines for the management of patients with decompensated cirrhosis. *J. Hepatol.* **2018**, *69*, 406–460. [CrossRef] [PubMed]
53. Niu, X.; Wang, X.; Niu, B.; Meng, Y.; He, H.; Wang, Y.; Li, G. Costunolide Loaded in pH-Responsive Mesoporous Silica Nanoparticles for Increased Stability and an Enhanced Anti-Fibrotic Effect. *Pharmaceuticals* **2021**, *14*, 951. [CrossRef] [PubMed]
54. Kim, S.; Chen, J.; Cheng, T.; Gindulyte, A.; He, J.; He, S.; Li, Q.; Shoemaker, B.A.; Thiessen, P.A.; Yu, B.; et al. PubChem in 2021: New data content and improved web interfaces. *Nucleic Acids Res.* **2021**, *49*, D1388–D1395. [CrossRef]
55. Daina, A.; Michielin, O.; Zoete, V. SwissTargetPrediction: Updated data and new features for efficient prediction of protein targets of small molecules. *Nucleic Acids Res.* **2019**, *47*, W357–W364. [CrossRef]
56. Wang, X.; Shen, Y.; Wang, S.; Li, S.; Zhang, W.; Liu, X.; Lai, L.; Pei, J.; Li, H. PharmMapper 2017 update: A web server for potential drug target identification with a comprehensive target pharmacophore database. *Nucleic Acids Res.* **2017**, *45*, W356–W360. [CrossRef]
57. UniProt, C. UniProt: The universal protein knowledgebase in 2021. *Nucleic Acids Res.* **2021**, *49*, D480–D489. [CrossRef]
58. Pinero, J.; Ramirez-Anguita, J.M.; Sauch-Pitarch, J.; Ronzano, F.; Centeno, E.; Sanz, F.; Furlong, L.I. The DisGeNET knowledge platform for disease genomics: 2019 update. *Nucleic Acids Res.* **2020**, *48*, D845–D855. [CrossRef]
59. Safran, M.; Rosen, N.; Twik, M.; BarShir, R.; Stein, T.I.; Dahary, D.; Fishilevich, S.; Lancet, D. The GeneCards Suite. In *Practical Guide to Life Science Databases*; Abugessaisa, I., Kasukawa, T., Eds.; Springer Nature: Singapore, 2021; pp. 27–56. [CrossRef]
60. Sherman, B.T.; Hao, M.; Qiu, J.; Jiao, X.; Baseler, M.W.; Lane, H.C.; Imamichi, T.; Chang, W. DAVID: A web server for functional enrichment analysis and functional annotation of gene lists (2021 update). *Nucleic Acids Res.* **2022**, *50*, W216–W221. [CrossRef]
61. Huang, D.W.; Sherman, B.T.; Lempicki, R.A. Systematic and integrative analysis of large gene lists using DAVID bioinformatics resources. *Nat. Protoc.* **2009**, *4*, 44–57. [CrossRef]
62. Zhou, Y.; Zhou, B.; Pache, L.; Chang, M.; Khodabakhshi, A.H.; Tanaseichuk, O.; Benner, C.; Chanda, S.K. Metascape provides a biologist-oriented resource for the analysis of systems-level datasets. *Nat. Commun.* **2019**, *10*, 1523. [CrossRef] [PubMed]
63. Szklarczyk, D.; Gable, A.L.; Lyon, D.; Junge, A.; Wyder, S.; Huerta-Cepas, J.; Simonovic, M.; Doncheva, N.T.; Morris, J.H.; Bork, P.; et al. STRING v11: Protein-protein association networks with increased coverage, supporting functional discovery in genome-wide experimental datasets. *Nucleic Acids Res.* **2019**, *47*, D607–D613. [CrossRef] [PubMed]
64. Shannon, P.; Markiel, A.; Ozier, O.; Baliga, N.S.; Wang, J.T.; Ramage, D.; Amin, N.; Schwikowski, B.; Ideker, T. Cytoscape: A software environment for integrated models of biomolecular interaction networks. *Genome Res.* **2003**, *13*, 2498–2504. [CrossRef] [PubMed]
65. *The PyMOL Molecular Graphics System*; Version 1.8; Schrodinger, LLC.: New York, NY, USA, 2015.
66. Morris, G.M.; Huey, R.; Lindstrom, W.; Sanner, M.F.; Belew, R.K.; Goodsell, D.S.; Olson, A.J. AutoDock4 and AutoDockTools4: Automated docking with selective receptor flexibility. *J. Comput. Chem.* **2009**, *30*, 2785–2791. [CrossRef] [PubMed]
67. Janet, H. Methods of Blood Collection in the Mouse. *Lab. Animal* **2000**, *29*, 47–53.

pharmaceuticals

MDPI

Article

Exploring the Effective Components and Mechanism of Action of Japanese *Ardisia* in the Treatment of Autoimmune Hepatitis Based on Network Pharmacology and Experimental Verification

Tian Fu [1,†], Yifei Chen [1,2,†], Junkui Li [3,4], Peili Zhu [3,4], Huajuan He [1], Wei Zhang [1], Ken Kin Lam Yung [3,4,*] and Wei Wu [1,*]

1 School of Pharmacy, Guilin Medical University, Guilin 541199, China
2 School of Traditional Chinese Medicine, Southern Medical University, Guangzhou 510515, China
3 Department of Biology, Hong Kong Baptist University, Hong Kong 999077, China
4 Golden Meditech Centre for NeuroRegeneration Sciences (GCNS), Hong Kong Baptist University, Kowloon Tong, Kowloon, Hong Kong 999077, China
* Correspondence: kklyung@hkbu.edu.hk (K.K.L.Y.); wuwei@glmc.edu.cn (W.W.)
† These authors contributed equally to this work.

Abstract: Japanese *Ardisia* is widely used as a hepatoprotective and anti-inflammatory agent in China. However, the active ingredients in Japanese *Ardisia* and their potential mechanisms of action in the treatment of autoimmune hepatitis (AIH) are unknown. The pharmacodynamic substance and mechanism of action of Japanese *Ardisia* in the treatment of AIH were investigated using network pharmacology and molecular docking technology in this study. Following that, the effects of Japanese *Ardisia* were evaluated using the concanavalin A (Con A)-induced acute liver injury rat model. The active ingredients and targets of Japanese *Ardisia* were searched using the Traditional Chinese Medicine Systems Pharmacology database, and hepatitis-related therapeutic targets were identified through GeneCards and Online Mendelian Inheritance in Man databases. A compound–target network was then constructed using Cytoscape software, and enrichment analysis was performed using gene ontology (GO) and Kyoto Encyclopedia of Genes and Genomes (KEGG) databases. Molecular docking technology was used to simulate the docking of key targets, and the AIH rat model was used to validate the expression of key targets. Nineteen active chemical components and 143 key target genes were identified. GO enrichment analysis revealed that the treatment of AIH with Japanese *Ardisia* mainly involved DNA-binding transcription factor binding, RNA polymerase II-specific DNA transcription factor binding, cytokine receptor binding, receptor-ligand activity, ubiquitin-like protein ligase binding, and cytokine activity. In the KEGG enrichment analysis, 165 pathways were identified, including the lipid and atherosclerotic pathway, IL-17 signaling pathway, TNF signaling pathway, hepatitis B pathway, and the AGE–RAGE signaling pathway in diabetic complications. These pathways may be the key to effective AIH treatment with Japanese *Ardisia*. Molecular docking showed that quercetin and kaempferol have good binding to *AKT1*, *IL6*, *VEGFA*, and *CASP3*. Animal experiments demonstrated that Japanese *Ardisia* could increase the expression of *AKT1* and decrease the expression of *CASP3* protein, as well as *IL-6*, in rat liver tissues. This study identified multiple molecular targets and pathways for Japanese *Ardisia* in the treatment of AIH. At the same time, the effectiveness of Japanese *Ardisia* in treating AIH was verified by animal experiments.

Keywords: Japanese *Ardisia*; autoimmune hepatitis; network pharmacology; molecular docking; underlying mechanism

Citation: Fu, T.; Chen, Y.; Li, J.; Zhu, P.; He, H.; Zhang, W.; Yung, K.K.L.; Wu, W. Exploring the Effective Components and Mechanism of Action of Japanese *Ardisia* in the Treatment of Autoimmune Hepatitis Based on Network Pharmacology and Experimental Verification. *Pharmaceuticals* **2022**, *15*, 1457. https://doi.org/10.3390/ph15121457

Academic Editor: Diana Roxana Pelinescu

Received: 9 October 2022
Accepted: 21 November 2022
Published: 24 November 2022

1. Introduction

Autoimmune hepatitis (AIH) is a common liver disease worldwide, seen in both men and women, but predominantly in women. According to epidemiological surveys, the incidence of AIH ranges from 0.67 to 2.0/100,000 people per year, and the prevalence ranges from 4.0 to 42.9/100,000 [1,2]. AIH has become the second most common inflammatory liver disease after viral hepatitis [3].

The pathogenesis of AIH is not completely understood. AIH is thought to be caused by genetic factors, molecular mimetic mechanisms, immune damage, and a variety of physical and chemical factors [4–11]. Prednisolone in combination with or without azathioprine (AZA) is generally recommended as the first-line drug for AIH [12], and second-generation alternatives, such as budesonide and tacrolimus, are recommended for this category of non-responders or intolerant patients, but these drugs have certain side effects. Traditional Chinese medicine has become increasingly important in the treatment of the disease in recent years.

Japanese *Ardisia*, known as *Ardisia japonica* or marlberry, is used as a medicinal plant in traditional Chinese medicine. It grows very slowly, and its leaves have a similar appearance to tea leaves. Bright red berries appear under the leaves in autumn, and therefore, it is also called 'aidicha' or 'yedizhu' in Chinese. Japanese *Ardisia* is mainly grown in the southern provinces of China, such as Hunan and Guangxi, where it is a popular medicinal herb used in Chinese folk medicine. The Chinese ancient medicine book 'Compendium of Materia Medica' records that *A. japonica* has the effect of 'detoxification and promoting blood circulation'. The pharmacodynamic components of *A. japonica* are saponins, coumarins, benzoquinones, and flavonoids [13–15]. It has pharmacological activities, such as relieving cough and asthma, protecting the liver, and anti-inflammatory, anti-viral, and anti-tumor activities [16,17]. In clinical practice, *A japonica* is commonly used to treat chronic bronchitis, pulmonary tuberculosis, tuberculous pleurisy, and acute icteric hepatitis. Meanwhile, *A. japonica* has shown remarkable curative effects in the treatment of chronic hepatitis [14]. However, its pharmacodynamic ingredients and mechanism of action remain unclear.

Cyberpharmacology can effectively reveal the material basis and mechanism of action of Chinese medicine by systematically and integrally exploring the relationship between drugs and diseases [18]. Thus, the aim of this study is to systematically elucidate the mechanism of Japanese *Ardisia* in the treatment of AIH through network pharmacology and molecular docking analysis of the interaction between drug molecules and AIH-related targets, and to provide a theoretical basis for clinical research. The specific flow chart is shown in Figure 1.

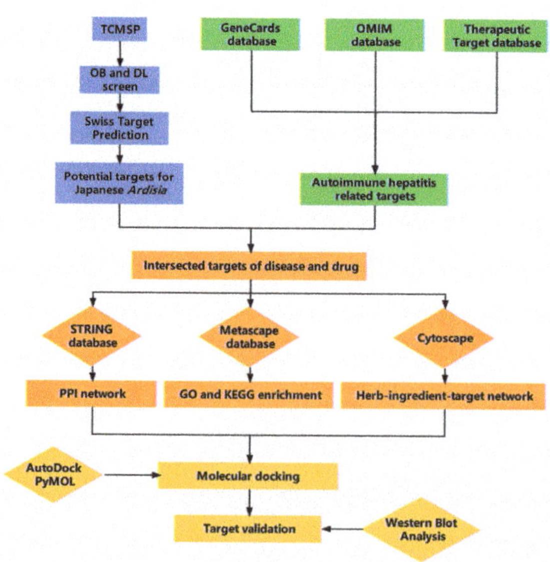

Figure 1. Network pharmacology analysis workflow.

2. Results

2.1. Screening of Active Compounds

Oral bioavailability (OB) is the fraction of an orally administered drug that reaches systemic circulation. This is an important consideration for bioactive molecules used as therapeutic agents. Drug likeness (DL) qualitatively assesses the capacity of a molecule to become an orally administered drug based on its bioavailability [19]. The main active components of Japanese *Ardisia* were obtained by searching the TCMSP database. Nineteen molecules with OB $\geq$ 30% and DL $\geq$ 0.18 were identified as bioactive compounds [20], as shown in Table 1.

Table 1. Basic information on the active compounds of Japanese *Ardisia*.

MOL ID	MOL Name	OB	DL
MOL010934	Ardisianoside K	31.98	0.63
MOL010953	Triterpenoid glycoside 1	34.11	0.63
MOL010964	Maesanin	42.77	0.35
MOL010973	Rapanone	34.15	0.24
MOL010974	Tri-O-methylnorbergenin	33.17	0.41
MOL010976	Triterpene glycoside 4	41.4	0.63
MOL010981	Triterpenoid glycoside 3	44.04	0.6
MOL010982	2,5-dihydroxy-3-[(10Z)-pentadec-10-en-1-yl][1,4] benzoquinone	34.74	0.6
MOL010983	2,5-Dihydroxy-3-[(10Z)-pentadec-10-en-1-yl] cyclohexa-2,5-diene-1,4-dione	37.3	0.32
MOL010985	2-hydroxy-5-methoxy-3-pentadecaenylbenzoquinone	41.61	0.32
MOL011002	5-ethoxy-2-hydroxy-3-[(10Z)-pentadec-10-en-1-yl][1,4] Benzoquinone	42.77	0.38
MOL011003	5-ethoxy-2-hydroxy-3-[(8Z)-tridec-8-en-1-yl][1,4] benzoquinone	43.23	0.3
MOL011019	Ardisianone A	44.22	0.25
MOL011020	Ardisianone B	60.9	0.2
MOL001663	(4aS,6aR,6aS,6bR,8aR,10R,12aR,14bS)-10-hydroxy-2,2,6a,6b,9,9,12a-heptamethyl-1,3,4,5,6,6a,7,8,8a,10,11,12,13,14b-tetradecahydropicene-4a-carboxylic acid	32.03	0.76
MOL002879	Diop	43.59	0.39
MOL000422	Kaempferol	41.88	0.24
MOL009278	Laricitrin	35.38	0.34
MOL000098	Quercetin	46.43	0.28

OB, oral bioavailability; DL, drug-likeness.

2.2. Compound Target Interaction Network

The information on the main bioactive components and corresponding targets in Japanese *Ardisia* were obtained from the TCMSP database (TCMSP, https://lsp.nwu.edu.cn/tcmsp.php, accessed on 14 November 2021). After screening to remove invalid gene IDs, the targets downloaded from the GeneCards (http://www.genecards.org, accessed on 14 November 2021), OMIM databases (OMIM, http://www.omim.org, accessed on 14 November 2021), and TTD (http://db.idrblab, accessed on 14 November 2021) were crossed with those from the TCMSP database to obtain potential targets for the treatment of AIH in Japanese *Ardisia*. The obtained data are presented as a Venn diagram (Figure 2). There were 5724 autoimmune hepatitis gene targets, of which 153 were potential targets related to the drug Japanese *Ardisia*. The drug-related genes and disease-specific targets were analyzed, and 143 key target genes were identified. The targets of the active ingredients of Japanese *Ardisia* are shown in Table 2.

2.3. Core Genes of the PPI Network

The common targets of Japanese *Ardisia* and autoimmune hepatitis were imported into the STRING protein interaction database (https://string-db.org, accessed on 15 November 2021) to construct the PPI network (Figure 3A). The top 30 target proteins analyzed by R software (https://www.r-project.org/ accessed on 15 November 2021) are shown in Figure 3B. The intersections of differential genes between Japanese *Ardisia* active ingredient targets and autoimmune hepatitis disease targets were imported into Cytoscape 3.7.0

(http://www.cytoscape.org, accessed on 15 November 2021) for topological analysis. The key targets were sorted according to the degree value, with higher degree values indicating nodes that were more central in the network and more important. Through the PPI protein interaction network, the disease-related targets are found, and the degree value is greater than the median by screening twice to find key targets. Thus, the key target genes for autoimmune hepatitis treatment mainly included *AKT1*, *IL6*, *VEGFA*, *CASP3*, *JUN*, *MYC*, etc. (Figure 3C).

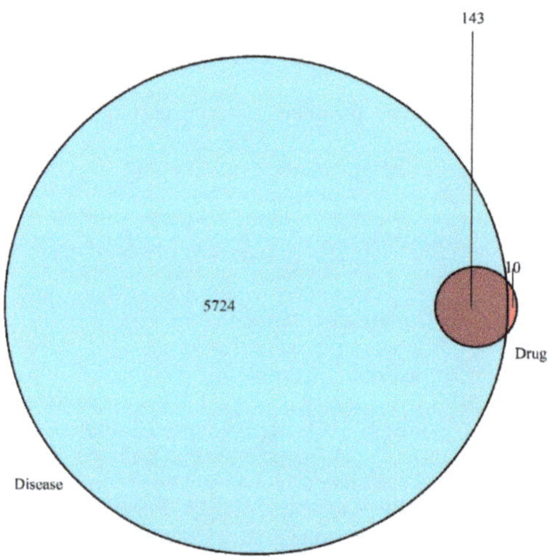

Figure 2. Intersection of drug targets and disease targets. Note: blue represents disease targets; red represents drug targets.

Table 2. Possible targets for each component.

MOL ID	Ingredients	Drug-Acting Targets of Disease
MOL010934	Ardisianoside K	NR3C1
MOL010964	Maesanin ($C_{23}H_{36}O_4$)	ACHE
MOL010974	Tri-O-methylnorbergenin	PRSS1
MOL011003	5-ethoxy-2-hydroxy-3-[(8Z)-tridec-8-en-1-yl][1,4] benzoquinone	ACHE
MOL011020	Ardisianone B	GABRA1, NCOA2
MOL002879	Diop	CHRM3
MOL000422	Kaempferol	PTGS1, AR, PPARG, NCOA2, PRSS1, PGR, CHRM1, ACHE, CHRM2, GABRA1, F7, RELA, IKBKB, BCL2, AHSA1, CASP3, MAPK8, PPARG, CYP3A4, CYP1A1, ICAM1, SELE, VCAM1, CYP1B1, ALOX5, GSTP1, AHR, PSMD3, SLC2A4, NR1I3, DIO1, GSTM1, GSTM2, AKR1C3
MOL009278	Laricitrin	ESR1, AR, PPARG, ESR2, GSK3B, PRSS1, PTGS1, NCOA2
MOL000098	Quercetin	PTGS1, AR, PPARG, NCOA2, AKR1B1, PRSS1, F7, ACHE, GABRA1, RELA, EGFR, VEGFA, CCND1, BCL2, FOS, EIF6, CASP9, PLAU, RB1, IL6, AHSA1, CASP3, TP63, ELK1, NFKBIA, POR, CASP8, RAF1, PRKCA, HIF1A, RUNX1T1, ERBB2, PPARG, ACACA, CYP3A4, CAV1, MYC, CYP1A1, ICAM1, SELE, VCAM1, PTGER3, BIRC5, DUOX2, NOS3, HSPB1, MGAM, CYP1B1, CCNB1, ALOX5, GSTP1, NFE2L2, NQO1, PARP1, AHR, PSMD3, SLC2A4, COL3A1, DCAF5, NR1I3, CHEK2, HSF1, CRP, RUNX2, RASSF1, CTSD, IGFBP3, IGF2, IRF1, ERBB3, PON1, DIO1, NPEPPS, HK2, RASA1, GSTM1, GSTM2

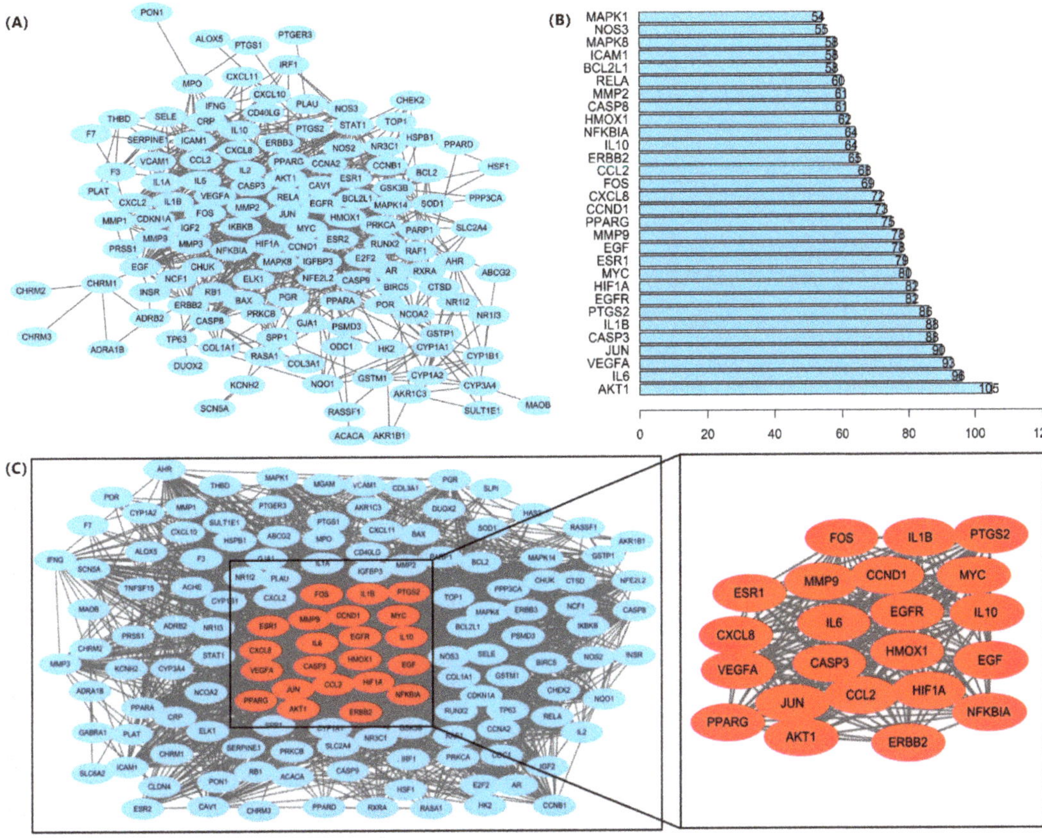

Figure 3. (**A**) PPI protein interaction network. Note: Network nodes represent proteins. Each node represents all proteins produced by a single protein-coding locus; the edge represents the protein-protein binding and promotes sharing function. (**B**) PPI network core genes. (**C**) Key targets of PPI network in the treatment of autoimmune hepatitis.

2.4. Network Pharmacology Visualization of Japanese Ardisia

The target PPI information obtained from the STRING protein interaction database was imported into Cytoscape 3.7.0 software (http://www.cytoscape.org, accessed on 15 November 2021), and the common targets between Japanese *Ardisia*, its active components, and autoimmune hepatitis were visualized. The data were merged into a component-target network diagram via the merge function in Cytoscape 3.7.0 to obtain the network diagram of 'Japanese *Ardisia*–component–gene–autoimmune hepatitis' (Figure 4).

2.5. GO Functional Enrichment Analysis

The 20 GO nodes with the greatest number of annotated proteins were selected for display. These nodes mainly involved DNA–binding transcription factor binding, RNA polymerase II-specific DNA binding, cytokine receptor binding, receptor-ligand activity, ubiquitin-like protein ligase binding, cytokine activity, ubiquitin–protein ligase binding, nuclear receptor activity, ligand-activated transcription factor activity, and kinase regulatory activity. The *p*-values were arranged from largest to smallest, and visual analysis was performed using an advanced bubble graph (Figure 5). DNA–binding transcription factor binding had the most obvious effect, and the greatest number of genes, followed by RNA polymerase II-specific DNA binding, cytokine receptor binding, receptor-ligand activity, ubiquitin-like protein ligase binding, cytokine activity, and ubiquitin–protein ligase binding. Meanwhile, nuclear receptor activity, ligand-activated transcription factor activity, kinase regulator activity, and other pathways indirectly affected a series of signaling pathways, eventually causing changes in biological processes.

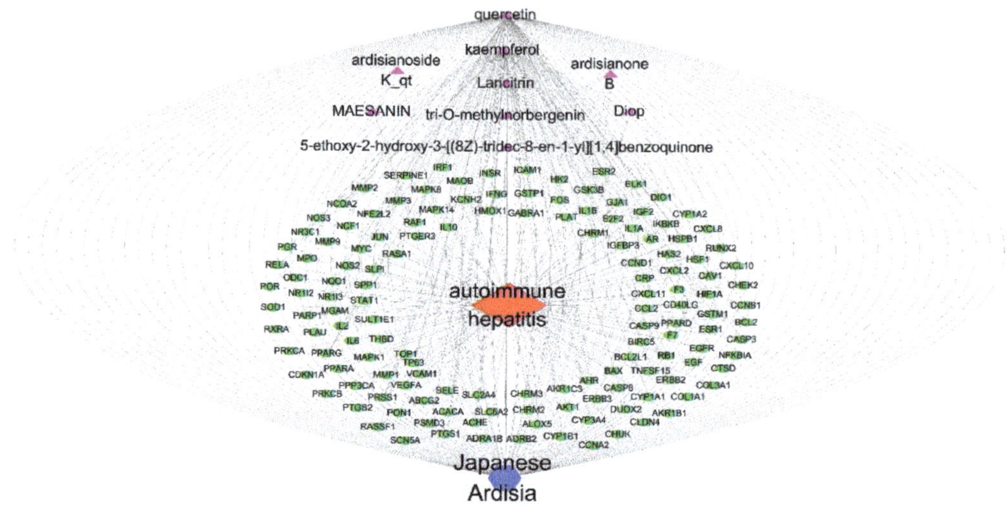

Figure 4. Network diagram of "Japanese *Ardisia*-component-gene-autoimmune hepatitis". Note: Japanese *Ardisia* is blue; components are purple; the gene is green; and autoimmune hepatitis is red.

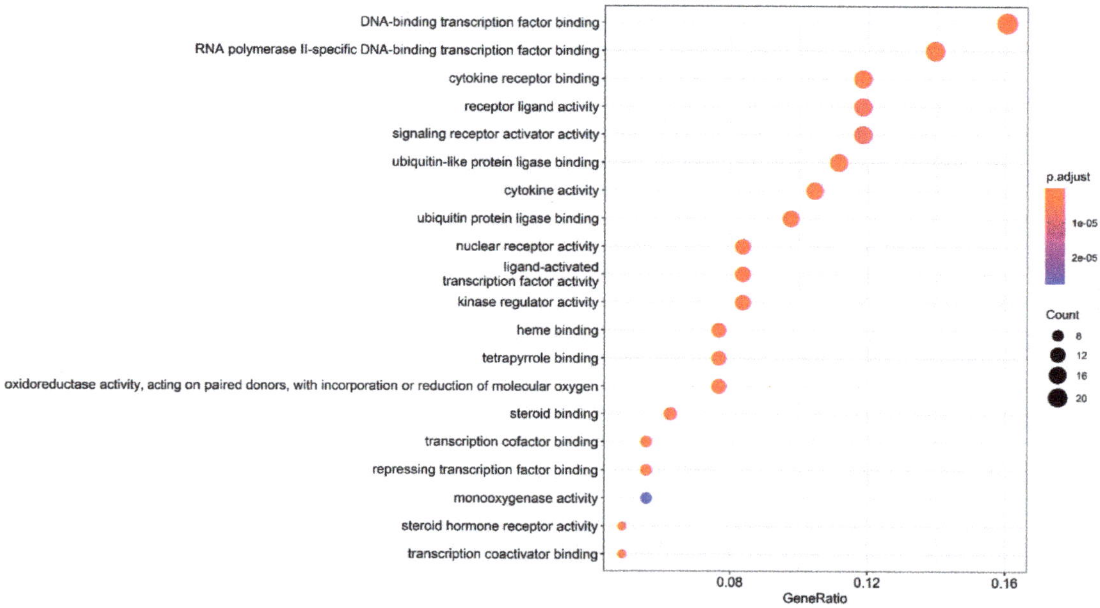

Figure 5. Bubble chart of GO enrichment analysis. Note: the abscissa is the gene ratio, the ordinate is the enriched pathway, the size of the point represents the number of genes, and the color represents the level of *p*-value.

2.6. KEGG Pathway Enrichment Analysis

One hundred and sixty-three pathways were identified in the KEGG enrichment analysis. The top 20 signaling pathways mainly involved the lipid and atherosclerosis pathway, Kaposi sarcoma-associated pathway, human cytomegalovirus infection pathway, IL-17 signaling pathway, TNF signaling pathway, hepatitis B pathway, and the AGE-RAGE signaling pathway were involved in diabetic complications (Figure 6). The results suggest that Japanese *Ardisia* can be used to treat autoimmune hepatitis through multi-target and multi-pathway regulation.

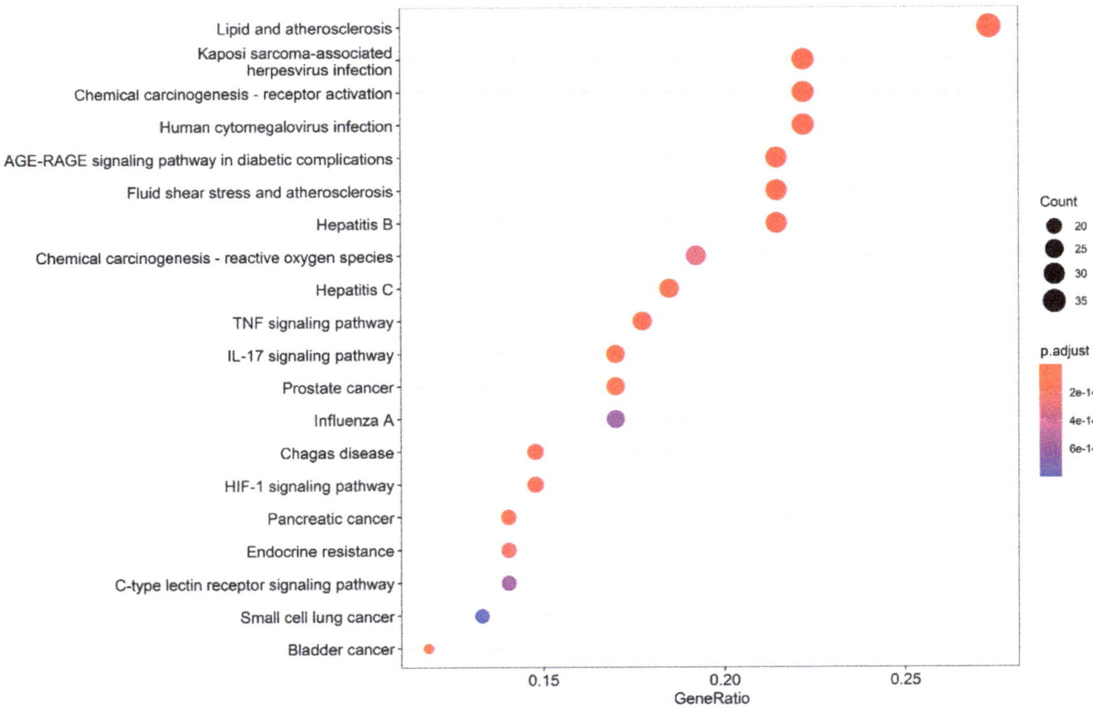

Figure 6. Bubble chart of KEGG pathway enrichment analysis. Note: the abscissa is the gene ratio, the ordinate is the enriched pathway, the size of the point represents the number of genes, and the color represents the level of *p*-value.

2.7. Molecular Docking

The X-ray crystal structures of the target molecules *AKT1*, *IL6*, *VEGFA*, and *CASP3* were obtained from the PDB protein structure database. PyMOL 2.5 was then used to remove water molecules and small molecules with ligand affinity. Subsequently, the protein receptor and ligand files were converted into PDBQT format using AutoDock Tools 1.5.6. AutoDock Vina 1.1.2 was used to characterize the molecular docking and calculate its affinity. The conformation with the highest affinity was selected as the final docking conformation, and PyMOL (https://pymol.org/2/ accessed on 16 November 2021) and AutoDock software (https://autodock.scripps.edu/ accessed on 16 November 2021) were used to visualize the docking results in the form of two-dimensional and three-dimensional diagrams (Figure 7). If the binding energy between the molecule and the target protein is negative, the ligand and receptor can spontaneously bind, and if the binding energy is less than −5 kcal/mol, a stable docking structure can be formed [21]. The docking binding energies between quercetin and *AKT1*, *IL6*, *VEGFA*, and *CASP3* were −4.91, −5.75, −4.86, and −5.89 kcal/mol, respectively. The docking binding energies between kaempferol and *AKT1*, *IL6*, *VEGFA*, and *CASP3* were −5.6, −6.42, −5.04, and −5.12 kcal/mol, respectively. The details are shown in Table 3. The Japanese *Ardisia* active ingredients quercetin and kaempferol had a good binding ability with the four key targets.

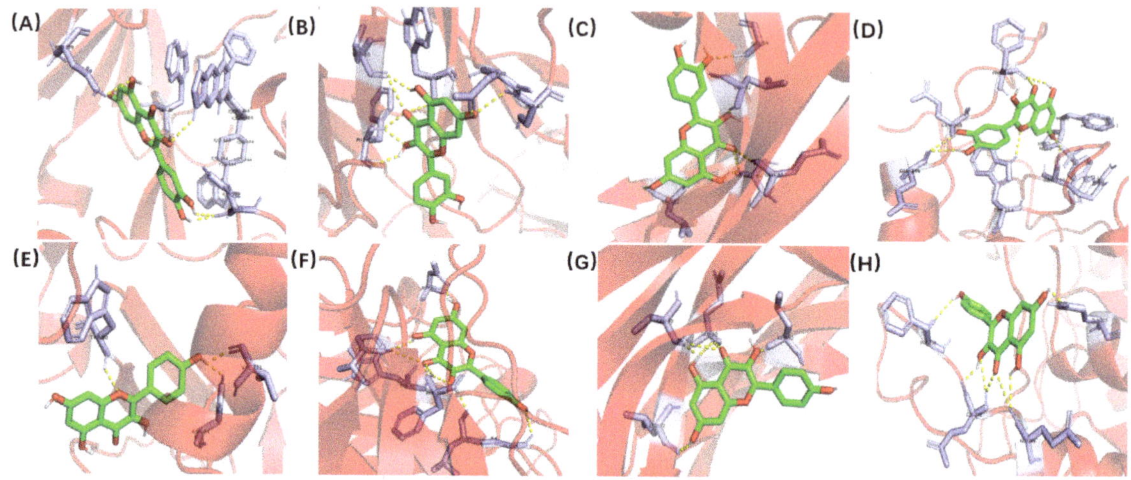

Figure 7. Molecular docking of quercetin and kaempferol with *AKT1, IL6, VEGFA* and *CASP3*. (**A**) Quercetin and *AKT1*. (**B**) Quercetin and *IL6*. (**C**) Quercetin and *VEGFA*. (**D**) Quercetin and *CASP3*. (**E**) Kaempferol and *AKT1*. (**F**) Kaempferol and *IL6*. (**G**) Kaempferol and *VEGFA*. (**H**) Kaempferol and *CASP3*.

Table 3. Docking and binding energy of main components of Japanese *Ardisia* and core targets (kcal·mol^{-1}).

Target Molecules	*AKT1*	*IL6*	*VEGFA*	*CASP3*
MOL000098 (Quercetin)	−4.91	−5.75	−4.86	−5.89
MOL000422 (Kaempferol)	−5.6	−6.42	−5.04	−5.12

2.8. Animal Experiments

2.8.1. Validation of the Therapeutic Effectiveness of Japanese *Ardisia*

To investigate the efficacy of Japanese *Ardisia* in the treatment of AIH, we established a Con A-induced immunological liver injury model (Figure 8A). Firstly, we found that Con A caused changes in body weight, liver weight, and liver coefficients in rats after ten days, with some reversal effect after preadministration of Japanese *Ardisia* (Figure 8B). The rats' liver tissue was then taken out and checked for cholestasis. Con A group showed signs of cholestasis and inflammation, and there was some relief from cholestasis after the drug was administered (Figure 8C). By testing serum markers of liver injury (ALT and AST), the results showed that Con A-induced acute liver injury resulted in a significant increase in ALT and AST levels. In contrast, preadministration of Japanese *Ardisia* was able to alleviate the altered biochemical levels. (Figure 8D).

Furthermore, HE staining of liver tissues showed that the model group had a large infiltration of inflammatory cells and a large amount of vacuolar-like degeneration of hepatocytes compared to the control group. Liver tissue damage was alleviated after preadministration of Japanese *Ardisia* (Figure 8E,F), indicating that the pre-administration of the drug was able to slow down the con A-induced histopathological damage to the liver.

2.8.2. The Effect of Japanese *Ardisia* on *AKT1*, *CASP3*, and *IL-6* Protein Levels

Western blot analyses are shown in Figure 9, which revealed that *CASP3* and *IL-6* protein expression levels were significantly higher ($p < 0.01$), and *AKT1* expression levels were significantly lower ($p < 0.01$) in the model group compared to the normal control group. *CASP3* protein expression was significantly lower ($p < 0.01$), *AKT1* protein expression was significantly higher ($p < 0.05$), and *IL-6* protein expression was significantly lower ($p < 0.05$) in the Japanese *Ardisia* administration group after 10 days compared to the model group.

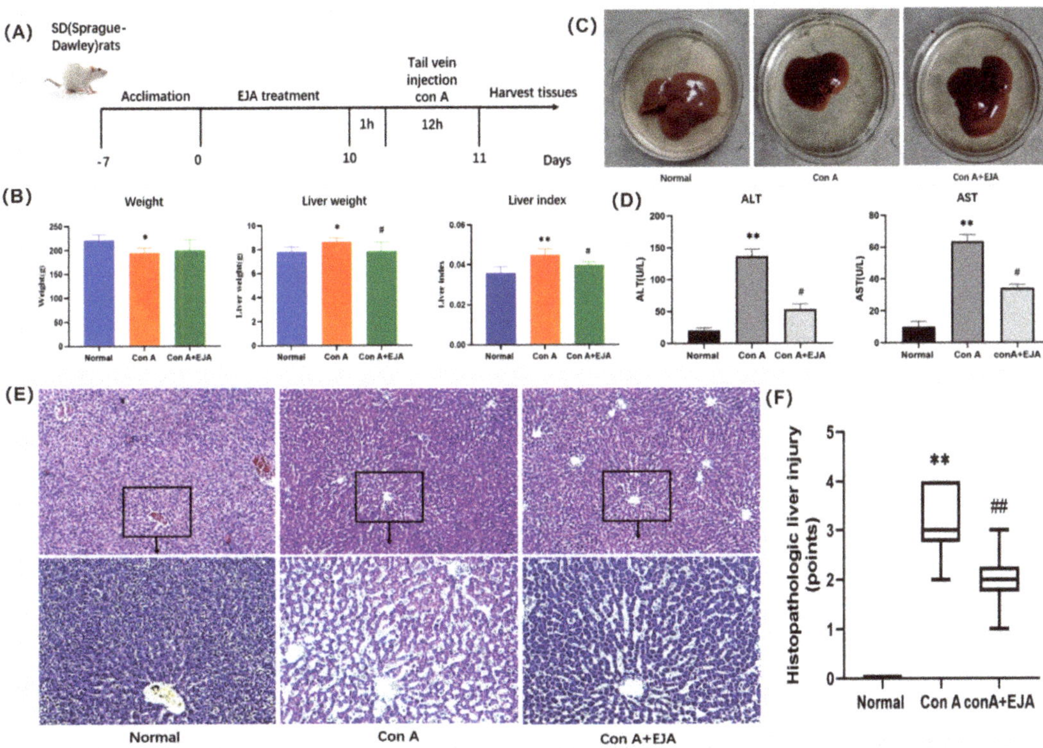

Figure 8. (**A**) Description of the rat model of acute immune liver injury used in this study. (**B**) Body weight, liver weight, and liver coefficient changes in rats 10 days after EJA treatment. Compared with normal group, ** $p < 0.01$, * $p < 0.05$; Compared with model group, # $p < 0.05$. Note: EJA is an extract of Japanese *Ardisia*. Data were shown as mean ± SEM ($n = 6$ per group). (**C**) Rat liver treated with EJA. (**D**) Effect of EJA on serum ALT and AST levels. Compared with normal group, ** $p < 0.01$; Compared with model group, # $p < 0.05$. Note: EJA: an extract of Japanese *Ardisia*. Data were shown as mean ± SEM ($n = 3$ per group). (**E**) Histopathological sections of rat liver (HE staining, 200× 400×). (**F**) Effect of Japanese *Ardisia* pretreatment on the scoring of liver pathological sections of rats with con A-induced liver injury. Compared with normal group, ** $p < 0.01$; compared with model group, ## $p < 0.01$. Note: EJA is an extract of Japanese *Ardisia*. Data were shown as mean ± SEM ($n = 6$ per group). Scoring criteria: 0—no necrosis; 1—individual cell necrosis; 2—less than 30% necrosis; 3—30–60% necrosis; 4—greater than 60% necrosis [22].

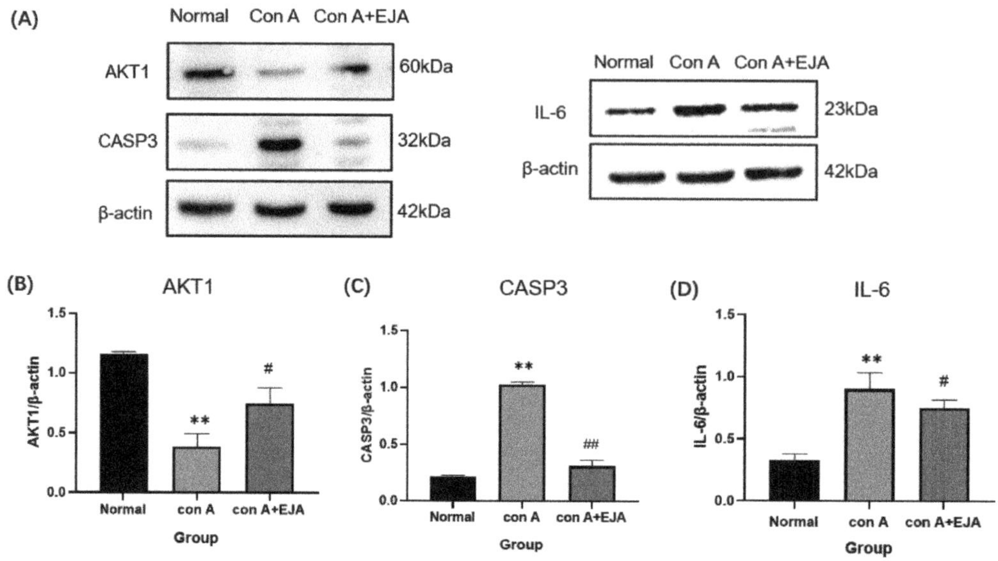

Figure 9. Western blot analyses. (**A**) Western blot analysis of *AKT1, CASP3,* and *IL-6*. (**B–D**) Relative expression of *CASP3, AKT1,* and *IL-6* in the liver tissue of SD rats. Compared with normal group, $^{**}p < 0.01$; Compared with model group, $^{\#\#}p < 0.05$, $^{\#}p < 0.05$. Data were shown as mean ± SEM (n = 3 per group). Note: EJA is an extract of Japanese *Ardisia*.

3. Discussion

AIH is a chronic inflammatory disease characterized by abnormally high levels of serum autoantibodies, hypergammaglobulinemia, and serum transaminases [23]. With the advancement of medical technology in recent years, AIH has received widespread attention and has gradually become a research hotspot.

Chinese medicine's multi-component and multi-target properties, and thus its holistic and systemic action characteristics, make it unique in its advantages and potential for complex diseases, but this complexity has also limited its application and development.

Cyberpharmacology analyzes drug action at the systemic level and reveals the synergistic mechanism of drug action on the human body, which fits well with the dialectical and holistic view of Chinese medicine theory, and it is expected to bring a breakthrough to Chinese medicine research characterized by a holistic approach, providing new methodological support for Chinese medicine to move from empirical to theoretical science [24–26]. We used network pharmacology to create a component-shared target network map in this study to systematically reveal the material basis and molecular mechanism of Japanese *Ardisia* for the treatment of AIH.

This study provides key information about the anti-hepatitis effect of Japanese *Ardisia*. The TCMSP database analysis resulted in the screening of 19 active ingredients. 143 effective targets for the treatment of autoimmune diseases were identified using databases, such as GeneCards and OMIM. According to the results of the component–target network analysis, 150 target genes were associated with the drug-disease target intersection. The core genes were *AKT1, IL6, VEGFA,* and *CASP3*. *AKT1*, also known as protein kinase B, regulates a wide variety of cellular functions, including cell proliferation, survival, metabolism, and angiogenesis, in both normal and malignant cells [27]. *IL6* is a pro-inflammatory cytokine with a wide variety of biological functions. It is involved in physiological activities such as the inflammatory response, cellular immunity, and hematopoietic regulation. *IL6* plays major roles in the differentiation of B cells into immunoglobulin-secreting cells and antibody production, the activation of T cell proliferation and differentiation, the immune response, and the promotion of inflammatory reactions [28]. *IL6* is rapidly synthesized in response to tissue injury or an inflammatory infection. This promotes the body's defense function by stimulating the acute immune response and the hematopoietic system. When the tissue recovers its homeostasis, *IL6* synthesis is discontinued [29]. The continuous activation of the *IL6* pathway is associated with liver injury and hepatocellular carcinoma [30]. *VEGFA* induces endothelial cell proliferation, promotes cell migration, inhibits apoptosis, and induces the permeabilization of blood vessels. It is essential for both physiological and pathological angiogenesis. *CASP3* is a cysteine aspartate protease that participates in the activation cascade of cysteine proteases. *CASP3* plays an important role in inflammation and tumor progression [31]. It is highly expressed in patients with hepatitis B [32], and it regulates cell proliferation and apoptosis [33] and tumor invasion and metastasis [34].

Japanese *Ardisia* has a therapeutic effect on hepatitis via a mechanism that may be related to key molecules involved in the regulation of autoimmune hepatitis. It inhibits associated inflammatory factors by regulating various signaling pathways, downregulating the concentrations of serum hyaluronic acid and tumor necrosis factor, protecting hepatocytes from injury, reducing liver inflammation, and protecting against lipid peroxidation [35]. Inflammatory stimulation is the cause of many chronic diseases [36]. GO enrichment analysis showed that the anti-hepatitis effect of Japanese *Ardisia* is related to the inflammatory response, cell cycle regulation, and hormone metabolism. In the KEGG enrichment analysis, 163 pathways were identified, including the lipid and atherosclerosis pathway, IL-17 signaling pathway, TNF signaling pathway, human cytomegalovirus infection pathway, fluid shear stress and atherosclerosis pathway, hepatitis B pathway, and AGE–RAGE signaling pathway in diabetic complications. It can be seen that the pathway is mainly related to oxidative stress, immune regulation, and inflammatory responses, with the AGE-RAGE signaling pathway being closely related to inflammation, which activates the MAPK and NF-KB pathways and interferes with immune and oxidative stress responses [37]. TNF is a key regulator of the inflammatory response, and its receptors TNFR1 and TNFR2 activate complex signaling pathways that lead to a series of inflammatory responses in the vascular endothelium, including thrombosis, leukocyte adhesion, and vascular leakage [38]. The IL-17 signaling pathway is involved in neutrophil infiltration and inflammatory responses and can be restricted by the ACE2 downregulation of the STAT3 pathway, thereby slowing down neutrophil infiltration and inflammation. These pathways may be the key to effective autoimmune hepatitis treatment with Japanese *Ardisia*, which affects the secretion of certain substances by acting on specific receptors or enzymes, to achieve its therapeutic effect.

The binding interaction between each compound and its receptor was scored by the molecular docking program. A lower score indicated more stable binding between the ligand and its receptor [39,40]. Through the molecular docking of quercetin and kaempferol, the effective components of Japanese *Ardisia*, with the key targets *AKT1*, *IL6*, *VEGFA*, and *CASP3*, the docking binding energy and the number of intermolecular hydrogen bonds were obtained. The results showed that quercetin and kaempferol had a good binding ability with the key targets and can spontaneously bind to form a stable binding conformation.

Animal studies have shown that Japanese *Ardisia* is effective in the treatment of AIH. Western blot analysis revealed that Japanese *Ardisia* could reduce the expression of *CASP3* and *IL-6* while increasing the expression of *AKT1*. These proteins are involved in the AGE–RAGE signaling pathway in diabetic complications. According to this, the Japanese *Ardisia* may be able to modulate these key targets to achieve therapeutic effects on immune liver injury.

Additionally, there are also limitations to the research at this stage, as network pharmacology techniques can only predict drug composition and targets qualitatively. High-performance liquid chromatography (HPLC) or ultraviolet spectrophotometry (UV) should be used to determine the plausibility of the screened active ingredients, and this should be combined with pharmacology, pharmacodynamics, and pharmacokinetics to make the screened active ingredients and mechanism of action more convincing.

4. Materials and Methods

4.1. Databases

The Traditional Chinese Medicine Systems Pharmacology (TCMSP) database, an analysis platform (http://tcmspw.com/tcmsp.php (accessed on 14 November 2021)), GeneCards (http://www.genecards.org (accessed on 14 November 2021)), Online Mendelian Inheritance in Man (OMIM, http://www.omim.org (accessed on 14 November 2021)), the Therapeutic Target Database (TTD; http://db.idrblab), the Search Tool for Retrieval of Interacting Genes (STRING) protein interaction database (https://string-db.org (accessed on 15 November 2021)), Cytoscape 3.7.0 (http://www.cytoscape.org (accessed on 15 November 2021)), and the Protein Data Bank (PDB) protein structure database (https://www.rcsb.org/pdb (accessed on 16 November 2021)) were used in this study.

4.2. Screening of Active Ingredients of Japanese Ardisia

The components of Japanese *Ardisia* were searched in TCMSP (TCMSP, https://lsp.nwu.edu.cn/tcmsp.php, accessed on 14 November 2021) and screened for active compounds by oral bioavailability (OB) and drug similarity (DL), as well as potential targets of action for activity, the active compounds obtained from the screening are presented in the form of a list.

4.3. Screening of Target Diseases

GeneCards, OMIM, and TTD were used to identify hepatitis-related gene targets. Specific search parameters, such as the keyword 'autoimmune hepatitis', were used for data collection. Using R 4.0.4 software (https://www.r-project.org/ accessed on 15 November 2021) to remove duplicate regions of gene targets, the intersection of the active ingredient and disease target was obtained and plotted as a Venn diagram.

4.4. Target Protein Localization and Interaction Analysis

The cross-targets of Japanese *Ardisia* in autoimmune hepatitis treatment and prevention were imported into the STRING protein interaction database for analysis. The PPI network was mapped using the Cytoscape 3.7.0 software. A protein–protein interaction (PPI) network was constructed according to proximity centrality, intermediate centrality, and degree value, and the key targets were screened using Cytoscape 3.7.0.

4.5. Construction of a Component-Target-Disease Interaction Network of Japanese Ardisia with Autoimmune Hepatitis

The previously acquired common genes and the associated active ingredients were visualized and analyzed using Cytoscape 3.7.0 software.

4.6. Gene Ontology and Kyoto Encyclopedia of Genes and Genomes Pathway Enrichment Analysis

The potential targets of Japanese *Ardisia* for autoimmune hepatitis treatment were imported into the Gene Ontology (GO) and Kyoto Encyclopedia of Genes and Genomes (KEGG) databases. GO was selected for enrichment analysis, KEGG was selected for pathway analysis, and the *p*-value threshold was set at <0.05 to determine the enrichment pathways of key targets. An advanced bubble diagram was constructed from the enrichment results: the smaller the *p*-value, the higher the enrichment; the larger the bubble, the richer the genes were.

4.7. Molecular Docking

The two-dimensional structures of quercetin, kaempferol, and laricitrin were obtained from the TCMSP database and saved as a Mol2 file. AKT serine/threonine kinase 1 (*AKT1*), interleukin 6 (*IL6*), vascular endothelial growth factor A (*VEGFA*), and caspase 3 (*CASP3*) were selected as the target proteins for molecular docking experiments. The three-dimensional structures of *AKT1*, *IL6*, *VEGFA*, and *CASP3* were downloaded from the PDB protein structure database and saved in PDB format [41]. Crystal structures with high resolution and corresponding bioactive ligand complexes were preferentially selected [42]. The structures of the protein macromolecules and small-molecule compounds were imported into AutoDock 4.2.6 software (https://autodock.scripps.edu/ accessed on 16 November 2021) for molecular docking, and the docking results were analyzed using the PyMOL (https://pymol.org/2/ accessed on 16 November 2021) visualization tool.

4.8. Animal Experiments

4.8.1. Drug

The whole herb of Japanese *Ardisia* was ground into a coarse powder and kept in a dry, cool environment. A weighed amount of coarse powder was soaked in ten times the amount of water for one night before being decocted twice for 1.5 h each time. The decoctions were combined and concentrated to 1 g/mL (1 mL equals 1 g raw material), the concentrate was centrifuged at 5000 rpm for 15 min, the supernatant was collected and concentrated to 1.5 g/mL (1 mL equals 1.5 g raw material), and the obtained concentrate was refrigerated at 4 °C [43,44].

4.8.2. Animal Grouping and Drug Administration

SPF-grade male SD (Sprague-Dawley) rats, weighing (200 ± 10) g, with six animals per cage were housed in specific pathogen-free facility with a 12 h light and 12 h dark cycle at 22 °C. All mice were randomly divided into three groups, control group (n = 6), con A group (n = 6), and extract of Japanese *Ardisia* (EJA) (n = 6) groups after being fed adaptively for a week. The control and model groups were given saline, while the administration group was given 36 g/kg of Japanese *Ardisia* by gavage once daily for 10 days. One hour after the last dose, all groups were given 25 mg/kg con A in the tail vein, except for the control group which was given an equal amount of saline in the tail vein, rats were euthanized 12 h after con A treatment, and their livers were dissected from the rats.

The kits purchased from Nanjing Jiancheng Institute of Biological Engineering were used, and the alanine transaminase (ALT) (Cat.No.C009-2-1) and aspartate transaminase (AST) (Cat.No.C010-1-1) measurements were performed according to the instructions provided by the kit supplier.

All the animals were provided by Hunan Sleek Jingda Laboratory Animal Co. License No. SCXK (Xiang) 2019-0005. Animal welfare and experimental procedures followed the regulations of the Animal Ethics Committee of Guilin Medical College.

4.8.3. Histopathological Section Analysis of Liver

Tissue removed from 2.8.2 was fixed in 4% paraformaldehyde, paraffin sections were embedded, and, finally, the pathological sections were observed by HE staining, and the pathological changes of liver tissues were observed under an electron microscope.

4.8.4. Western Blot

To verify the protein expression level, we extracted protein from rat liver tissue, weighed a certain amount of liver tissue, fully lysed it with RIPA lysis solution (Solarbio, Beijing, China), used SDS-PAGE electrophoresis, and then transferred it to PVDF (Solarbio, Beijing, China) membrane. We then blotted and closed them at room temperature for one hour, and then we incubated them with *AKT1* (60203-2, proteintech, Wuhan, China), *CASP3* (ab184787, Abcam, Cambridge, UK), *IL-6* (ab259341, Abcam, Cambridge, UK), β-actin antibody (66,009, proteintech, Wuhan, China), and incubated them overnight at 4 °C. Then, we incubated them with goat anti-rabbit IgG (H + L) and goat anti-mouse Ig G (H + L) for 1 h at room temperature. The protein bands were visualized using an ECL chemiluminescence kit (Beyotime, Shanghai, China) and quantified under Image J system, version 6.0.

4.9. Statistical Analysis

The experimental results are presented as mean ± SEM for each group, with at least three independent experiments. The differences between the treatment and normal groups were analyzed by a one-way analysis of variance using GraphPad Prism 8.0 (GraphPad Software Inc., San Diego, CA, USA). $p < 0.05$ was used to indicate statistical significance.

5. Conclusions

In summary, we first performed a network pharmacology and molecular docking approach to elucidate the anti-AIH effects and potential mechanisms of Japanese *Ardisia*, and finally determined the effect of Japanese *Ardisia* anti-AIH on the expression of key target proteins by Western blotting analysis. Japanese *Ardisia* tends to work through the IL-17 signaling pathway, the TNF signaling pathway, the AGE–RAGE signaling pathway in diabetic complications, and other targets such as *AKT1*, *IL-6*, and *CASP3* to exert drug effects. The potential signaling pathways uncovered in this study lay the theoretical groundwork and point the way for future experimental validation.

Author Contributions: T.F.: Writing—Methodology, Investigation, and Original Draft; Y.C.: Methodology, Software, and Visualization; J.L. and P.Z.: Software and Validation; H.H. and W.Z.: Investigation; K.K.L.Y. and W.W.: Conceptualization, Writing—Review, and Editing. All authors have read and agreed to the published version of the manuscript.

Funding: This work is supported by the National Natural Science Foundation of China (No. 81760704), the Natural Science Foundation of Guangxi (No. 2018GXNSFAA281332), and the Scientific Research Project of Traditional Chinese medicine Bureau of Guangdong Province (No. 20222138).

Institutional Review Board Statement: The animal experimental protocol was approved by the Animal Experimentation Ethics Committee of Guilin Medical University (Application Approval No. GLMC-IACUC-2022012).

Informed Consent Statement: Not applicable.

Data Availability Statement: All data generated or analyzed during this study are included in this published article.

Acknowledgments: The authors thank the Cultivation Plan for Thousands of Young and Middle-aged Key Teachers in Colleges and Universities in Guangxi for supporting this work.

Conflicts of Interest: The authors declare no conflict of interests.

References

1. Van Gerven, N.M.F.; Verwer, B.J.; Witte, B.I.; van Erpecum, K.J.; van Buuren, H.R.; Maijers, I.; Visscher, A.P.; Verschuren, E.C.; van Hoek, B.; Coenraad, M.J.; et al. Epidemiology and clinical characteristics of autoimmune hepatitis in the Netherlands. *Scand. J. Gastroenterol.* **2014**, *49*, 1245–1254. [CrossRef] [PubMed]
2. Ngu, J.H.; Bechly, K.; Chapman, B.A.; Burt, M.J.; Barclay, M.L.; Gearry, R.B.; Stedman, C.A.M. Population-based epidemiology study of autoimmune hepatitis: A disease of older women? *J. Gastroenterol. Hepatol.* **2010**, *25*, 1681–1686. [CrossRef] [PubMed]
3. Medina, J.; Garcia-Buey, L.; Moreno-Otero, R. Hepatitis C virus-related extra-hepatic—Aetiopathogenesis and management. *Aliment. Pharmacol. Ther.* **2004**, *20*, 129–141. [CrossRef] [PubMed]
4. Zachou, K.; Muratori, P.; Koukoulis, G.K.; Granito, A.; Gatselis, N.; Fabbri, A.; Dalekos, G.N.; Muratori, L. Review article: Autoimmune hepatitis—Current management and challenges. *Aliment. Pharmacol. Ther.* **2013**, *38*, 887–913. [CrossRef]
5. Longhi, M.S.; Ma, Y.; Mieli-Vergani, G.; Vergani, D. Aetiopathogenesis of autoimmune hepatitis. *J. Autoimmun.* **2010**, *34*, 7–14. [CrossRef]
6. Vergani, D.; Mieli-Vergani, G. Aetiopathogenesis of autoimmune hepatitis. *World J. Gastroenterol.* **2008**, *14*, 3306–3312. [CrossRef]
7. Kirstein, M.M.; Seibel, E.; Manns, M.P.; Vogel, A. Prediction of short- and long-term outcome in patients with autoimmune hepatitis. *J. Hepatol.* **2014**, *60*, S200. [CrossRef]
8. Taubert, R.; Hardtke-Wolenski, M.; Noyan, F.; Lalanne, C.; Jonigk, D.; Schlue, J.; Krech, T.; Lichtinghagen, R.; Falk, C.S.; Schlaphoff, V.; et al. Hyperferritinemia and hypergammaglobulinemia predict the treatment response to standard therapy in autoimmune hepatitis. *PLoS ONE* **2017**, *12*, 15. [CrossRef]
9. Suriawinata, A.A.; Thung, S.N. Acute and chronic hepatitis. *Semin. Diagn. Pathol.* **2006**, *23*, 132–148. [CrossRef]
10. Torgutalp, M.; Efe, C.; Babaoglu, H.; Kav, T. Relationship between serum adenosine deaminase levels and liver histology in autoimmune hepatitis. *World J. Gastroenterol.* **2017**, *23*, 3876–3882. [CrossRef]

11. Sun, Q.; Xu, X.; Yang, X.; Weng, D.; Wang, J.S.; Zhang, J.F. Salecan protected against concanavalin A-induced acute liver injury by modulating T cell immune responses and NMR-based metabolic profiles. *Toxicol. Appl. Pharmacol.* **2017**, *317*, 63–72. [CrossRef] [PubMed]
12. Lohse, A.W.; Chazouilleres, O.; Dalekos, G.; Drenth, J.; Heneghan, M.; Hofer, H.; Lammert, F.; Lenzi, M.; European Association for the Study of the Liver. EASL Clinical Practice Guidelines: Autoimmune hepatitis. *J. Hepatol.* **2015**, *63*, 1543–1544.
13. Yu, K.Y.; Wu, W.; Li, S.Z.; Dou, L.L.; Liu, L.L.; Li, P.; Liu, E.H. A new compound, methylbergenin along with eight known compounds with cytotoxicity and anti-inflammatory activity from Ardisia japonica. *Nat. Prod. Res.* **2017**, *31*, 2581–2586. [CrossRef] [PubMed]
14. Yu, K.Y.; Gao, W.; Li, S.Z.; Wu, W.; Li, P.; Dou, L.L.; Wang, Y.Z.; Liu, E.H. Qualitative and quantitative analysis of chemical constituents in Ardisiae Japonicae Herba. *J. Sep. Sci.* **2017**, *40*, 4347–4356. [CrossRef]
15. Chen, Y.; Du, K.Z.; Li, J.; Bai, Y.; An, M.R.; Tan, Z.J.; Chang, Y.X. A Green and Efficient Method for the Preconcentration and Determination of Gallic Acid, Bergenin, Quercitrin, and Embelin from Ardisia japonica Using Nononic Surfactant Genapol X-080 as the Extraction Solvent. *Int. J. Anal. Chem.* **2018**, *2018*, 1707853. [CrossRef]
16. Li, Q.; Li, W.; Hui, L.P.; Zhao, C.Y.; He, L.; Koike, K. 13,28-Epoxy triterpenoid saponins from Ardisia japonica selectively inhibit proliferation of liver cancer cells without affecting normal liver cells. *Bioorg. Med. Chem. Lett.* **2012**, *22*, 6120–6125. [CrossRef]
17. Lee, I.S.; Cho, D.H.; Kim, K.S.; Kim, K.H.; Park, J.; Kim, Y.; Jung, J.H.; Kim, K.; Jung, H.J.; Jang, H.J. Anti-inflammatory effects of embelin in A549 cells and human asthmatic airway epithelial tissues. *Immunopharmacol. Immunotoxicol.* **2018**, *40*, 83–90. [CrossRef]
18. Hopkins, A.L. Network pharmacology. *Nat. Biotechnol.* **2007**, *25*, 1110–1111. [CrossRef]
19. Daina, A.; Michielin, O.; Zoete, V. SwissADME: A free web tool to evaluate pharmacokinetics, drug-likeness and medicinal chemistry friendliness of small molecules. *Sci. Rep.* **2017**, *7*, 42717. [CrossRef]
20. Ma, C.; Wang, L.R.; Xie, X.Q. GPU Accelerated Chemical Similarity Calculation for Compound Library Comparison. *J. Chem. Inf. Model.* **2011**, *51*, 1521–1527. [CrossRef]
21. Lanini, S.; Ustianowski, A.; Pisapia, R.; Zumla, A.; Ippolito, G. Viral Hepatitis Etiology, Epidemiology, Transmission, Diagnostics, Treatment, and Prevention. *Infect. Dis. Clin. N. Am.* **2019**, *33*, 1045–1062. [CrossRef] [PubMed]
22. Zhu, J.L.; Wang, J.F.; Sheng, Y.; Zou, Y.; Bo, L.L.; Wang, F.; Lou, J.S.; Fan, X.H.; Bao, R.; Wu, Y.P.; et al. Baicalin Improves Survival in a Murine Model of Polymicrobial Sepsis via Suppressing Inflammatory Response and Lymphocyte Apoptosis. *PLoS ONE* **2012**, *7*, e35523. [CrossRef] [PubMed]
23. Lohse, A.W.; Wiegard, C. Diagnostic criteria for autoimmune hepatitis. *Best Pract. Res. Clin. Gastroenterol.* **2011**, *25*, 665–671. [CrossRef] [PubMed]
24. Hao, D.C.; Xiao, P.G. Network Pharmacology: A Rosetta Stone for Traditional Chinese Medicine. *Drug Dev. Res.* **2014**, *75*, 299–312. [CrossRef] [PubMed]
25. Zhang, R.Z.; Zhu, X.; Bai, H.; Ning, K. Network Pharmacology Databases for Traditional Chinese Medicine: Review and Assessment. *Front. Pharmacol.* **2019**, *10*, 123. [CrossRef] [PubMed]
26. Zhou, Z.C.; Chen, B.; Chen, S.M.; Lin, M.Q.; Chen, Y.; Jin, S.; Chen, W.Y.; Zhang, Y.Y. Applications of Network Pharmacology in Traditional Chinese Medicine Research. *Evid. Based Complement. Altern. Med.* **2020**, *2020*, 1646905. [CrossRef]
27. Harrison, B.C.; Leinwand, L.A. Fighting fat with muscle: Bulking up to slim down. *Cell Metab.* **2008**, *7*, 97–98. [CrossRef]
28. Narazaki, M.; Tanaka, T.; Kishimoto, T. The role and therapeutic targeting of IL-6 in rheumatoid arthritis. *Expert Rev. Clin. Immunol.* **2017**, *13*, 535–551. [CrossRef]
29. Tanaka, T.; Kishimoto, T. The Biology and Medical Implications of Interleukin-6. *Cancer Immunol. Res.* **2014**, *2*, 288–294. [CrossRef]
30. Shakiba, E.; Ramezani, M.; Sadeghi, M. Evaluation of serum interleukin-6 levels in hepatocellular carcinoma patients: A systematic review and meta-analysis. *Clin. Exp. Hepatol.* **2018**, *4*, 182–190. [CrossRef]
31. Zhang, H.Z.; Liu, L.S.; Jiang, C.M.; Pan, K.Q.; Deng, J.; Wan, C.Y. MMP9 protects against LPS-induced inflammation in osteoblasts. *Innate Immun.* **2020**, *26*, 259–269. [CrossRef] [PubMed]
32. Pop, C.; Salvesen, G.S. Human Caspases: Activation, Specificity, and Regulation. *J. Biol. Chem.* **2009**, *284*, 21777–21781. [CrossRef] [PubMed]
33. Kuo, W.T.; Shen, L.; Zuo, L.; Shashikanth, N.; Ong, M.; Wu, L.C.; Zha, J.M.; Edelblum, K.L.; Wang, Y.T.; Wang, Y.M.; et al. Inflammation-induced Occludin Downregulation Limits Epithelial Apoptosis by Suppressing Caspase-3 Expression. *Gastroenterology* **2019**, *157*, 1323–1337. [CrossRef] [PubMed]
34. Perry, C.J.; Munoz-Rojas, A.R.; Meeth, K.M.; Kellman, L.N.; Amezquita, R.A.; Thakral, D.; Du, V.Y.; Wang, J.X.; Damsky, W.; Kuhlmann, A.L.; et al. Myeloid-targeted immunotherapies act in synergy to induce inflammation and antitumor immunity. *J. Exp. Med.* **2018**, *215*, 877–893. [CrossRef] [PubMed]
35. Yang, Y.X.; Zhang, P.; Wang, Y.Y.; Wei, S.Z.; Zhang, L.; Wang, J.B.; Lu, X.H.; Zhou, H.Q.; Li, R.S.; Wen, J.X.; et al. Hepatoprotective Effect of San-Cao Granule on Con A-Induced Liver Injury in Mice and Mechanisms of Action Exploration. *Front. Pharmacol.* **2018**, *9*, 624. [CrossRef]
36. Furman, D.; Campisi, J.; Verdin, E.; Carrera-Bastos, P.; Targ, S.; Franceschi, C.; Ferrucci, L.; Gilroy, D.W.; Fasano, A.; Miller, G.W.; et al. Chronic inflammation in the etiology of disease across the life span. *Nat. Med.* **2019**, *25*, 1822–1832. [CrossRef]
37. de Medeiros, M.C.; Frasnelli, S.C.T.; Bastos, A.D.; Orrico, S.R.P.; Rossa, C. Modulation of cell proliferation, survival and gene expression by RAGE and TLR signaling in cells of the innate and adaptive immune response: Role of p38 MAPK and NF-KB. *J. Appl. Oral Sci.* **2014**, *22*, 185–193. [CrossRef]

38. Bradley, J.R. TNF-mediated inflammatory disease. *J. Pathol.* **2008**, *214*, 149–160. [CrossRef]
39. Morris, C.J.; Della Corte, D. Using molecular docking and molecular dynamics to investigate protein-ligand interactions. *Mod. Phys. Lett. B* **2021**, *35*, 2130002. [CrossRef]
40. Caballero, J. The latest automated docking technologies for novel drug discovery. *Expert Opin. Drug Discov.* **2021**, *16*, 625–645. [CrossRef]
41. Berman, H.M.; Battistuz, T.; Bhat, T.N.; Bluhm, W.F.; Bourne, P.E.; Burkhardt, K.; Iype, L.; Jain, S.; Fagan, P.; Marvin, J.; et al. The Protein Data Bank. *Acta Crystallogr. Sect. D Struct. Biol.* **2002**, *58*, 899–907. [CrossRef] [PubMed]
42. Tanaka, Y.; Mola, E.M. IL-6 targeting compared to TNF targeting in rheumatoid arthritis: Studies of olokizumab, sarilumab and sirukumab. *Ann. Rheum. Dis.* **2014**, *73*, 1595–1597. [CrossRef] [PubMed]
43. Yu, X.A.; Azietaku, J.T.; Li, J.; Wang, H.; Zheng, F.; Hao, J.; Chang, Y.X. Simultaneous Quantification of Gallic Acid, Bergenin, Epicatechin, Epicatechin Gallate, Isoquercitrin, and Quercetin-3-Rhamnoside in Rat Plasma by LC-MS/MS Method and Its Application to Pharmacokinetics after Oral Administration of Ardisia japonica Extract. *Evid. Based Complement. Altern. Med.* **2018**, *2018*, 4964291. [CrossRef] [PubMed]
44. Feng, S.X.; Han, X.X.; Zhao, D.; Li, R.R.; Liu, X.F.; Tian, Y.G.; Li, J.S. Simultaneous quantitation of 31 bioactive components in different parts of Ardisiae Japonicae Herba from different regions by UPLC-Orbitrap Fusion MS. *J. Liq. Chromatogr. Relat. Technol.* **2021**, *44*, 649–662. [CrossRef]

pharmaceuticals

MDPI

Article

The First Anti-Snakebite and Hepatoprotective Characterization of a Trypsin Kunitz-like Inhibitor (EcTI) from the Plant *Enterolobium contortisiliquum;* A Case of Two Soul Mates Meeting

Caroline R. C. Costa [1,2], Mariana N. Belchor [1,2], Airam Roggero [2], Laila L. Moraes [2], Ricardo Samelo [2], Isabelly Annunciato [2], Camila R. Bonturi [3], Maria L. V. Oliva [3], Sergio F. Sousa [4], Marcos A. de Oliveira [1,2] and Marcos H. Toyama [1,2,*]

[1] Center of Natural and Human Sciences, Federal University of ABC (UFABC), Santo André 09210-580, SP, Brazil
[2] Biosciences Institute of Paulista Coast Campus (IB/CLP), University of São Paulo State (UNESP), São Vicente 11330-900, SP, Brazil
[3] National Institute of Pharmacology (INFAR), Federal University of São Paulo (UNIFESP), São Paulo 04044-020, SP, Brazil
[4] Unit of Applied Biomolecular Sciences (UCIBIO), REQUIMTE-BioSIM-Medicine Faculty, Porto University, 4050-345 Porto, Portugal
* Correspondence: marcoshikaritoyama@gmail.com; Tel.: +55-019-9911-14885

Citation: Costa, C.R.C.; Belchor, M.N.; Roggero, A.; Moraes, L.L.; Samelo, R.; Annunciato, I.; Bonturi, C.R.; Oliva, M.L.V.; Sousa, S.F.; de Oliveira, M.A.; et al. The First Anti-Snakebite and Hepatoprotective Characterization of a Trypsin Kunitz-like Inhibitor (EcTI) from the Plant *Enterolobium contortisiliquum;* A Case of Two Soul Mates Meeting. *Pharmaceuticals* **2023**, *16*, 632. https://doi.org/10.3390/ph16040632

Academic Editors: Diana Roxana Pelinescu and Daniela De Vita

Received: 28 February 2023
Revised: 6 April 2023
Accepted: 17 April 2023
Published: 21 April 2023

Abstract: Snake venom serine protease (SVSP) interferes with the regulation and control of important biological reactions in homeostasis and can be classified as an activator of the fibrinolytic system and platelet aggregation. Our group has recently isolated a new serine protease from *Crotalus durissus terrificus* total venom (Cdtsp-2). This protein exhibits edematogenic capacity and myotoxic activity. A Kunitz-like EcTI inhibitor protein with a molecular mass of 20 kDa was isolated from *Enterolobium contortisiliquum* and showed high trypsin inhibition. Thus, the objective of this work is to verify the possible inhibition of the pharmacological activities of Cdtsp-2 by the Kutinz-type inhibitor EcTI. To isolate Cdtsp-2 from total *C. d. terrificus* venom, we used three-step chromatographic HPLC. Using the mice paw edema model, we observed an edematogenic effect, myotoxicity and hepatotoxicity caused by Cdtsp-2. In vitro and in vivo experiments showed that the alterations in hemostasis caused by Cdtsp-2 are crucial for the development of marked hepatotoxicity and that EcTI significantly inhibits the enzymatic and pharmacological activities of Cdtsp-2. Kunitz-like inhibitor may be a viable alternative for the development of ancillary treatments against the biological activities of venoms.

Keywords: serine protease; Cdtsp-2; Kunitz-type inhibitor; *Crotalus durissus terrificus*

1. Introduction

Snake venoms exhibit several toxic enzymes, including serine proteases, which catalysis the cleavage of peptide covalent bonds in proteins. These macromolecules play key roles in several biological processes ranging from digestion to control and regulation of blood coagulation, immune system modulation, besides exhibiting an essential control of inflammation. Serine proteases from snake venoms are trypsin-like enzymes with highly conserved S1 substituents, as we can see in Figure 1. They show high selectivity towards molecular substrates, such as blood coagulation factors, while also being glycosylated molecules; this allows them to interact with other molecules [1,2]. Several experimental data show that these proteins exhibit various pharmacological roles, including a clear pro-inflammatory action [2–4]. Recent studies with Gyroxin, a venom serine protease, have shown that this enzyme can act on PAR (protease activated receptor) receptors, which activate the phospholipids breakdown in the membrane and generate IP3 upon proteolytic cleavage [5–7].

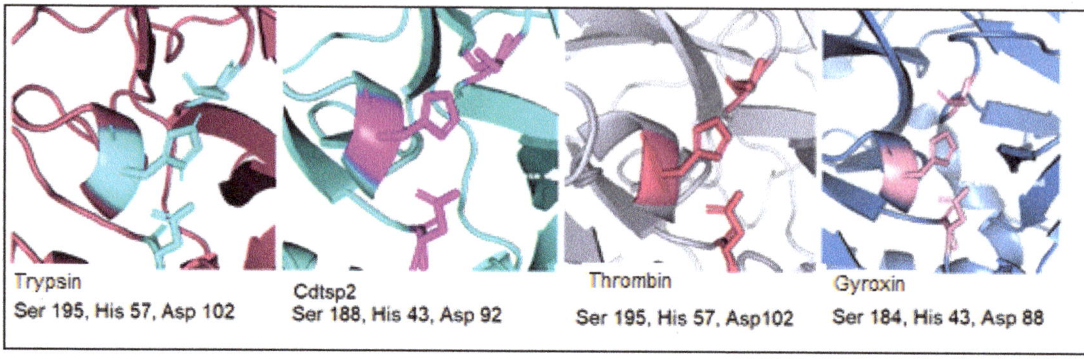

Figure 1. The active sites of the serine proteases are highly conserved regardless of their origin.

Costa and collaborators isolated Cdtsp-2 (*Crotalus durissus terrificus* serine protease 2) with a 98% purity and characterized this protein as another serine protease different from Gyroxin. This new macromolecule causes edema by enzymatic cleavage of PAR1 receptors, PAR2 and G proteins, activating PLC and PKC, which raise cPLA2 activity by increasing arachidonic acid metabolism and its interaction with oxidative stress. Thus, the evidence suggests that this toxin is an essential and still neglected aggravating factor for the action of snake venom and the venom of other animals. Reviews on clinical manifestations induced by the bite of *Crotalus durissus* venom show that besides the neurotoxic, myotoxic and hemorrhagic action, there are also reports on the hepatotoxic potential of rattlesnake venom. Several data from experimental animal studies indicate that components of the coagulation cascade, particularly coagulation factor Xa and thrombin, lead to profibrogenic events, leading to liver fibrosis and acute liver inflammation [1–3,7–12].

Antivenoms cannot neutralize the pharmacological actions of serine proteases, therefore the search for inhibitors is fundamental to allow for adequate treatment following ophidian accidents [13,14]. Serine protease inhibitors (SPIs) are widely distributed in living organisms, such as bacteria, fungi, plants and humans, and their main activity is to regulate proteolytic activity. In plants, SPIs are involved in the control of endogenous proteolytic processes, such as in the regulation of proteases in seeds. Batista et al. [15] isolated a Kunitz-type inhibitor named EcTI (*Enterolobium Contortisiliquum* trypsin inhibitor), with 97% purity, from seeds, using a plant extract with saline and acetone solution, with four chromatographic steps. This protein exhibited 173 amino acids, four of which were cysteine, forming two disulfide bridges. As could be observed in this study, EcTI was not efficient at inhibiting all the serine proteases to which it was exposed; it managed to block the activity of trypsin, but did not have the same success with thrombin. The literature shows that EcTI has anti-inflammatory activity and acts in modulating cytokines, inducing apoptosis of breast cancer cells [16]. In addition, numerous studies reveal that EcTI exhibits anti-inflammatory activity, including in lung tissues, modulates the metabolism of various tumor cells, including in brain tumors, considerably reducing their activities [2,12–15].

Hence, this work aimed to characterize the Cdtsp-2 hepatic inflammatory effect performing the first set of experimental and in silico data of the EcTI inhibitor action on the enzymatic, edematogenic, myonecrotic and hepatotoxic effects induced by serine protease Cstsp2 from *Crotalus durissus terrificus* venom.

2. Results

Firstly, enzymatic activity of Cdtsp-2 incubated with the Kunitz-like inhibitor EcTI, was evaluated. The result shown in Figure 2 demonstrate that EcTI abolished the enzymatic activity of the serine protease Cdtsp-2 (positive control). The second step was an evaluation of the anti-inflammatory capacity of EcTI through the edematogenic (paw edema) assay against Cdtsp-2 (positive control). In this test we used pure EcTI and Cdtsp-2 and all samples had their concentrations adjusted to 1 mg/mL. The EcTI effect was performed in two situations: applied 10 min before Cdtsp-2 injection and 30 min before Cdtsp-2 application. In the first edematogenic assay, 10 min after injecting Cdtsp-2 in the right paw, 50 μL EcTI was injected into the peritoneum of the mice, simulating a parenteral treatment. Under these conditions, we can observe in Figure 3 that EcTI inhibited the inflammatory process in a subtle way (Figure 3B). However, to verify the protective capacity against the inflammation caused by Cdtsp-2, we tested again the paw edema model, in which the inhibitor was injected into the peritoneum 30 min before Cdtsp-2 application in the paw. For control, saline solution (0.9% NaCl) was used in both analyses and all the concentrations were adjusted to 1 mg/mL. In Figure 3A, we can clearly see that EcTI has a significant anti-inflammatory protective capacity. In Figure 4, we did the same test as in Figure 3, but we used the anti-inflammatory Dexamethasone (DXP), a Phospholipase A 2 (PLA) inhibitor, and we can see that DXP was a protective molecule for paw edema as EcTI.

To analyse the myotoxic and protective capacities of the Kunitz-like inhibitor EcTI and Cdtsp-2, we performed the Creatine Kinase (CK) breakdown assay, with the concentrations adjusted to 1 mg/1 mL. Myonecrosis induced by Cdtsp-2 is shown both in the test 10 min after (Figure 5B) and in the test 30 min before (Figure 5A). The results with the Kunitz-type inhibitor showed a myoprotective capacity in both tests, significantly minimizing CK levels in both situations, which used the inhibitor 30 min before Cdtsp-2 application (Figure 5A) or applied it 10 min after the serine protease (Figure 5A). Due to these results we chose to investigate the hepatotoxicity of Cdtsp-2 and the EcTI hepatoprotective capacity against serine protease.

Acetaminophen (paracetamol, N-acetyl-p-aminophenol) was used as a positive control for hepatotoxicity in all tests. In Figure 6A, the first hepatotoxic marker tested was LDH (Lactate dehydrogenase), and in this test we can see that pure EcTI is not hepatotoxic when compared to paracetamol; however, Cdtsp-2 shows LDH levels above paracetamol. Another result we can see in Figure 6A is that EcTI was not able to mitigate LDH release against serine protease. Still in reference to the hepatotoxicity, we used five different hepatotoxicity marker tests, of which the second was the Gamma-Gt presented in Figure 6B. In the graph, we observe that the Kunitz-type inhibitor was not able to mitigate the release of the y-GT molecule against Cdtsp-2. Figure 6C exhibits the release of C-reactive protein (CRP), revealing that the inhibitor was able to reduce the damage caused by serine protease. AST and ALT were also evaluated, and Figure 6 shows that, in both tests, EcTI revealed no hepatotoxic activity itself, but could not significantly reduce hepatotoxicity caused by Cdtsp-2. However, neither infiltrated inflammatory or signals of necrosis can be seen with the samples present normal aspect, as shown in Figure 7.

The circular dichroism data exhibited in Figure 8B reveals that EcTI interacts so strongly with Cdtsp-2 that it significantly modifies its three-dimensional structure, changing the percentages of the secondary structures of the serine protease. A theoretical model of the *Crotalus durissus terrifucus* serine protease (Cdtsp-2) (Figure 9) was generated through homology of the *Agkistrodonhalys* thrombin-similar protein (identity 72.77%) coordinates of the enzyme (pdbcoordinates 4e7n) [12]. The structures represented were generated using PyMOL. It is possible to verify the estimated structures by dichroism, whose random coil is the most expressive. The structure was validated using MolProbity software, generating the Ramachandra graph. The quality validation of the theoretical model by Ramachandra resulted in the presentation of only three amino acids in unfavorable positions, thus the theoretical model generated by our group is viable for studies in bioinformatics interactions.

In the interaction of Cdtsp-2 with the EcTI inhibitor simulation (Figure 10D) we found a completely different interaction from the one shown by Zhou [17]. The interaction of Cdtsp-2 with EcTI shows bonds of all types, van der Walls, hydrogen bridges and even covalent bonds in the active site, as we show in Figure 10, whose distances were from 0.9 Å to 3.5 Å. However, the interactions between the protein and the inhibitor occurred through all the amino acids of both proteins, not only between the active site and the reactive loop, as shown by Zhou [17]. It is possible to see all the interactions and alignments that occurred between the amino acid residues of the proteins.

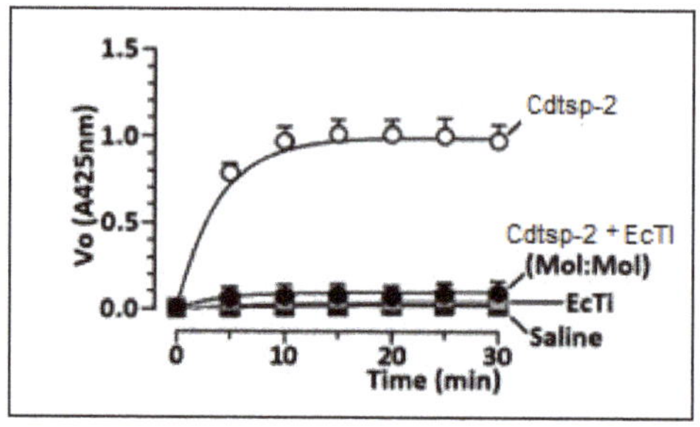

Figure 2. Graph of the enzymatic assay of Cdtsp-2 incubated with pure compounds and with EcTI protein (velocity × time). Prasa et al.'s [18] protocol was used, where the wavelength for detection was 405 nm and the protein substrate BapNa. The treatments were controlled, with substrate only, Cdtsp-2 incubated with substrate, Cdtsp-2 with protein, which were pipetted in microplate and incubated at 37 °C. The readings were performed on SPECTRA MAX (Molecular Devices, San Jose, CA, USA). It is observed that a decrease in the enzymatic activity of the protein was evident in the presence of EcTI.

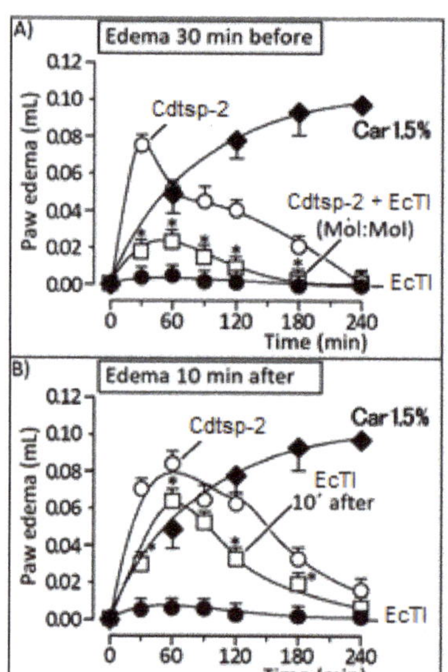

Figure 3. (**A**): Graph of paw edema volume (μL) of Swiss mice during inflammation, induced by the presence of Cdtsp-2 venom or the toxin with EcTi, which was injected 30 min before protein application. (**B**): Plot of the paw edema volume (μL) of Swiss mice during inflammation, induced by the presence of Cdtsp-2 or the toxin with EcTi injected 10 min after the protein application. Both assays were performed using 10 μg of protein and 50 μg of EcTi. Results were expressed as mean ± standard deviation (n = 5), and treatments with significant difference marked with (*) (Two-way ANOVA, with Bonferroni as a posteriori test, F = 150.0 and $p < 0.001$).

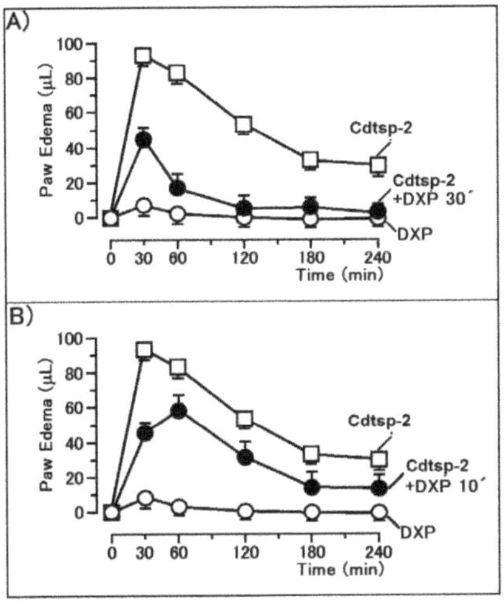

Figure 4. (**A**): Graph of paw edema volume (µL) of Swiss mice during inflammation, induced by the presence of Cdtsp-2 venom or the toxin with DXP (Dexametasone), which was injected 30 min before protein application. (**B**): Plot of the paw edema volume (µL) of Swiss mice during inflammation, induced by the presence of Cdtsp-2 or the toxin with DXP injected 10 min after the protein application. Both assays were performed using 10 µg of protein and 50 µg of DXP. Results were expressed as mean ± standard deviation ($n = 5$), and treatments with significant difference marked with (*) (Two-way ANOVA, with Bonferroni as a posteriori test, $F = 150.0$ and $p < 0.001$).

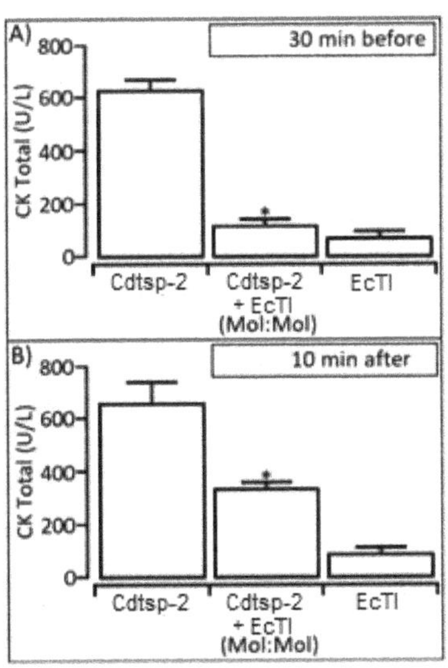

Figure 5. (**A**): Creatine kinase breakdown assay, revealing a significant decrease in the myotoxity of the protein when it was applied EcTI 30 min after Cdtsp-2. (**B**): There was no significant decrease in the myotoxicity of the protein when it was applied 10 min after the pure compounds. However, there was a decrease in myotoxicity when Cdtsp-2 was exposed to EcTI. Data were expressed as mean and standard deviation, and treatments with significant difference marked with (*) and ANOVA was used as the statistical test, with Dunnet as the a posteriori test, $F = 9.797$, $p = 0.0017$.

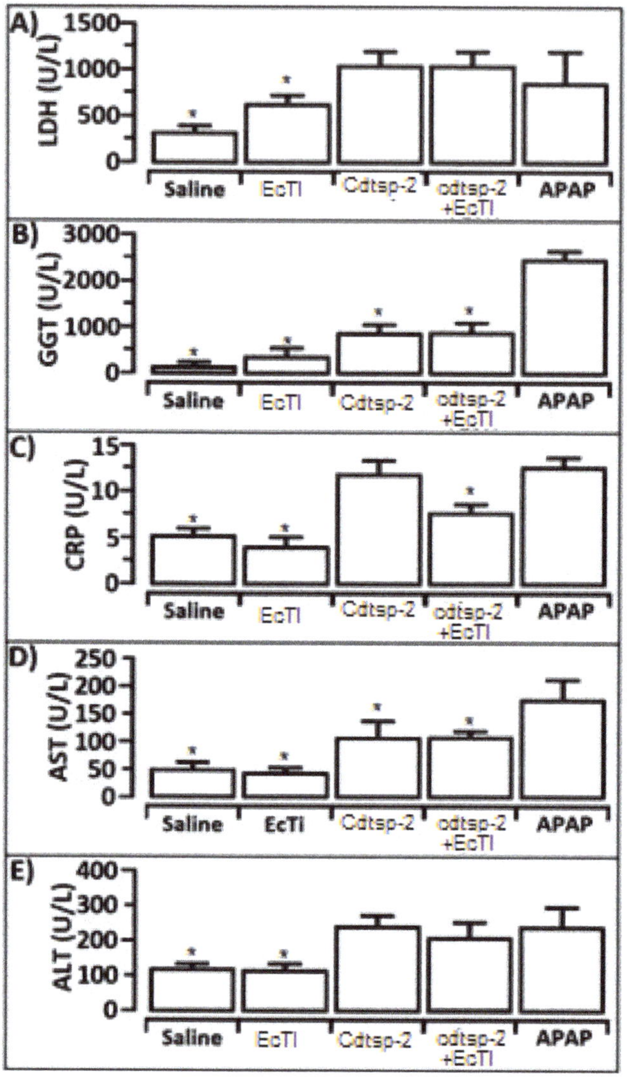

Figure 6. (**A**): It is possible to see an increase in LHD excretion in the presence of Cdtsp-2 protein when it was applied to EcTI 30 min after Cdtsp-2. (**B**): There was a significant decrease in Hepatotoxicity in relation to y-GT excretion caused by the protein when it was applied to EcTI 30 min after Cdtsp-2. (**C**): There was a significant decrease in liver inflammation caused by Cdtsp-2 protein when the same was applied to EcTI 10 min after Cdtsp-2. (**D**,**E**): There was no significant decrease in hepatotoxicity of the protein when it was applied to EcTI 10 min after Cdtsp-2. Data were expressed as mean and standard deviations, and treatments with significant difference marked with (*) and ANOVA was used as a statistical test, with Dunnet as a posteriori test, F = 9.797, p = 0.0017.

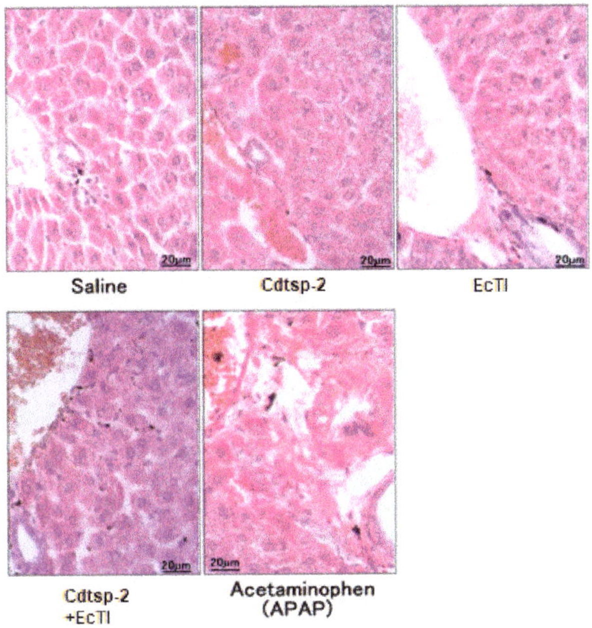

Figure 7. All samples present normal aspects, with neither infiltrated inflammatory or signals of necrosis. The pieces were processed according to the method described by Medeiros et al. for tests [19] and histopathological evaluations of the liver were made qualitatively with the aid of a light microscope in a blinded trial, based on the guidelines issued by the Society for Toxicologic Pathology [20] and the National Toxicology Program Health and Human Services [21].

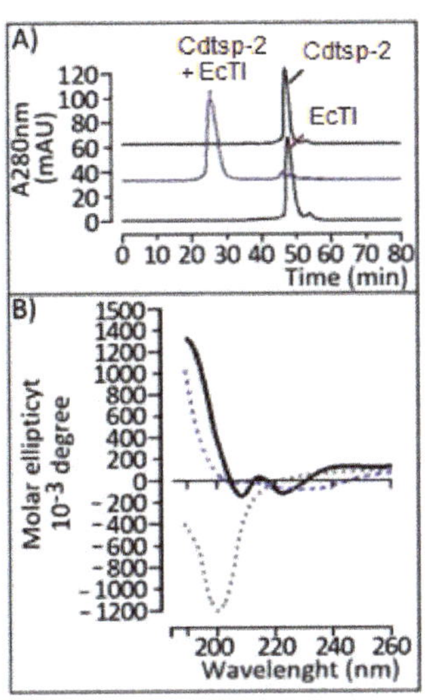

Figure 8. (**A**): Analysis of interaction through HPLC molecular exclusion. (**B**): Analysis on bench circular dichroism, Jasco, with monitoring at 190 to 260 nm in 0.1%TFA and 66% ACN buffer and the protein with concentration 0.02 mg/mL. Profile analysis (recording the differentiated absorption of polarized light) was given by Spectra manager program: "Randon coil" about 61% of the structure, 17.84% of the structure is organized in alpha-helices; 21.16% in sheet.

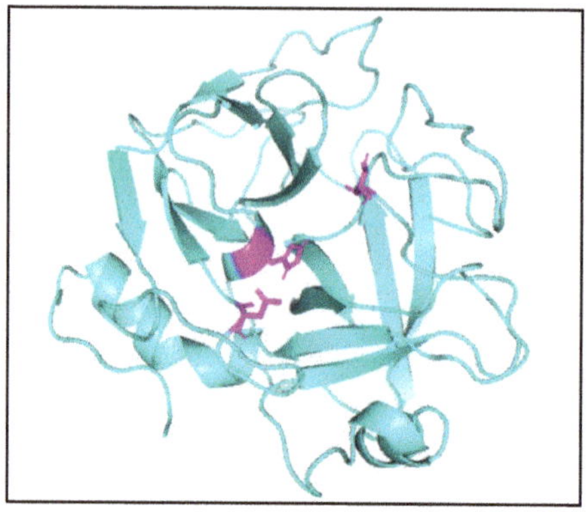

Figure 9. The theoretical model of Cdtsp-2 generated by PyMol. The amino acid residues referring to the active site are represented in purple.

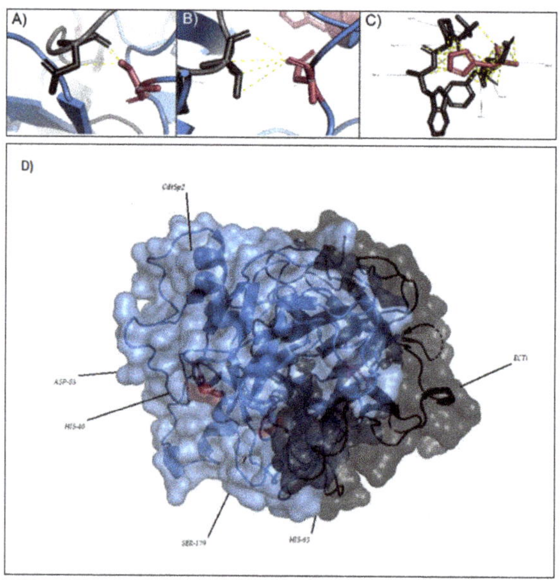

Figure 10. Interaction between EcTI with the amino acid residues of the Active Site of Cdtsp-2. The bonds are 0.9 Å to 3.5 Å. (**A**): His 188. (**B**): Asp 92. (**C**): His 43. (**D**): Interaction between EcTI and Cdtsp-2.

3. Discussion

Cdtsp-2 is a protein that closely resembles human Gyroxin and thrombin, and its active sites are highly preserved [6] (Figure 1). The protein can act in the coagulation cascade and in the breakdown of fibrinogen into fibrin monomers. However, Cdtsp-2 is not as glycosylated as Gyroxin. Therefore, it is plausible that protein glycosylations are crucial for its function and to inhibit the action of other elements. Another significant difference between Thrombin and Cdtsp-2 is the type of clot formation; the former produces firm clots, while the latter produces only loose clots.

Batista [15], when isolating EcTI, found that it was possible to inhibit factor XIIa of the coagulation cascade, plasma kallikrein and plasmin, chymotrypsin and trypsin, but it was not possible to inhibit factor Xa and thrombin. However, this result is not seen when compared to Cdtsp-2. Although the active sites are highly preserved in the thrombin mimics, structural changes in the allosteric regions may account for the differences in inhibition of enzyme activity [22,23]. More tests were done to evaluate the pharmacological characteristics of the Cdtsp-2 protein and its possible actions in in vivo tests. As Cdtsp-2 is a recently discovered serine protease [12], not much is known about its activities. In particular, it is not known whether it is hepatotoxic or not. For the in vivo studies, 25 g to 35 g female Swiss mice were used. The paw model described by DI ROSA [24] was used, injecting 10μg of Cdtsp-2 protein into the right paw, and saline (0.9% NaCl) into the intraperitoneal region 10 min later, to mimic a parenteral treatment. The edema test aims to observe the effectiveness of protective treatments, such as the injection of saline, the pure compounds and the inhibitor, 30 min before exposure to Cdtsp-2.

Our group's study [12] paw edema assay revealed that the serine protease Cdtsp-2 was not only able to induce damage to the coagulation cascade as a thrombin similar, but was also able to produce significant edema, i.e., the protein in question also triggers the inflammatory process. This result, shown in Figure 3, is consistent with Menaldo et al. [25], who showed a serine protease purified from the venom of a snake of the genus Bothrops, which can also cause edema.

The EcTI inhibitor showed a significant ability in reducing edema (Figure 3A). This Kunitz-like inhibitor also acts by decreasing key inflammatory cytokines, such as TGF-α, IL-6, IL-8 and MCP-1, in addition to NFκB, a transcription factor, responsible for aggressive and metastatic characteristics [16]. Cdtsp-2, indeed, presents the ability to cause myonecrosis, as published by Costa and collaborators in 2018; we can observe this in Figure 5. Its ability to cause edema and cell death is quite particular when compared to the other enzymes of the same type, but very similar to the effects caused by sPLA2, already described in the literature [26], and we can see that a inhibitor of sPLA2 the DXPshowed a significant ability in reducing edema from Cdtsp-2 (Figure 4). Besides, CK results showed that there was an expressive reduction in the myotoxicity caused by Cdtsp-2 when exposed to EcTI (Figure 5). The induction of myonecrosis by Cdtsp-2 was very similar to the myonecroses presented by other classes of proteins.

Accidents with *Crotalus durissus terrificus* can be fatal, since the venom has neurotoxic, myotoxic, coagulant and hepatotoxic actions [9,11]. Therefore, tests with Cdtsp-2 related to hepatotoxicity are necessary, and the inhibitory capacity of EcTI against these parameters is essential for the development of a possible adjuvant treatment. JR Mitchell and coworkers [22] observed in their studies that glutathione depletion by acetaminophen was increased by treatments that potentiate liver necrosis and covalent binding produced by the toxic metabolite of acetaminophen. Due to this fact, many studies now use paracetamol (acetaminophen) as a gold standard for hepatotoxicity. In our work, we also used it as a positive standard.

Lactate dehydrogenase (LDH) is an enzyme present in almost all tissues [27]. Increases can be caused by diseases such as hepatitis, anemia, heart attack, fractures, muscle trauma, cancer and infections such as encephalitis, meningitis and HIV [28]. Since LDH is nonspecific and routine isoenzyme analysis is usually not available in clinical laboratories, LDH measurements provide incomplete data and other tests, such as CK for muscle, ALT for liver, troponin for heart disease, etc. are required. Also, LDH activity is impaired when there is hemolysis in the blood sample. Since red blood cells (RBCs) have the LDH protein, hemolysis leads to false-positive results [29].

Our samples had an elevated concentration of LDH, which despite being a nonspecific test, corroborates with the results we found for CK and all the other hepatitis and inflammatory markers. These data are in line with the results found in the studies of Barravieira and collaborators in [9], France [11] and Al-Quraishy [30]. Still regarding the LDH results, the EcTI inhibitor was unable to reverse the damage already caused by Cdtsp-2 (Figure 5A). Another non-specific marker for hepatotoxicity used was Gamma GT (Gamma-Glutamyltransferase); this marker can be produced in organs such as liver, kidneys, seminal vesicles, spleen, pancreas, heart and brain [31]. It may be increased in renal failure, pancreatic disease, myocardial infarction, diabetes and other diseases besides liver disease.

GGT levels increase markedly in the setting of bile duct obstruction, and the ratio of alanine aminotransferase (ALT) to GGT can guide the clinician in deciding between obstructive jaundice (low ALT/medium-high GGT) versus hepatitis (high ALT/low-medium GGT) during the investigation of a patient with jaundice [32]. Alanine aminotransferase is the most widely used biological indicator for liver health [33]. ALT, as the name suggests, is linked to the transamination of alanine and is concentrated in greater amounts in the liver than in other organs. The release of ALT by the hepatocyte into the bloodstream occurs after hepatocellular injury, and it is eliminated with a plasma half-life of approximately 42 h in humans and 24 h in mice [34]. Elevation of ALT can indicate liver injury, but is not liver-specific. ALT can be caused by muscle, skeletal or cardiac damage and drugs that increase ALT gene expression [35].

Aspartate aminotransferase is a transaminase enzyme which catalyzes the reaction between aspartate and alpha-ketoglutarate to form oxaloacetate and glutamate. The enzyme AST, also called serum glutamate oxalate transaminase (SGOT), is present in all tissues except bone and levels are highest in the liver and skeletal muscles. The concentration of AST is increased after injury, trauma, necrosis, infection or neoplasia in the liver or muscle [36]. Another liver marker, C-reactive protein (CRP), was discovered by researchers Tillett and Francis [37] when they identified a substance in the serum of patients with acute inflammation. Our results suggest that there was no bile duct obstruction, since in the results found in Figure 6 the GGT is low-medium, and both aspartate aminotransferase (AST) and alanine aminotransferase (ALT) in Figure 6 are at high levels, thus suggesting hepatotoxicity.

CRP is a protein composed of five subunits synthesized by the liver which reacts to the acute phase of an inflammatory/infection process, mainly due to the action of IL-6 on the gene that controls CRP transcription [38]. In Figure 6C, we show that Cdtsp-2 generated a significant increase in CRP, as much as paracetamol when compared to the untreated animals. Another interesting result shown in the graph was that the Kunitz-like inhibitor EcTI was able to reduce the damage caused by the serine protease (Cdtsp-2); these results corroborate with the studies of T Shimomura and co-workers [39].

Barravieira et al. [7,9] showed in his works that the liver presents alterations after 6 h of snake-venom exposure. However, our samples were exposed to Cdtsp-2 for 4 h, and because of this we cannot see any alterations in histopatology analysis.

Shimomura, in his studies, revealed that the hepatocyte growth factor (HGF) activator is a serine protease produced by the liver that circulates inactive in the blood.

Shearer and colleagues [40] described in their work that by suppressing PAR2 there was an effective inhibition to in induced fibrosis, inflammation, steatosis and hepatocellular necrosis, thus suggesting a new multifaceted approach to treat severe liver diseases.

Cdtsp-2 is a serine protease that depends on PAR1 and PAR2 to trigger its pharmacological effects [12], thus, our results demonstrate that the EcTI inhibitor was able to strongly interact with Cdtsp-2, not allowing the protease to activate PAR 2 in the liver, thus preventing the hepatobiliary inflammatory cascade (Figures 6–8 and Scheme 1).

Tissue injury activates HGF, which is converted to the active form by proteolysis. The activator of HGF activates an inactive protein which is transformed into a biologically active heterodimer in the injured tissue. Activated HGF may be involved in the regeneration of injured tissue. The inhibitor (HGF) is a member of the Kunitz family of serine protease inhibitors, as is EcTI [41].

It is crucial to understand the secondary and tertiary structures of a protein to fully understand its physiological functions. Thus, the circular dichroism assay was performed with the Cdtsp-2 fraction. This experiment shows us that 17.84% of the structure is composed of alpha-helices; 21.16% is in beta sheets and the other amino acids are in random structures (Randon coil), about 61% of the structure (Figure 8A,B). The results that we found through circular dichroism correlate with the structures found in studies conducted by [7,9,22]. The inhibition found by Rodrigues C.F. et al. [23] was associated with the structural changes of Thrombin and the modification in its activity, which occurred by changes in its catalytic site or allosteric regions, which did not occur with Cdtsp-2, possibly due to the absence of glycosylations.

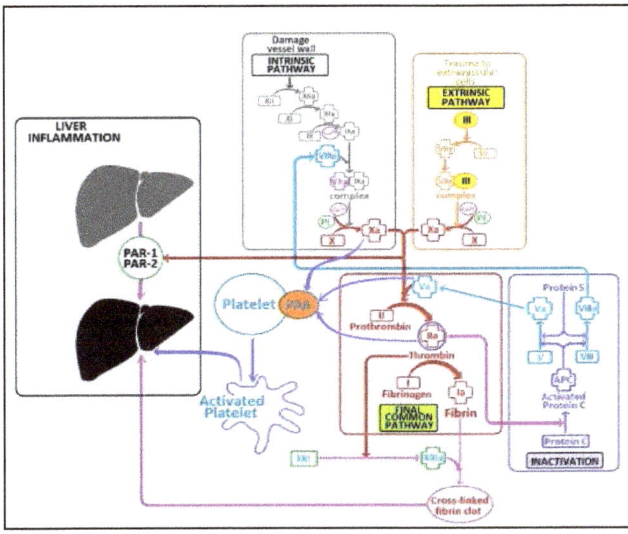

Scheme 1. In summary, this scheme shows that Cdtsp-2 is involved with perturbation of coagulation cascade, in activation of platelet and liver inflammation through activation of PAR's receptors.

The model of Cdtsp-2 is consistent with sequencing and circular dichroism, which shows a predominance of the secondary structure random coil, different from the serino-protease already described for *Crotalus durissus terrificus* venom. The percentage of serine proteases in the venom increased from 2.5% to about 7% of the total [12], increasing the problematic nature of the antivenom serum, since serine proteases are not fully neutralized in their enzymatic and pharmacological activity by the antivenom serum [42].

Typically, inhibition of trypsin-like serine proteases is through the interaction of the reactive loop with the active site cleft [43]. The Kunitz-like inhibitor (EcTI), when exposed to serine protease trypsin, forms a complex with the standard inhibitory mechanism in which the reactive loop of the inhibitor is anchored to the active site of the trypsin, with Arg64 and Ile65 side chains occupying S1 and S1′ pockets, respectively [17]. However, when we study interactions with thrombin simile-type proteins, the forms of interaction and inhibition are different.

Otlewski and co-workers [44] also summarized the types of interactions that occur between inhibitors and thrombin simile-type serine proteases. One relevant example is direct active site blockade. This type of inhibition is given by non-canonical inhibitors of serine proteases. These inhibitors insert their N-terminal tail into the active site of the enzyme, forming a small β-sheet parallel with the enzyme residues. Non-canonical inhibitors were created in hematophagous animals as anticoagulants to inhibit thrombin or factor X a. The paradigmatic example is the recognition of thrombin by the leech hirudin inhibitor. The N-terminal end of the globular domain of hirudin establishes contact with the active site through the parallel β-sheet, while the C-terminal acidic end is recognized by the anionic fibrinogen recognition exosite [44].

According to Richardson and co-workers (2000), the hemedin of a terrestrial leech makes interaction through the N-terminal in a similar way to that cited by Gutter [44], but the acidic C-terminal segment interacts with the heparin binding surface, which differs from the two interaction models [45]. In both examples, the inhibitors form a monomer-like protein complex, as found in the protein-protein docking shown in Figure 8D, i.e., Cdtsp-2 interacted with the EcTI inhibitor like thrombin interacts with inhibitors from blood-phage animals; this result is compatible with those found in the strong enzymatic and pharmacological inhibitions against Cdtsp-2.

4. Materials and Methods

4.1. Purification of Cdtsp-2

The purification of the Serine Portease Cdtsp-2 was done following the methodology of Fonseca et al. [46] and Costa et al. [12].

4.2. Reagentes

We obtained commercially from Sigma-Aldrich (St. Louis, MO, USA) all other chemicals, reagents and kits were purchased from Sigma-Aldrich®, Bio-Rad (Hercules, CA, USA), Cayman Chemical (Ann Arbor, MI, USA).

4.3. Enzyme Assay of Cdtsp-2 Inhibition

The inhibition potential of the EcTI inhibitor was checked against Cdtsp-2. The enzyme was dissolved in 0.9% NaCl saline solution, with a final concentration of 1 mg/mL. The samples were incubated with the pure compounds for 20 min at room temperature, then the solution was incubated on a microplate together with the substrate BapNa, solubilized in water at 37 °C and buffer 50 mM Tris-HCl, pH 7.4, 100 mM NaCl, plus a control (blank − substrate + saline + buffer) and positive control. The samples were read at sequential intervals (5 min, with a total time of 90 min) in a spectrophotometer for reading (λ = 405 nm) (adapted from Prasa et al. [18]) and all absorbance data were transformed in velocity using this formula: V_0 = nm.258 mmol/Tempo(min).

4.4. Circular Dichroism

Polarized light is used in the distal ultraviolet (UV) range (from 180–260 nm). This technique allows the evaluation of the structural integrity of proteins, conformational changes, processes of denaturation (unfolding) and renaturation (folding), making it possible to estimate the composition of the elements in the secondary structure of this macromolecule. The circular dichroism analysis was performed with 0.02 mg/mL of Cdtsp-2 monitored using wavelengths of 190–260 nm. The analysis was performed to 8 convolutions, and data were treated by Spectra manager.

4.5. Evaluation of Paw Edema

In vivo experimental models were performed to evaluate inhibition of acute inflammation caused by purified Cdtsp-2 using randomly chosen Swiss mice (~25 g, n = 5). These assays were performed only after in vitro investigation of inhibition or interaction of the compounds with the enzymes used in this work. Animals were treated with 50 μL (0.5 μg/μg per animal/10 μg per animal) of the compounds via peritoneal or intravenous injections. A saline solution (0.9% NaCl) was used for the controls. There were 5 groups: Saline, Cdtsp-2, Cdtsp-2 previously incubated with EcTI, EcTI and carrageenin (1.5% in solution, with positive control). Edema volume monitoring was achieved using a digital plethysmometer for about four hours. After the tests, the mice were anaesthetized and sacrificed via cervical dislocation. In vivo experiments were performed according to the institutional rules and were approved by the ethics committee from UNESP, number 10/2018-CEUA.

4.6. Hepatotoxicity

The hepatotoxicity study was performed using the methodology according to Hewedy (2021) [47], where hepatic injury was induced by intraperionial injection using as positive gold standard the acetaminophen molecule (paracetamol, N-acetyl-p-aminophenol).

Fifty µg/µL of the EcTI inhibitor, previously incubated for 30 min with Cdtsp-2 (1:1 *w*/*w*), were injected into the right peritoneum (n = 5). After 30 min on average, blood from the animals was collected from the tail in a heparinized tube, which was centrifuged and frozen for further testing. Control animals (n = 5) were subjected to the same procedure as treated animals, but were only inoculated with saline, pure compounds and inhibitor. The observation time of the animals was 4 h. After testing, the mice were anesthetized and sacrificed via cervical dislocation. With the separated serum we used commercial kits where we followed the manufacturers' guidelines for the determination of CRP (C-reactive protein), AST (aspartate aminotransferase), ALT (alanine aminotransferase, LDH (lactate dehydrogenase) and gamma-GT (gamma glutamyl transferase).

Histopathology

Livers were collected and their wet weights (absolute and relative to body weight) were recorded and destined for histological procedures. Samples were fixed in buffered formaldehyde solution (10%) for 24 h and lately transferred to 70% alcohol. The pieces were processed according to the method described by Medeiros et al. [19] and histopathological evaluations of the livers were made qualitatively with the aid of a light microscope in a blinded trial, based on the guidelines issued by the Society for Toxicologic Pathology [20] and the National Toxicology Program Health and Human Services [21]. The following parameters were observed for hepatic tissue: presence of hydropic or fat degeneration, defense cells foci, mitotic figures and binucleated hepatocytes and hyaline casts. All parameters were classified according to the frequency: absent, mild (<25% of the histological section), moderate (25% to 50% of the histological section) and severe (occurring between 50% and 100% of the histological section).

4.7. CK Level Measurement

In total, 20 µL (2.5 µg) of the compounds were added into the peritoneum 10 min after Cdtsp-2 (n = 5) was injected into the gastrocnemius muscle. After an average of 30 min, blood from the animals was collected from the tail in a heparinized tube, which was centrifuged and frozen for later analysis. Control groups (n = 5) underwent the same procedure as the treated animals, but were inoculated with 20 µL of 0.9% NaCl, 20 µL (2.5 µg) of the purified compounds or 20 µL (10 µg) of isolated Cdtsp-2. After testing, the mice were sacrificed by cervical dislocation. Serum creatine kinase (CK) levels were determined according to the kit manufacturer (Bioliquid, Pinhais, Brazil).

4.8. Statistical Analysis

Data are expressed as mean ± standard deviations. The results were analysed through an analysis of variance (ANOVA) of one or two routes followed by the Bonferroni and t-test with statistical variance $p < 0.05$.

4.9. Preparation of Ligands

Thrombin (1doj), Trypsin (6yiw), Cdtsp-2 (Gka1) and EcTI (4j2k) proteins were found in PDB (Protein Data Bank-https://www.rcsb.org accessed on 18 September 2022) and NCBI (NCBI https://www.ncbi.nlm.nih.gov accessed on 18 September 2022) database and coding was used in this analysis to find the amino acid sequence of all the proteins. The crystallographic model was chosen as the best model for building the theoretical structural model from human models.

The structure information of the compounds was taken from the PubChem platform (https://pubchem.ncbi.nlm.nih.gov accessed on 20 September 2022) associated with the Molinspiration and SwissADME platforms were used for better visualization of the structure of the compounds [40]. The ligands were in 3D format with the SDF extension. After downloading, the ligands and files were converted to Mol2. format and the PDB models were generated by RCBS, to be inserted into the SwissDock platform for further analysis in the Chimera 1.14 tool, where we observed the sites that have the highest interactions between the compound and the protein, among the 250 positions generated by SwissDock in this preliminary docking [42].

The Swiss ADME platform (www.swissadme.ch accessed on 20 January 2023) was created by the Swiss bioinformatics institute. The platform was used to estimate individual types of ADME biological behavior, prior to chemical synthesis and biological testing, to approximate the biological activity profiles of the simulated molecules [46].

4.10. Molecular Docking

In this study, the molecular docking experiments were performed by the SwissDock platform [48], subsequently we used the CavityPlus platform [49], following the idea of XU et al. [50] that the association of using the platforms generated better results to generate a Grid Box more faithful to reality, using the Autodock Vina software [49,50]. Several results from various fitting runs were generated and summarized in a table for further analysis [45,47].

The SwissDock and Cavity Plys platform (https://swissmodel.expasy.org, accessed on 16 April 2023) was used to assemble the structural molecular model of the protein and evaluate the possibilities of binding the chosen protein.

After a prior study of the mechanisms and essential residues of these proteins, molecular docking experiments were performed by Autodock Vina with two Grid Boxes: one larger, covering more site residues, and the other with the $40 \times 40 \times 40$ parameters focused on the site of largest interactions with 100 exhaustive assays per sample [51].

Coupling calculations using the Lamarckian genetic algorithm and standard procedures for fitting a versatile ligand to a rigid protein were performed with AutoDock 5.6. The docking calculations were performed on the binding of each protein target and catalytic site.

Once the possible binding sites were identified, the coupling of compounds to these sites was performed to decide the most likely and most energetically desirable binding conformations [43].

Autodock Vina 1.1.2 was used to obtain robust docking simulations involving a grid box at the identified binding site. A 32 Å matrix with a grid spacing of 0.375 Å was bounded by an active site. Affinity scores were obtained and ranked based on the free energy binding theory provided by AutoDock Vina (in kcal/mol). To verify the binding relationships, we used Discovery Studio Visualizer 2.5 (http://3dsbiovia.com/products/ accessed on 16 April 2023), PyMOL and LigPlot+ [52,53], evaluating the binding energies, distances and orientations of the molecules in the microenvironment of the active site of the enzymes with the binding compounds [50].

5. Conclusions

The Kunitz-like inhibitor purified from *Enterolobium contortisiliquum* EcTI was efficient against the damage caused by serine protease. These results were important pointers for studies on the mechanism of interaction between Cdtsp-2 and EcTI proteins. The EcTI inhibitor showed strong interactions not only with the triad of the catalytic site, but also with all other amino acids of the protein, forming a complex and inactivating the enzymatic activities of Cdtsp-2. Therefore, we can see that Kunitz inhibitor is a plausible option for developing ancillary treatments against the biological activities of venoms.

Author Contributions: C.R.C.C., M.N.B., A.R., M.A.d.O., R.S., I.A., L.L.M. and C.R.B., M.L.V.O., S.F.S. and M.H.T. are the authors responsible for all the biochemical, in vivo, in vitro and in silico experimental assays; the authors C.R.C.C., M.N.B., M.A.d.O. and M.H.T. assisted in analysis of the results and wrote and edited this manuscript; M.H.T. is responsible for obtaining the laboratory and financial resources of the project and the work, and was the coordinator of the group that carried out the work. All authors have read and agreed to the published version of the manuscript.

Funding: The presented project was financially support by the FAPESP, Process No. 2017/20291-0 under the responsibility of Marcos Hikari Toyama, as well as resources received from the FAPESP

Process No. p2017/19942-7 of Marcos Antônio Oliveira, to CNPq Process No. 309271/2022-3 of Marcos Hikary Toyama and UNESP and FAPESP Process No. 2003/10516-2 of Maria Luiza Vilela Oliva.

Institutional Review Board Statement: The animal study protocol was approved by the Institutional Review Board (or Ethics Committee) of Instituto de Biociências-UNESP (protocol code CEUA 10/2018 and 19 March 2019).

Informed Consent Statement: Not applicable.

Data Availability Statement: Data are in the article.

Acknowledgments: The presented project was financially support by the FAPESP, Process No. 2017/20291-0 under the responsibility of Marcos Hikari Toyama, as well as resources received from the FAPESP Process No. p2017/19942-7 of Marcos Antônio Oliveira, to CNPq Process No. 309271/2022-3 of Marcos Hikary Toyama and UNESP and FAPESP Process No. 2003/10516-2 of Maria Luiza Vilela Oliva. Our most sincere thanks to Professor Juliana Perobelli jperobelli@unifesp.br and her team, who were responsible for providing resources for histopathological analysis.

Conflicts of Interest: The authors declare no conflict of interest.

References

1. Da Silva, G.M.; De Souza, D.H.B.; Waitman, K.B.; Ebram, M.C.; Fessel, M.R.; Zainescu, I.C.; Portaro, F.C.; Heras, M.; De Andrade, S.A. Design, synthesis, and evaluation of Bothrops venom serine protease peptidic inhibitors. *J. Venom. Anim. Toxins Incl. Trop. Dis.* **2021**, *27*, e20200066. [CrossRef]
2. Moio Da Cunha, E.; Martins, O.A. Principais Compostos Químicos Presente nos Venenos de Cobras dos Gêneros Bothrops e Crotalus-Uma Revisão. **2012**, *2*, 21–26. Available online: www.scielo.br (accessed on 12 January 2023).
3. Kang, D.H.; Lee, D.J.; Lee, K.W.; Park, Y.S.; Lee, J.Y.; Lee, S.-H.; Koh, Y.J.; Koh, G.-Y.; Choi, C.; Yu, D.-Y.; et al. Peroxiredoxin II Is an Essential Antioxidant Enzyme that Prevents the Oxidative Inactivation of VEGF Receptor-2 in Vascular Endothelial Cells. *Mol. Cell* **2011**, *44*, 545–558. [CrossRef]
4. Waheed, H.F.; Moin, S.I.; Choudhary, M. Snake Venom: From Deadly Toxins to Life-saving Therapeutics. *Curr. Med. Chem.* **2017**, *24*, 1874–1891. [CrossRef]
5. Picolo, G.; Chacur, M.; Gutiérrez, J.M.; Teixeira, C.F.P.; Cury, Y. Evaluation of Antivenoms in the Neutralization of Hyperalgesia and Edema Induced by Bothrops Jararaca and Bothrops Asper Snake Venoms. *Braz. J. Med. Biol. Res.* **2002**, *35*, 1221–1228. Available online: http://www.scielo.br/scielo.php?script=sci_arttext&pid=S0100-879X2002001000016&lng=en&tlng=en (accessed on 15 March 2023). [CrossRef]
6. Miyagui, C.; Brando Prieto da Silva lvaro R de Santana, G. Serine proteases—Cloning, Expression and Potential Applications. In *An Integrated View of the Molecular Recognition and Toxinology—From Analytical Procedures to Biomedical Applications*; IntechOpen: London, UK, 2013.
7. Barraviera, B.; Júnior, J.C.B.; Arakaki, D.; Domingues, M.A.C.; Pereira, P.C.M.; Mendes, R.P.; Machado, J.M.; Meira, D.A.; Arkaki, D. A Retrospective Study of 40 Victims of Crotalus Snake Bites: Analysis of the Hepatic Necrosis Observed in One Patient. *Rev Soc Bras Med. Trop* **1989**, *22*, 5–12. Available online: http://www.scielo.br/scielo.php?script=sci_arttext&pid=S0037-86821989000100002&lng=en&tlng=en (accessed on 15 March 2023). [CrossRef]
8. Vital Brazil, O.; Fontana, M.D.; Vital Brazil, O.; Fontana Toxins, M.D. Toxins as tools in the study of sodium channel distribution in the muscle fibre membrane. *Toxicon Off. J. Int. Soc. Toxinology* **1993**, *31*, 1085–1098. [CrossRef]
9. Barraviera, B. Acute-phase response in snakebite. *Rev. Inst. Med. Trop. São Paulo* **1994**, *36*, 479. [CrossRef]
10. Calvete, J.J.; Sanz, L.; Angulo, Y.; Lomonte, B.; Gutiérrez, J.M. Venoms, venomics, antivenomics. *FEBS Lett.* **2009**, *583*, 1736–1743. [CrossRef]
11. Ral, O.; Acg, P.J. Acute hepatotoxicity of *Crotalus durissus terrificus* (South American rattlesnake) venom in rats. *J. Venom. Anim.Toxins Incl. Trop. Dis. V* **2008**, *13*, 61–78.
12. Costa, C.R.C.; Belchor, M.N.; Rodrigues, C.F.B.; Toyama, D.D.O.; De Oliveira, M.A.; Novaes, D.P.; Toyama, M.H. Edema Induced by a *Crotalus durissus terrificus* Venom Serine Protease (Cdtsp 2) Involves the PAR Pathway and PKC and PLC Activation. *Int. J. Mol. Sci.* **2018**, *19*, 2405. [CrossRef]
13. Pant, A.; Kopec, A.K.; Luyendyk, J.P. Role of the blood coagulation cascade in hepatic fibrosis. *Am. J. Physiol. Gastrointest. Liver. Physiol.* **2018**, *315*, G171–G176. [CrossRef]
14. Costa, C.R.C. Avaliação de Compostos Polifenólicos da Laguncularia Racemosa Sobre a Atividade Enzimática e Farmacológica de Serino Proteases de *Crotalus durissus terrificus*. 22 February 2018. Available online: https://repositorio.unesp.br/handle/11449/153268 (accessed on 14 March 2023).
15. Batista, I.F.C.; Nonato, M.C.; Bonfadini, M.R.; Beltramini, L.M.; Oliva, M.L.V.; Sampaio, M.U.; Sampaio, C.A.; Garratt, R.C. Preliminary crystallographic studies of EcTI, a serine proteinase inhibitor from Enterolobium contortisiliquum seeds. *Acta Crystallogr. D Biol. Crystallogr.* **2001**, *57*, 602–604. [CrossRef]

16. Lobo, Y.A.; Bonazza, C.; Batista, F.P.; Castro, R.A.; Bonturi, C.R.; Salu, B.R.; Sinigaglia, R.D.C.; Toma, L.; Vicente, C.M.; Pidde, G.; et al. EcTI impairs survival and proliferation pathways in triple-negative breast cancer by modulating cell-glycosaminoglycans and inflammatory cytokines. *Cancer Lett.* **2020**, *491*, 108–120. [CrossRef]
17. Zhou, D.; Lobo, Y.A.; Batista, I.F.C.; Marques-Porto, R.; Gustchina, A.; Oliva, M.L.V.; Wlodawer, A. Crystal Structures of a Plant Trypsin Inhibitor from Enterolobium contortisiliquum (EcTI) and of Its Complex with Bovine Trypsin. *PLoS ONE* **2013**, *4*, e62252. [CrossRef]
18. Prasa, D.; Svendsen, L.; Stürzebecher, J. Inhibition of thrombin generation in plasma by inhibitors of factor Xa. *Thromb Haemost* **1997**, *78*, 1215–1220. [CrossRef]
19. da Cunha de Medeiros, P.; Samelo, R.R.; Silva, A.P.G.; da Silva Araujo Santiago, M.; Duarte, F.A.; Perobelli, J.E. Prepubertal exposure to low doses of sodium arsenite impairs spermatogenesis and epididymal histophysiology in rats. *Environ Toxicol.* **2019**, *34*, 83–91. [CrossRef]
20. Society of Toxicology Pathology. Standardized System of Nomenclature and Diagnostic Criteria (SSNDC) Guides. Available online: https://www.toxpath.org/ssndc.asp (accessed on 10 December 2018).
21. National Toxicology Program Health and Human Services. Guides. Available online: https://ntp.niehs.nih.gov/nnl/hepatobiliary/liver/fatty_change/index.htm#group-5 (accessed on 10 December 2018).
22. Mitchell, J.R.; Jollow, D.J.; Potter, W.Z.; Davis, D.C.; Gillette, J.R.; Brodie, B.B. Acetaminophen-induced hepatic necrosis. I. Role of drug metabolism. *J. Pharmacol. Exp. Ther.* **1973**, *187*, 185–194.
23. Rodrigues, C.F.B.; Gaeta, H.H.; Belchor, M.N.; Ferreira, M.J.P.; Pinho, M.V.T.; Toyama, D.D.O.; Toyama, M.H. Evaluation of Potential Thrombin Inhibitors from the White Mangrove (*Laguncularia racemosa* (L.) C.F. Gaertn). *Mar. Drugs* **2015**, *13*, 4505–4519. [CrossRef]
24. DI ROSA, M. Biological Properties of Carrageenan. *J. Pharm. Pharmacol.* **2011**, *24*, 89–102. Available online: https://academic.oup.com/jpp/article/24/2/89/6200741 (accessed on 14 March 2023). [CrossRef]
25. Menaldo, D.L.; Bernardes, C.P.; Pereira, J.C.; Silveira, D.S.; Mamede, C.C.; Stanziola, L.; de Oliveira, F.; Pereira-Crott, L.S.; Faccioli, L.H.; Sampaio, S.V. Effects of two serine proteases from Bothrops pirajai snake venom on the complement system and the inflammatory response. *Int. Immunopharmacol.* **2013**, *15*, 764–771. [CrossRef]
26. Yadav, R.; Liu, Y.; Kwok, S.; Hama, S.; France, M.; Eatough, R.; Pemberton, P.; Schofield, J.; Siahmansur, T.J.; Malik, R.; et al. Effect of extended-release niacin on high-density lipoprotein (HDL) functionality, lipoprotein metabolism, and mediators of vascular inflammation in statin-treated patients. *J Am Heart Assoc* **2015**, *4*, e001508. [CrossRef]
27. Schumann, K.; Mauch, C.; Klespe, K.; Loquai, C.; Nikfarjam, U.; Schlaak, M.; Akçetin, L.; Kölblinger, P.; Hoellwerth, M.; Meissner, M.; et al. Real-World Outcomes Using PD-1 Antibodies and BRAF + MEK Inhibitors for Adjuvant Melanoma Treatment from 39 Skin Cancer Centers in Germany, Austria and Switzerland. *J. Eur. Acad. Dermatol. Venereol.* **2022**, *37*, 894–906. Available online: https://onlinelibrary.wiley.com/doi/full/10.1111/jdv.18779 (accessed on 14 March 2023). [CrossRef] [PubMed]
28. Drent, M.; Cobben, N.A.M.; Henderson, R.F.; Wouters, E.F.M.; Van Dieijen-Visser, M. Usefulness of Lactate Dehydrogenase and Its Isoenzymes as Indicators of Lung Damage or Inflammation. *Eur. Respir. J.* **1996**, *9*, 1736–1742. Available online: https://erj.ersjournals.com/content/9/8/1736 (accessed on 14 March 2023). [CrossRef] [PubMed]
29. Farhana, A.; Lappin, S.L. Biochemistry, Lactate Dehydrogenase. *StatPearls* **2022**. Available online: https://www.ncbi.nlm.nih.gov/books/NBK557536/ (accessed on 14 March 2023).
30. Al-Quraishy, S.; Dkhil, M.A.; Abdel Moneim, A.E. Hepatotoxicity and oxidative stress induced by Naja hajecrude venom. *J. Venom. Anim. Toxins Incl. Trop. Dis.* **2014**, *20*, 42. [CrossRef]
31. Naftalin, L.; Sexton, M.; Whitaker, J.F.; Tracey, D. Clinica chimica acta 293 a routine procedure for estimating serum y-glutamyl-transpeptidase activity. *Clin. Chim. Acta* **1997**, *258*, 21–30.
32. Aronson, S.J.; Junge, N.; Trabelsi, M.; Kelmemi, W.; Hubert, A.; Brigatti, K.W.; Fox, M.D.; de Knegt, R.J.; Escher, J.C.; Ginocchio, V.M.; et al. Disease burden and management of Crigler-Najjar syndrome: Report of a world registry. *Liver Int.* **2022**, *42*, 1593–1604. [CrossRef] [PubMed]
33. Yigit, M.; Sogut, O.; Tataroglu, Ö.; Yamanoglu, A.; Yigit, E.; Guler, E.M.; Ozer, O.F.; Kocyigit, A. Oxidative/antioxidative status, lymphocyte DNA damage, and urotensin-2 receptor level in patients with migraine attacks. *Neuropsychiatr. Dis. Treat* **2018**, *14*, 367–374. [CrossRef]
34. Gwaltney-Brant, S.M. Veterinary Forensic Toxicology. *Vet Pathol.* **2016**, *53*, 1067–1077. [CrossRef]
35. Yang, Y.; Pham, T.X.; Wegner, C.J.; Kim, B.; Ku, C.S.; Park, Y.K.; Lee, J.Y. Astaxanthin Lowers Plasma TAG Concentrations and Increases Hepatic Antioxidant Gene Expression in Diet-Induced Obesity Mice. *Br. J. Nutr.* **2014**, *112*, 1797–1804. Available online: https://www.cambridge.org/core/journals/british-journal-of-nutrition/article/astaxanthin-lowers-plasma-tag-concentrations-and-increases-hepatic-antioxidant-gene-expression-in-dietinduced-obesity-mice/49F447C0D80B03FB07FAD952E805E990 (accessed on 15 March 2023). [CrossRef]
36. McGill, M.R. The Past and Present of Serum Aminotransferases and the Future of Liver Injury Biomarkers. *EXCLI J.* **2016**, *15*, 817–828. Available online: http://www.ncbi.nlm.nih.gov/pubmed/28337112 (accessed on 12 March 2023).
37. Thomas Francis, J. *Biographical Memoirs: Volume 44-National Academy of Sciences-Google Livros*; Rock & Read Books Publishing Company: Shanghai, China, 1936.
38. Nehring, S.M.; Goyal, A.; Bansal, P.; Patel, B.C. C Reactive Protein. *StatPearls* **2017**, *65*, 237–244. Available online: http://europepmc.org/books/NBK441843 (accessed on 15 March 2023).

39. Shimomura, T.; Denda, K.; Kitamura, A.; Kawaguchi, T.; Kito, M.; Kondo, J.; Kagaya, S.; Qin, L.; Takata, H.; Miyazawa, K.; et al. Hepatocyte Growth Factor Activator Inhibitor, a Novel Kunitz-Type Serine Protease Inhibitor. *J. Biol. Chem.* **1997**, *272*, 6370–6376. Available online: http://www.jbc.org/article/S0021925818411520/fulltext (accessed on 15 March 2023). [CrossRef]
40. Shearer, A.M.; Rana, R.; Austin, K.; Baleja, J.D.; Nguyen, N.; Bohm, A.; Covic, L.; Kuliopulos, A. Targeting Liver Fibrosis with a Cell-Penetrating Protease-Activated Receptor-2 (PAR2) Pepducin. *J. Biol. Chem.* **2016**, *291*, 23188–23198. [CrossRef]
41. Oliva, M.L.V.; Silva, M.C.C.; Sallai, R.C.; Brito, M.V.; Sampaio, M.U. A novel subclassification for Kunitz proteinase inhibitors from leguminous seeds. *Biochimie* **2010**, *92*, 1667–1673. [CrossRef]
42. Kuniyoshi, A.K.; Rocha, M.; Carvalho, D.C.; Juliano, M.A.; Neto, L.J.; Tambourgi, D.V.; Portaro, F.C.V. Angiotensin-degrading serine peptidase: A new chymotrypsin-like activity in the venom of Bothrops jararaca partially blocked by the commercial antivenom. *Toxicon* **2012**, *59*, 124–131. [CrossRef]
43. Otlewski, J.; Jelen, F.; Zakrzewska, M.; Oleksy, A. The Many Faces of Protease–Protein Inhibitor Interaction. *EMBO J.* **2005**, *24*, 1303–1310. Available online: https://onlinelibrary.wiley.com/doi/full/10.1038/sj.emboj.7600611 (accessed on 15 March 2023). [CrossRef]
44. Grütter, M.G.; Priestle, J.P.; Rahuel, J.; Grossenbacher, H.; Bode, W.; Hofsteenge, J.; Stone, S.R. Crystal Structure of the Thrombin-Hirudin Complex: A Novel Mode of Serine Protease Inhibition. *EMBO J.* **1990**, *9*, 2361–2365. Available online: https://onlinelibrary.wiley.com/doi/full/10.1002/j.1460-2075.1990.tb07410.x (accessed on 15 March 2023). [CrossRef]
45. Richardson, L. Turning Points in Qualitative Research: Tying Knots in a Handkerchief-Google Livros. Available online: https://books.google.com.br/books?hl=pt-BR&lr=&id=8aXWAQAAQBAJ&oi=fnd&pg=PA379&dq=Writing+a+method+of+inquiry&ots=lhcYI-FNKx&sig=XXfeR0Yi77HNo2OuMuzujUSvKas&redir_esc=y#v=onepage&q=Writing%20a%20method%20of%20inquiry&f=false (accessed on 15 March 2023).
46. Xu, Y.; Wang, S.; Hu, Q.; Gao, S.; Ma, X.; Zhang, W.; Shen, Y.; Chen, F.; Lai, L.; Pei, J. CavityPlus: A web server for protein cavity detection with pharmacophore modelling, allosteric site identification and covalent ligand binding ability prediction. *Nucleic Acids Res.* **2018**, *46*, W374–W379. [CrossRef]
47. Hewedy, W.A. Effects of treatment with sitagliptin on hepatotoxicity induced by acetaminophen in mice. *Braz. J. Pharm. Sci.* **2021**, *56*, e18482. Available online: https://www.scielo.br/j/bjps/a/RjzXGxphxnzrZ3MxxwKdCWD/?lang=en# (accessed on 16 March 2023). [CrossRef]
48. Goodsell, D.S.; Morris, G.M.; Olson, A.J. Automated Docking of Flexible Ligands: Applications of AutoDock. *J. Mol. Recognit.* **1996**, *9*, 2–6. [CrossRef]
49. Fonseca, F.V.; Antunes, E.; Morganti, R.P.; Monteiro, H.S.A.; Martins, A.M.C.; Toyama, D.O.; Marangoni, S.; Toyama, M.H. Charcterization of a new platelet aggregating factor from crotoxin *Crotalus durissus cascavella* venom. *Protein J.* **2006**, *25*, 183–192. [CrossRef]
50. Halgren, T.A. Identifying and characterizing binding sites and assessing druggability. *J. Chem. Inf. Model.* **2009**, *49*, 377–389. [CrossRef]
51. Pettersen, E.F.; Goddard, T.D.; Huang, C.C.; Couch, G.S.; Greenblatt, D.M.; Meng, E.C.; Ferrin, T.E. UCSF Chimera—A visualization system for exploratory research and analysis. *J. Comput. Chem.* **2004**, *25*, 1605–1612. [CrossRef]
52. Trott, O.; Olson, A.J. AutoDock Vina: Improving the speed and accuracy of docking with a new scoring function, efficient optimization, and multithreading. *J. Comput. Chem.* **2009**, *31*, 455–461. [CrossRef]
53. Wallace, A.C.; Laskowski, R.A.; Thornton, J.M. LIGPLOT: A Program to Generate Schematic Diagrams of Protein-Ligand Interactions. *Protein Eng. Des. Sel.* **1995**, *8*, 127–134. Available online: https://academic.oup.com/peds/article/8/2/127/1561050 (accessed on 15 March 2023). [CrossRef]

pharmaceuticals

Article

In Silico Evaluation of Quercetin Methylated Derivatives on the Interaction with Secretory Phospholipases A2 from *Crotalus durissus terrificus* and *Bothrops jararacussu*

Mariana Novo Belchor [1,2], Caroline Ramos da Cruz Costa [1,2], Airam Roggero [2], Laila L. F. Moraes [2], Ricardo Samelo [2], Isabelly Annunciato [2], Marcos Antonio de Oliveira [1,2], Sergio F. Sousa [3] and Marcos Hikari Toyama [1,2,*]

[1] Center of Natural and Human Sciences, Federal University of ABC (UFABC), Santo André 09210-580, SP, Brazil; belchor.novo@ufabc.edu.br (M.N.B.)
[2] Biosciences Institute of Paulista Coast Campus (IB/CLP), University of São Paulo State (UNESP), São Vicente 11330-900, SP, Brazil
[3] Unit of Applied Biomolecular Sciences (UCIBIO), REQUIMTE-BioSIM-Medicine Faculty, Porto University, 4050-345 Porto, Portugal
* Correspondence: marcoshikaritoya@gmail.com

Abstract: Quercetin derivatives have already shown their anti-inflammatory potential, inhibiting essential enzymes involved in this process. Among diverse pro-inflammatory toxins from snake venoms, phospholipase A2 is one of the most abundant in some species, such as *Crotalus durissus terrificus* and *Bothrops jararacussu* from the Viperidae family. These enzymes can induce the inflammatory process through hydrolysis at the sn-2 position of glycerophospholipids. Hence, elucidating the main residues involved in the biological effects of these macromolecules can help to identify potential compounds with inhibitory activity. In silico tools were used in this study to evaluate the potential of quercetin methylated derivatives in the inhibition of bothropstoxin I (BthTX-I) and II (BthTX-II) from *Bothrops jararacussu* and phospholipase A2 from *Crotalus durissus terrificus*. The use of a transitional analogous and two classical inhibitors of phospholipase A2 guided this work to find the role of residues involved in the phospholipid anchoring and the subsequent development of the inflammatory process. First, main cavities were studied, revealing the best regions to be inhibited by a compound. Focusing on these regions, molecular docking assays were made to show main interactions between each compound. Results reveal that analogue and inhibitors, Varespladib (Var) and p-bromophenacyl bromide (BPB), guided quercetins derivatives analysis, revealing that Leu2, Phe5, Tyr28, glycine in the calcium-binding loop, His48, Asp49 of BthTX-II and Cdtspla2 were the main residues to be inhibited. 3MQ exhibited great interaction with the active site, similar to Var results, while Q anchored better in the BthTX-II active site. However, strong interactions in the C-terminal region, highlighting His120, seem to be crucial to decreasing contacts with phospholipid and BthTX-II. Hence, quercetin derivatives anchor differently with each toxin and further in vitro and in vivo studies are essential to elucidate these data.

Keywords: natural compounds; inflammation; toxins; snake venoms; molecular docking

Citation: Belchor, M.N.; Costa, C.R.d.C.; Roggero, A.; Moraes, L.L.F.; Samelo, R.; Annunciato, I.; de Oliveira, M.A.; Sousa, S.F.; Toyama, M.H. In Silico Evaluation of Quercetin Methylated Derivatives on the Interaction with Secretory Phospholipases A2 from *Crotalus durissus terrificus* and *Bothrops jararacussu*. *Pharmaceuticals* **2023**, *16*, 597. https://doi.org/10.3390/ph16040597

Academic Editor: Diana Roxana Pelinescu

Received: 6 March 2023
Revised: 7 April 2023
Accepted: 10 April 2023
Published: 15 April 2023

1. Introduction

Flavonoids, compounds that are secondary metabolites from plants, have a nucleus which consists of A, B and C rings. A series of modification reactions, such as hydroxylation, glycosylation, prenylation, and methylation, can enhance multiple physiological functions corresponding to both their structural diversity and tissue specificities. The *O*-methylation of aglycone flavonoids, such as Quercetin (Q), results in the reduction of the molecular activity of a hydroxyl fraction and the consequent increase in lipophilicity, which modifies its intracellular compartmentalization. Furthermore, *O*-methylation provides a branch point in the biosynthesis of several metabolic pathways, including production of modified flavonoids with increased antimicrobial properties [1]. Considered a post-modification product, these derivatives are formed through the methyl group fixation with oxygen at the flavonoid hydroxyl moiety. Due to the diverse hydroxyl groups in the flavonoid core, flavonoids' methylation positions are diverse and provide multiple health benefits, such as increased bioavailability compared to flavonoid precursors [2]. Therefore, these methylated flavonoids are potential candidates for use as anti-inflammatories by decreasing the enzymatic activity of enzymes such as cyclooxygenases (COX), pro-inflammatory interleukins, reactive oxygen species (ROS) and nitrogen (RNS) production [2–4]. Previous analyzes using ChEMBL and SwissTargetPrediction tools reveals that both Rhamnetin (Rhm; 7-*O*-Methylquercetin); 3-*O*-Methylquercetin (3MQ; 3-*O*-Methylquercetin) are compounds with an antioxidant capacity, besides to exhibits anti-inflammatory potential by strongly decreasing the cyclooxygenase (COX) and Lipoxygenase (LOX) activity. Rhamnazin (Rhz; 7,3′-Di-*O*-methylquercetin) exhibits two methylations, which seem to considerably increase the ability to sequester free radicals in cells. Methylated derivatives of Q are also found in some plant species, such as *Coriandrum sativum*, *Achyrocline satureioides*, and *Rhamnus petiolaris* with Rhm, 3MQ, and Rhz compounds, respectively, which have already shown different responses against BthTX-II activities [5].

Snake venoms consist of a complex mixture of biologically active molecules, and the phospholipase A2 (PLA2) group is one of the most studied toxins. There are two main groups of PLA2 (E.C. 3.1.1.4) in snake venoms: phospholipase A2-like (PLA2-like), such as the Lys49-PLA2 and the classic phospholipase A2, Asp49-PLA2. These proteins belong to the secreted PLA2 (sPLA2), a subgroup found in diverse secretions, body fluids and venom of bees, scorpions, and snakes [6]. Therefore, these macromolecules hydrolyze the sn-2 position of glycerophospholipids, leading to a release of fatty acids, such as arachidonic acid (AA) and lysophospholipids, triggering the inflammatory process [7,8]. Bothropstoxin II (BthTX-II) is an Asp49-PLA2 from *Bothrops jararacussu* (Bj), which belongs to the Viperidae family. This toxin is considered a PLA2-like due to some characteristics, such as the Ca^{2+} binding loop distortion, leading to changes in the C-terminal region. In this way, this protein shows a low phospholipase A2 activity besides to exhibit myotoxic, edematogenic, and hemolytic effects [9,10]. Bothropstoxin I (BthTX-I) is a Lys49-PLA2, which reveals high myotoxic activity with a lack of enzymatic activity. The myotoxic mechanism of these enzymes has already been demonstrated and includes mainly some amino acids from the C-terminal region, which interacts with membrane [8].

Crotalus durissus terrificus (Cdt) belongs to the same family and has an essential heterodimeric complex named Crotoxin in its venom, which is a potent β-neurotoxin with a phospholipase A2 activity. This toxin consists of two subunits: a basic pla2 with a weak neurotoxicity, named crotoxin B (CB) or Cdt PLA2, associated with a small acidic, nontoxic and nonenzymatic protein named crotapotin or crotoxin A (CA) [11]. CB presents four isoforms, CBa2, CBb, CBc, and CBd, performing 16 different CA-CB complexes. To date, there are three crystal structures from these isomers: one of them includes CA2 and CBb, the other with CBd (tetramer), and the structure used in this study consists of isoforms CBa2 and CBc (tetramer) [12].

Commercial inhibitors of PLA2, such as p-bromophenacyl bromide (BPB) and Varespladib (Var), have already shown their potential against PLA2 from diverse snake venoms. The classical inhibitor BPB has been used since 1970 to inhibit the catalytic PLA2 once it binds specifically with the His48 residue. However, this compound has already been shown to inhibit myotoxic activity of PrTX-I, a Lys49-PLA2 from *Bothrops pirajai*, through a covalent binding to His48, leading to a Ca^{2+}-binding loop distortion and then, a C-terminus rearrangement [13]. Var (LY315920) is a synthetic molecule which was clinically tested to block the inflammatory cascade initiated by secreted PLA2. There are several studies that reveal the efficacy of this compound in the treatment of phospholipase A2-rich snake venoms [14]. In snake venoms, this compound has already been shown to inhibit the cytotoxic e myotoxic effect of MjTX-II from *Bothrops moojeni* through the physical blockage of its allosteric activation [15]. In addition to this, a synergic effect of this compound with the antivenom was observed to decrease neuromuscular blockage induced by crotamine, highlighting the broad-spectrum effect of this drug [16].

Due to PLA2 role in the inflammatory process, the abundance of this toxin in the Cdt and Bj venoms, and the fact that these proteins show similar structures with sPLA2 from mammals [12], it is essential to find new inhibitors of this enzyme. Hence, in this study, we performed some steps to better understand how these compounds could interact with these three phospholipases A2. First, we aim to elucidate the main protein's regions to anchor with an inhibitor using CavityPlus. Afterwards, using the phosphonate transition-state analogue from 1POB crystal structure, named L-1-0-octyl-2-heptylphos-phonyl-sn-glycero-3-phosphoethanolamine (Analogue) [17], we focus on the elucidation of the main residues that interact with each toxin. Moreover, two commercial inhibitors were used for comparison purposes to identify the main residues involved in the anchoring with all toxins. Finally, we intend to compare the compounds' interactions with phospholipase A2 from *Bothrops jararacussu* and *Crotalus durissus terrificus* with the analogue and the inhibitors. Thus, we analyze if compounds anchor equally with all toxins, and if not, we try to elucidate main differences and the reasons for that.

2. Results

2.1. Properties of Compounds

Firstly, compounds were evaluated concerning physical–chemical features to better understand their biological activity. Data obtained from SwissADME, Mollinspiration and ChEMBL support these results. Q reveals a molecular weight (MW) of 302.24 g/mol, Rhm and 3MQ show the same value, and Rhz, with two methylations, exhibits a higher MW. The topological polar surface area (TPSA) is higher in Q, the same in Rhm and 3MQ, and Rhz exhibits the smallest value (Table 1). A classical descriptor evaluated lipophilicity: the partition coefficient between *n*-octanol and water, which is essential to analyzing physicochemical properties for pharmacokinetics drug discovery [18].

Table 1. Physical–chemical characteristics from Q, Rhm, 3MQ and Rhz.

	Q	Rhm	3MQ	Rhz
MW (g/mol)	302.24	316.26	316.26	330.29
Num. rotatable bonds	1	2	2	3
Num. H-bond acceptors	7	7	7	7
Num. H-bond donors	5	4	4	3
TPSA	131.36 Å^2	120.36 Å^2	Å^2	109.36 Å^2
Lipophilicity (Consensus Log Po/w)	1.23	1.63	1.75	2.02

Each compound shows specific alerts related to chemical structures which reveals a high tendency to interact with a great number of molecules and macromolecules. Moreover, the physicochemical properties are exhibited in the second column, revealing specific differences between them, such as a higher amount of unsaturation in Q than the others, besides being more polar. In general, these compounds show that enzymes are their main target (Figure 1).

2.2. Molecular Docking

2.2.1. Cavity Analysis

The cavities of each toxin were evaluated using the CavityPlus web server to detect potential binding sites on the protein surface. Figure 2 shows two cavities of each protein with the first and the second highest values of druggability. Figure 2a exhibits the first cavity of BthTX-I between chains A and B and the second (Figure 2b) in the hydrophobic channel, which includes residues such as Gly30, Tyr22, Lys7, Tyr52, Lys69, Ala19, Cys45, Lys69, Lys49, His48, and Ile9. Figure 2 reveals the binding region in the active site of

BthTX-II (Figure 2c) and in the interface between chains A and B (Figure 2d). On the other hand, the highest values of druggability found in CdtsPLA2 were in the tetramer interface, including the active sites of the protein (Figure 2e,f).

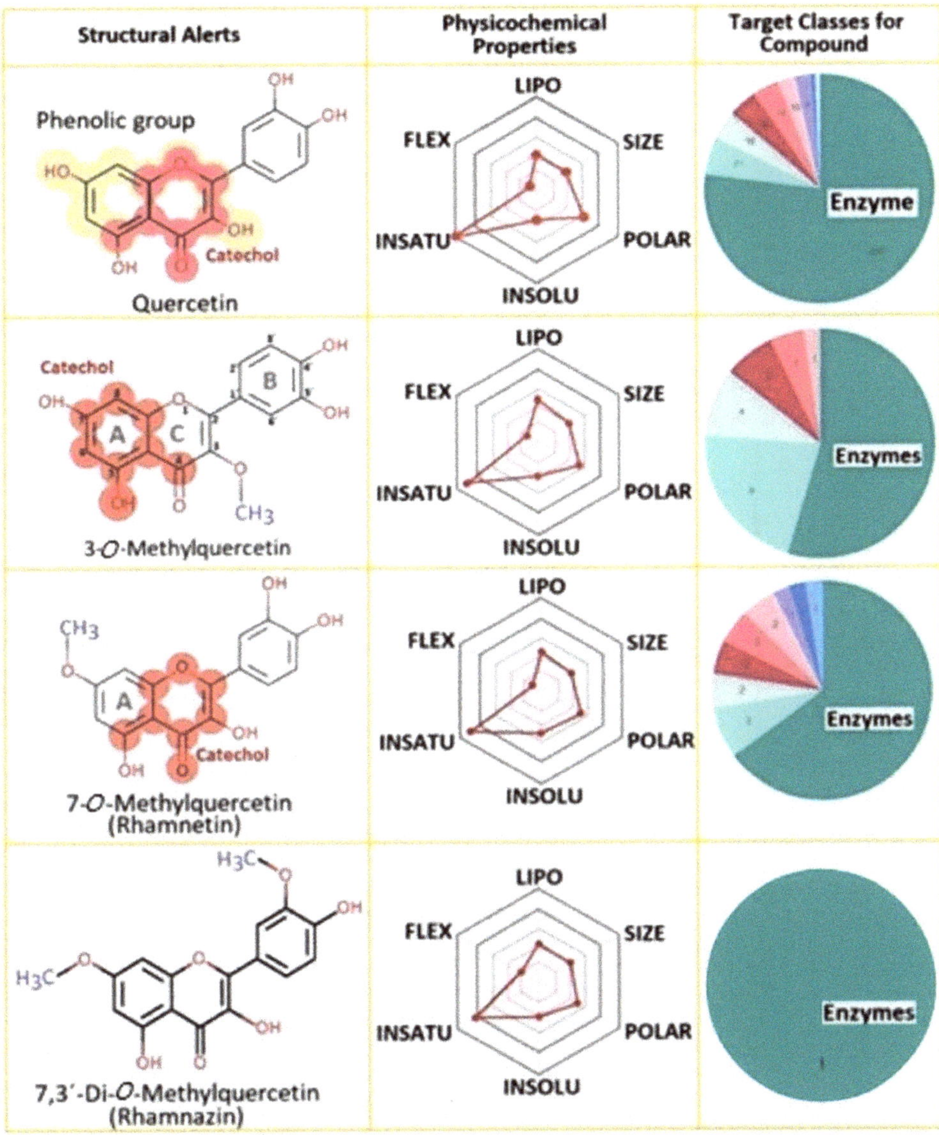

Figure 1. Structural alerts, physicochemical properties of possible targets of Q, Rhm, 3MQ and Rhz. Physicochemical properties in the second column are abbreviated and mean: Lipo: lipophilicity, Flex: Flexibility, Insatu: Insaturations, Insolu: insolubility, Polar: Polarity.

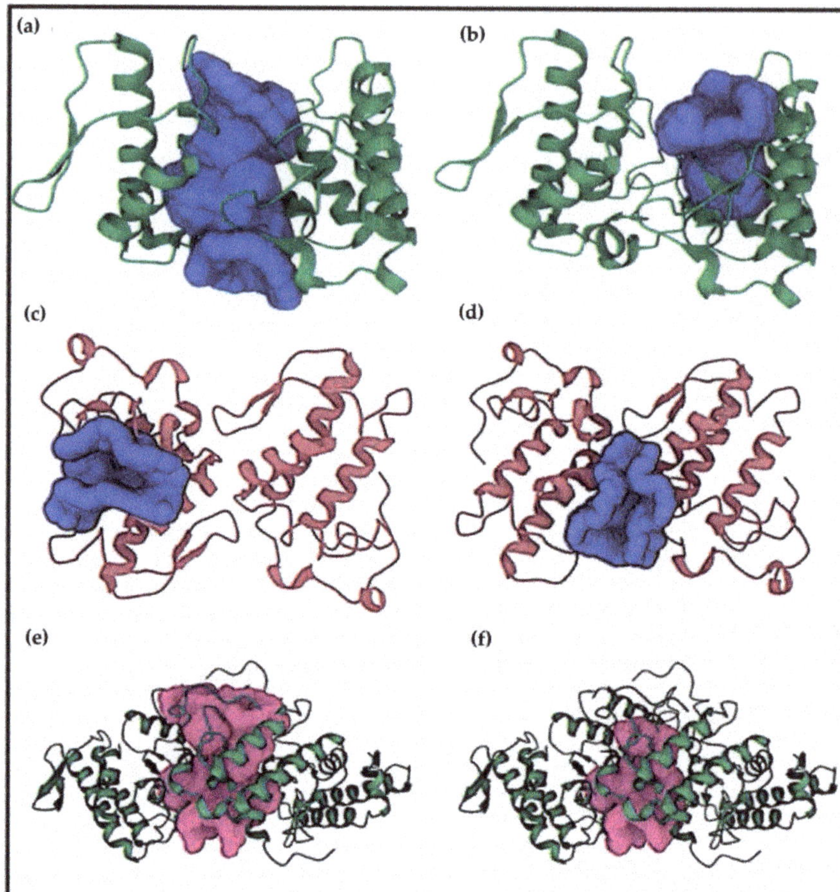

Figure 2. Two best cavities of each protein according to the highest druggability values. (**a**) First cavity observed in BthTX-I between chain A and B and the second region exhibited in the chain A hydrophobic channel (**b**). In (**c**), the highest value of druggability shows cavity in the chain B active site, and the second, in the interface between chain A and B (**d**). In (**e**), CdtsPLA2 reveal a huge area in the interface, and a small one, also in the interface (**f**).

2.2.2. Docking Analysis with a Transitional Analogous

Figure 3a shows that the molecule establishes hydrogen bonds with Gly30, Lys49, and Tyr22 of chain B of BthTX-I. Hydrophobic contacts with Leu2, Leu5, and Gly6 of the same chain are observed, besides essential contacts with Lys20 and His120 of chain A. BthTX-II shows hydrogen bonds with Gly32–33 and Thr23, despite presenting hydrophobic contacts with Asp49, Tyr28 and Cys29 (Figure 3b). Glycine is also essential in CdtsPLA2 in its interaction with the molecule, once hydrogen bonds were observed with Gly32 with diverse interactions with calcium ion, besides the hydrogen bond with His48 (Figure 3c). Hydrophobic contacts with Cys29, Gly30, Trp31 and Lys69 are also observed. Hence, these are the main residues that must be inhibited to prevent phospholipid hydrolysis and its entrance.

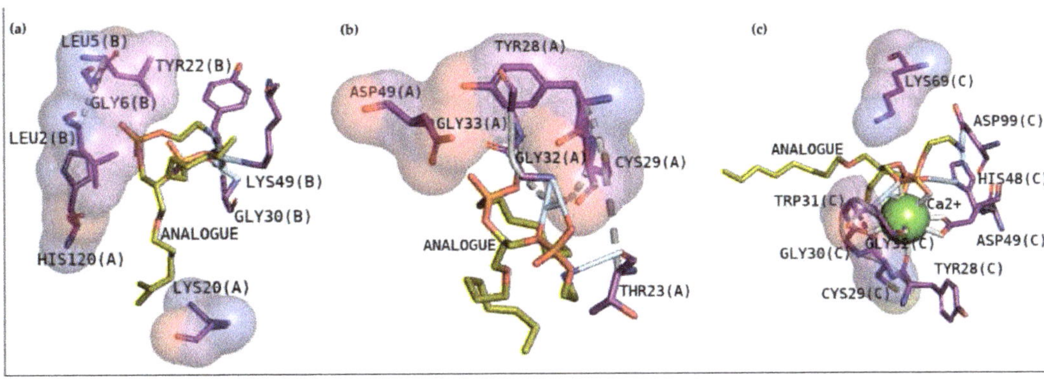

Figure 3. Interactions between BthTX-I (**a**), BthTX-II (**b**) and CdtsPLA2 (**c**) with a phosphonate transition-state analogue.

2.2.3. Classical Inhibitors BPB and Var

The compound BPB was used as a guide once it became a classic inhibitor of PLA2 [19]. It mainly shows contacts involved in the catalytic activity of these enzymes or in the channel responsible for phospholipid entrance in BthTX-I. Figure 4 exhibits that all toxins present great interactions with residues essential to the phospholipid fitting (BthTX-I, a and b) or in the active site (BthTX-II (c and d)) and CdtsPLA2 (e and f). Table 2 shows affinity values and rmsd of each docking analysis.

Table 2. Binding free energy (affinity) and rmsd values of each toxin with BPB.

	1° Cluster		2° Cluster	
	Affinity (kcal/mol)	**Rmsd (l.b.; u.b.)**	**Affinity (kcal/mol)**	**Rmsd (l.b.; u.b.)**
BthTX-I:BPB	−5.6	0; 0	−5.4	2.086; 4.796
BthTX-II:BPB	−5	0; 0	−5	0.102; 1.460
CdtsPLA2:BPB	−5.5	0; 0	−5.4	1.129; 2.875

Results with the inhibitor Var are exhibited in Figure 5. BthTX-I shows hydrogen bonds with Gly30, Asn28, Lys49, and His48 in the first cluster, with a binding free energy of −8.7 kcal/mol (Table 3). Similarly, hydrogen bonds were observed with the same residues, except Ans28 in the second cluster with −8.5 kcal/mol of affinity (Figure 5a,b). Results with BthTX-II revealed less affinity than the other two toxins, and both clusters show −7.1 and −6.7 kcal/mol, referring to clusters 1 and 2, respectively (Figure 5c,d). Residues involved in the anchoring observed in Figure 4c,d were Cys45, Asp49, and Tyr28, also revealing a lower number of hydrogen bonds when compared with BthTX-I and CdtsPLA2. Figure 4e,f present the same residues involved in the CdtsPLA2 interaction with Var, with hydrogen bonds with residues present in the calcium-binding loop, such as Tyr28, Gly30 and 32, besides with Asp49 and His48, residues involved in the protein-catalytic network (Figure 5e,f). In this case, binding free energy revealed −8.3 and −8.2 kcal/mol in clusters 1 and 2, respectively.

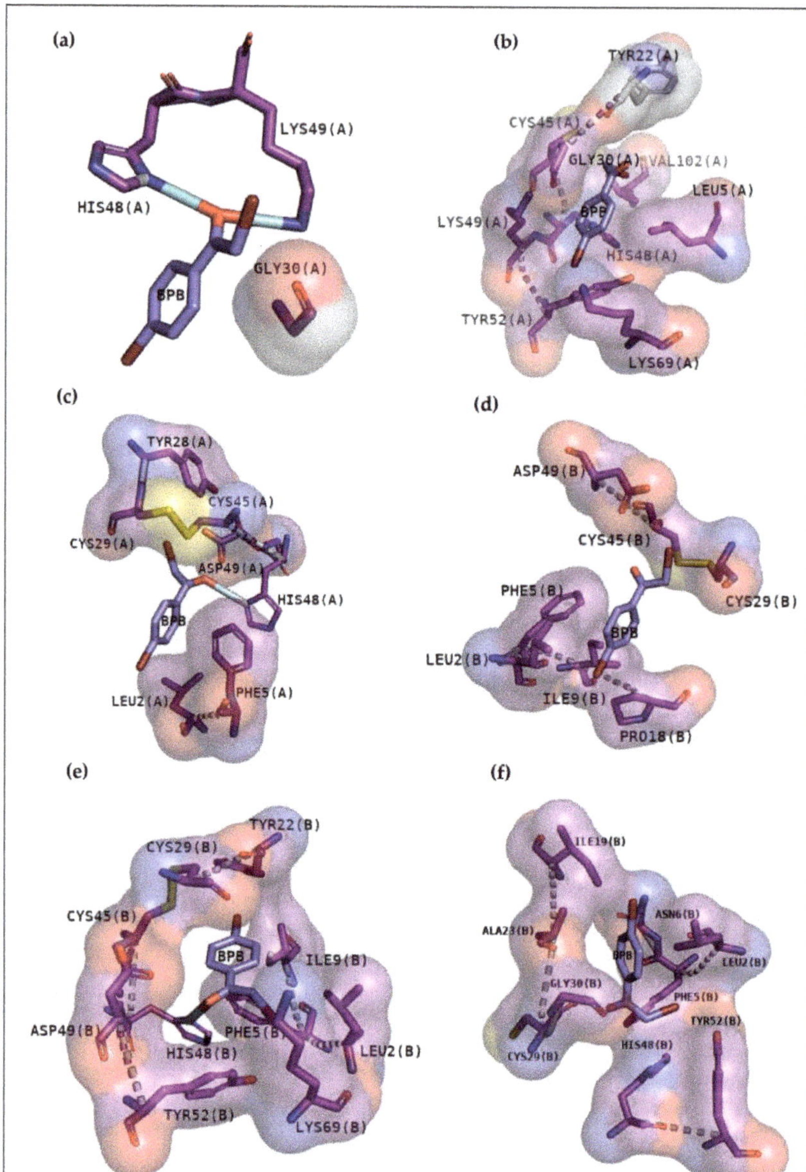

Figure 4. BPB docking results with BthTX-I, BthTX-II and CdtsPLA2. BthTX-I exhibits in the first cluster hydrogen bonds with His48 and Lys49, besides to exhibit hydrophobic contacts with Gly30 (**a**), while second cluster shows hydrophobic contacts with N-terminal residues, Gly30, His48, Lys49, Tyr52 and Lys69 (**b**). Hydrogen bond with His48 is also observed in BthTX-II in the first cluster, besides to show hydrophobic contacts with N-terminal region (**c**), and hydrophobic contacts in the second cluster (**d**). Similarly, BPB with CdtsPLA2 reveal hydrogen bond with His 48 and hydrophobic contacts with essential amino acids (**e**), and (**f**) exhibits just hydrophobic contacts. Purple and red colors around residues represent hydrophobic contacts with the compound.

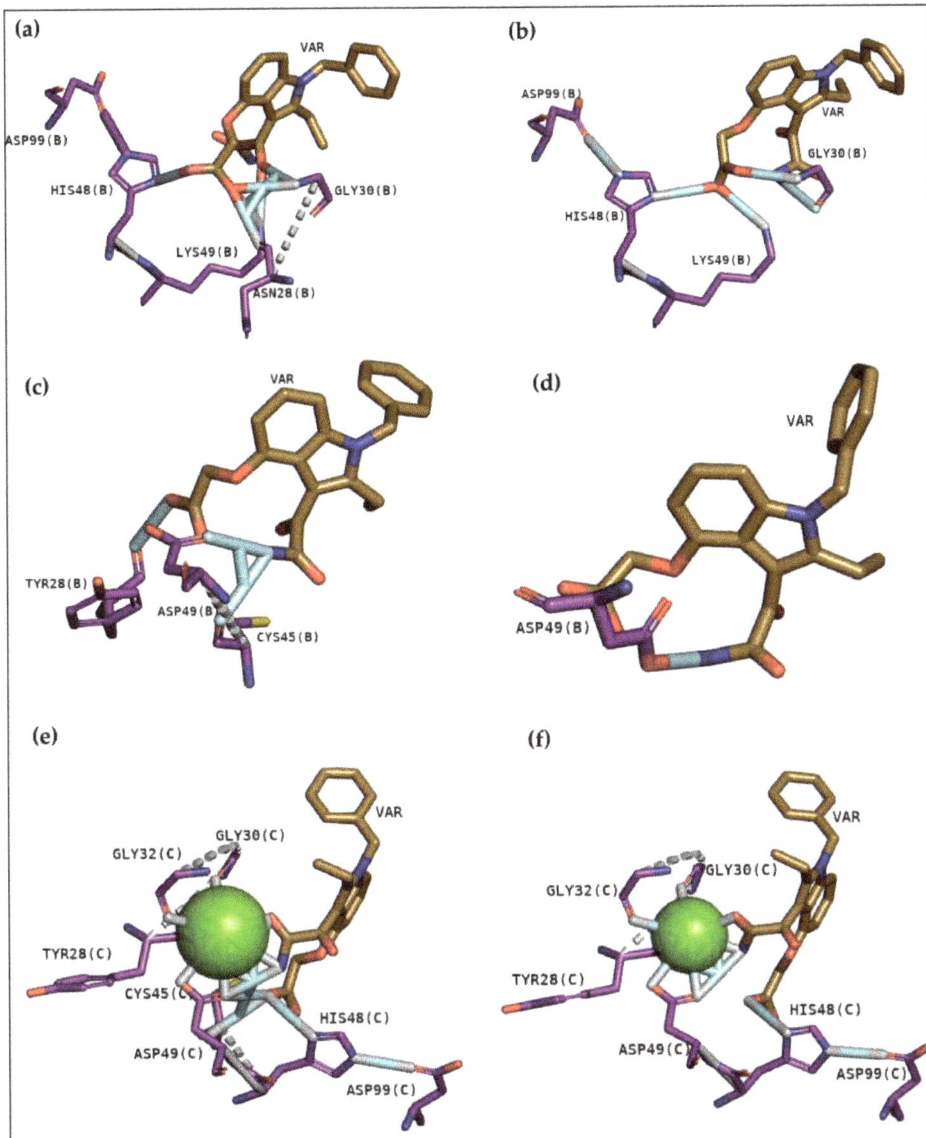

Figure 5. Var docking results with BthTX-I in the first cluster, (**a**) and second cluster (**b**) revealing hydrogen bonds with Gly30, His48 and Lys49. BthTX-II-Var in the first cluster (**c**), with hydrogen bonds with Asp49 and Tyr28, and just with Asp49 in the second cluster (**d**). CdtsPLA2 anchoring with Var, exhibiting strong interactions with the active site and calcium binding loop in the first (**e**) and second (**f**) cluster. Blue cylinders represent hydrogen bonds between residues and the compound.

Table 3. Binding free energy (affinity) and rmsd values of each toxin with Var.

	1° Cluster		2° Cluster	
	Affinity (kcal/mol)	Rmsd (l.b.; u.b.)	Affinity (kcal/mol)	Rmsd (l.b.; u.b.)
BthTX-I:Var	−8.7	0; 0	−8.5	1.871; 2.797
BthTX-II:Var	−7.1	0; 0	−6.7	2.444; 3.928
CdtsPLA2:Var	−8.3	0; 0	−8.2	1.942; 3.649

2.2.4. BthTX-I

Docking analysis with BthTX-I reveals that quercetin derivatives can interact with the hydrophobic channel responsible for anchoring with phospholipids. Hydrogen bonds with His48 and Val31, or Ala19 and Lys20, were observed in Figure 6a,b, which corresponds to clusters 1 and 2, respectively, of Q and BthTX-I. Rhm exhibits great interactions between the chains in the first cluster, with hydrogen bonds in Ala19 (A) and Lys20(A), besides the hydrophobic contacts with residues of both protein chains. Figure 6d also shows that Rhm can exhibit hydrogen bonds with residues essential to the phospholipid anchoring, such as His48, Gly30, and Lys69 (Figure 6c,d). Hydrogen bonds with His48, Lys49, Lys69, and His120 were found in the two clusters of 3MQ with the toxin, besides presenting hydrophobic contacts with essential residues from both chains. Similar to Rhm, Rhz exhibits great contact with both chains, revealing hydrogen bonds with Ala19, Lys20, His48, Lys49, and Lys69. Furthermore, hydrophobic contacts with essential amino acids, such as Gly30 and Val31, were observed. Table 4 shows the values of affinity and rmsd from each analysis.

Table 4. Binding free energy (affinity) and rmsd values of BthTX-I with each quercetin derivative.

	1° Cluster		2° Cluster	
	Affinity (kcal/mol)	Rmsd (l.b.; u.b.)	Affinity (kcal/mol)	Rmsd (l.b.; u.b.)
Q	−8.2	0; 0	−7.9	2.655; 3.657
Rhm	−8.2	0; 0	−8.1	1.625; 2.395
3MQ	−7.9	0; 0	−7.9	2.817; 4.116
Rhz	−8.2	0; 0	−8	2.465; 3.041

2.2.5. BthTX-II

Cluster 1 of BthTX-II:Q exhibits hydrogen bonds with His48, Cys29, and Cys45 and hydrophobic contacts with Asp49, Thr23, and Leu2 of chain A, with a binding energy of −7.4 kcal/mol (Figure 7a). The second cluster shows hydrogen bonds with Cys29, Cys45, and His48, besides hydrophobic contact with Leu2 (Figure 7b). Rhm was the unique compound which exhibits great interactions in the interface between monomer A and B, showing hydrogen bonds with Arg111(B), Leu110(B), Arg107(B), Gly80(A), Arg77(A), Arg77(B), and Glu12(B) with a binding energy of −7.7 kcal/mol (Figure 7c). The second cluster reveals similar results, with contacts between chains A and B (Figure 7d). Comparable with Q, 3MQ exhibits hydrogen bonds with Cys29, Cys45, and His48. However, the energy affinity was lower, −6.9 kcal/mol, and there were no hydrophobic contacts. Otherwise, the second cluster is between the chains with lower affinity (Figure 7e,f). On the other hand, Rhz shows a hydrogen bond with Asp49 and Lys69, despite presenting hydrophobic interactions with Leu2, Phe5, Ile9, Pro18, Tyr22, Thr23, Cys29, Cys45, and Tyr52, with an energy of −7.3 kcal/mol. Hydrogen bonds with Asp49 and Lys69 were observed in the second cluster; additionally, hydrophobic contacts with Phe5 and Tyr52 were exhibited. Table 5 shows the affinity values and their respective rmsd.

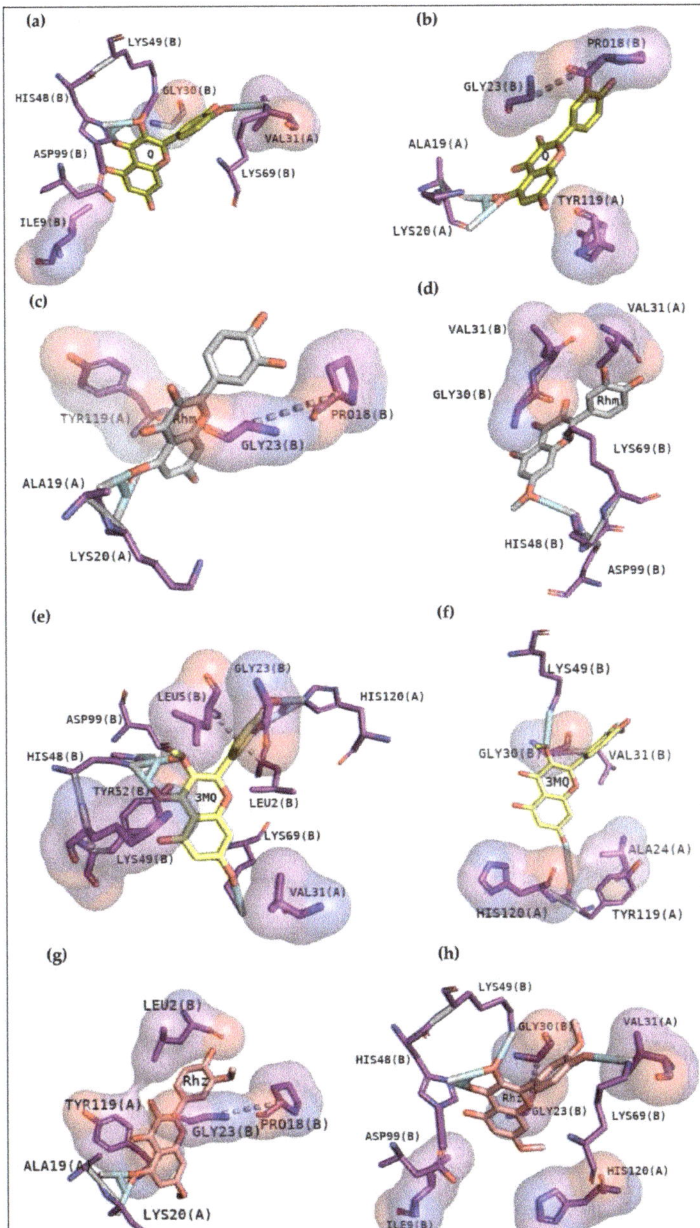

Figure 6. Docking results of BthTX-I with Q in the first cluster (**a**) and the second cluster (**b**), revealing great interactions with His48, Lys49 and Lys69. BthTX-I-Rhm anchoring, emphasizing Lys20, Tyr119 in the first cluster (**c**) and His48, Lys49 and Gly30 in the second (**d**). 3MQ exhibit in the first image (**e**) to anchor strongly with His48 plus Lys69, and in the second with Gly30, Lys49 and the C-terminal region (**f**). Lys20 and Tyr 119 also appears anchoring with Rhz (**g**) and Lys49 and His48 are the strong interactions in the second cluster (**h**). Hydrogen bonds are represented by blue cylinders, and hydrophobic contacts by purple and red colors around the residues.

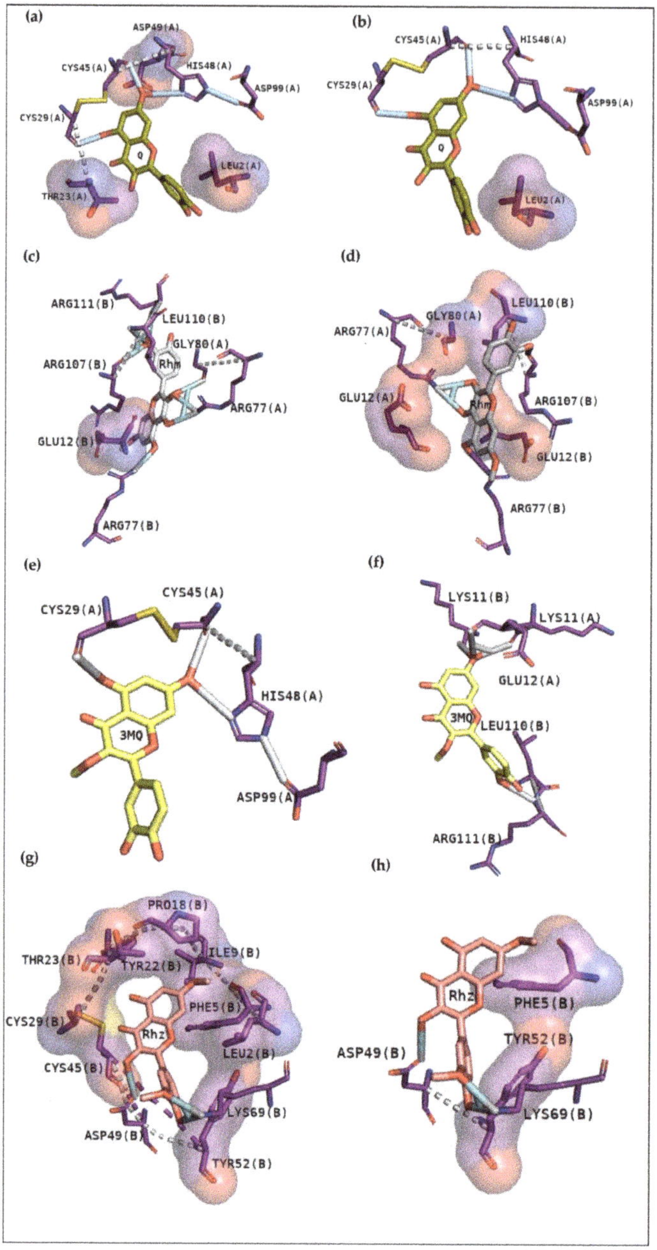

Figure 7. Docking results of BthTX-II with quercetin derivatives. Q reveals to anchor with the active site in the first (**a**) and also in the second cluster (**b**), while Rhm fits in the interface between the chains, emphasizing Gly80, Leu110 and Arg 111 in the first (**c**) and second clusters (**d**). First, 3MQ reveals hydrogen bond with His48 and Cys45 (**e**), and in the second cluster, it was observed contacts with N-terminal and C-terminal region with both chains (**f**). Rhz first shows to anchor with active site and the N-terminal region with mostly hydrophobic contacts (**g**), and in the last cluster, similar residues are shown (**h**).

Table 5. Binding free energy (affinity) and rmsd values of BthTX-II with each quercetin derivative.

	1° Cluster		2° Cluster	
	Affinity (kcal/mol)	Rmsd (l.b.; u.b.)	Affinity (kcal/mol)	Rmsd (l.b.; u.b.)
Q	−7.4	0; 0	−7.3	1.688; 2.723
Rhm	−7.7	0; 0	−7.5	0.629; 1.395
3MQ	−6.9	0; 0	−6.7	1.908; 2.325
Rhz	−7.3	0; 0	−6.9	1.826; 6.889

2.2.6. CdtsPLA2 Molecular Docking

Q mostly interacts with the interface of the chains B and D or B, C and D and does not exhibit contact with the active site in the first cluster, which exhibited an affinity of −9.7 kcal/mol, with hydrogen bonds with residues from the N- terminal region and Gly26, Cys29, and Tyr120. Hydrophobic contacts between chains B and D were also observed (Figure 8a). The second cluster presents less affinity (−8.0 kcal/mol) (Table 6) and shows a hydrophobic contact with an amino acid which is fundamental to the catalytic activity: Gly30. In addition to this, it shows hydrogen bonds with Lys69, Trp31, Phe24, Gly26, and Tyr120 (Figure 8b). Rhm showed the first cluster at the interface with chains A and C, exhibiting hydrogen bonds with Cys29, Tyr25, Gly26, Tyr120, and Asp122. Hydrophobic contacts were observed with the N-terminal region and Gly30 (Figure 8c), with an affinity of −9.4kcal/mol. Figure 8d shows hydrogen bonds with Lys69, Trp31, Phe24, and Gly26, and hydrophobic contacts were observed with Gly30, Tyr120, and Cys27.

Table 6. Binding free energy and rmsd values of Cdt sPLA2 molecular docking with quercetin derivatives.

	1° Cluster		2° Cluster	
	Affinity (kcal/mol)	Rmsd (l.b.; u.b.)	Affinity (kcal/mol)	Rmsd (l.b.; u.b.)
Q	−9.7	0; 0	−8	1.878; 2.823
Rhm	−9.4	0; 0	−8.2	0.530; 1.426
3MQ	−9.4	0; 0	−8.5	1.908; 2.325
Rhz	−8.9	0; 0	−8.8	1.807; 4.896

3MQ reveals an energy affinity of −9.4 kcal/mol with hydrogen bonds with Asn6, Ala18, Gly26 and Cys29. Diverse hydrophobic contact between chains B and D were observed around the molecule (Figure 8e). The second cluster showed great interactions with CdtsPLA2 active site, with an energy affinity of −8.5 kcal/mol (Table 6). Figure 8f also highlights the hydrogen bonds with Gly32, an essential amino acid in the calcium bind loop. Furthermore, all residues are connected with calcium ions, which is essential to catalytic activity. In addition, hydrophobic contacts with N-terminal amino acids and Cys-29 were observed. The first cluster of Rhz shows interactions between monomers B, C, and D, with a bind free energy of −8.9 kcal/mol. Similarly, with −8.8 kcal/mol, the second cluster exhibits interactions between monomers B, C, and D. Rhz does not show contact with residues directly involved in the catalytic activity of CdtPLA2.

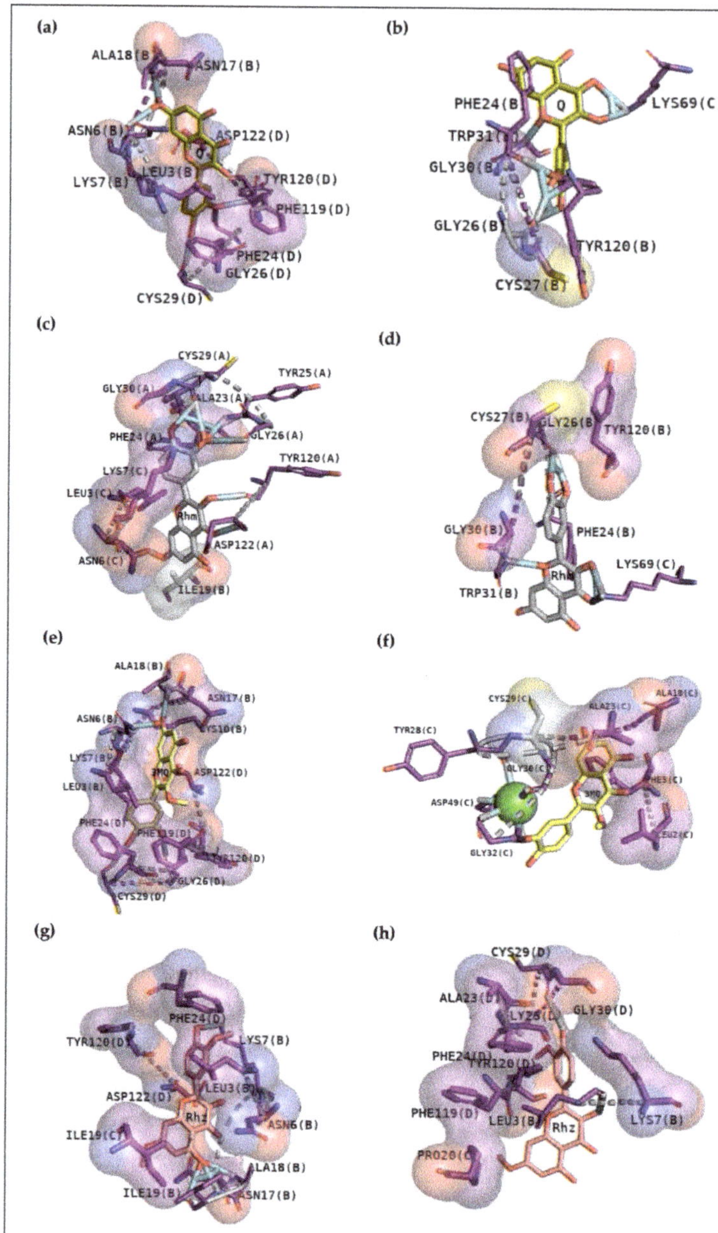

Figure 8. Docking results of CdtsPLA2 with Quercetin derivatives. (**a**) First cluster of Q revealing interactions between chain B and D, and the second, between B and C (**b**). Chain A, B and C were involved in the first Rhm anchoring with the protein (**c**), however, just B and C were observed in the second (**d**). Differently, with a high number of hydrophobic contacts, the first cluster of 3MQ anchors between chains B and D (**e**), and the second one fits in the protein active site (**f**). Similar residues are involved in the Rhz-CdtsPLA2 anchoring, with hydrophobic contacts in the first (**g**) and second cluster (**h**).

3. Discussion

To date, diverse in silico tools help to evaluate molecular structures, mainly to select compounds which exhibit great possibility to become an effective drug [13]. These tools also have an essential role in identifying the interaction of target compounds [20]. The TPSA is an important descriptor to understand the specific regions of the protein-compound's interactions, and Rhz shows the smallest value. Rhm and 3MQ show a similar value, and Q exhibits the highest polar area, which exceeds the range (20–130 Å^2) [13]. Therefore, lipophilicity could be another descriptor to support the results obtained in the molecular docking analysis. Diverse natural compounds, such as flavonoids, exhibit an inhibitory potential against sPLA2 activity that seems to be dependent on the 5-hydroxyl group, besides the double bond and the double-bonded oxygen in the oxane ring and the hydroxyl groups at the 3′ and 4′ position [21]. Q is widely found as a secondary metabolite in fruits, vegetables, and flowers. Its structure consists of three rings, consisting of a basic nucleus of a phenyl benzo (γ) pyrone and the side groups usually are hydroxyl, glycosyl, or methoxyl [22].

In recent studies, the search to provide an alternative or complementary treatment to antivenom therapy using synthetic and natural compounds has been the subject of investigation. Var is a synthetic compound known for its inhibitory potential against human-secreted groups IIA PLA2. Due to their high structural homology with PLA2 from snake venoms, studies were made to verify its inhibitory potential [10]. Furthermore, this compound has already been shown to inhibit myotoxins revealing great contacts with Gly30, Lys49 and His48 of MjTX-I and MjTX-II from *Bothrops moojeni* [15]. Additionally, Var also shows to anchor with other myotoxins, such as PrTX-I from *Bothrops pirajai* and BthTX-I from *Bothrops jararacussu* [10]. Herein, results with inhibitor Var reveal that the compound shows higher affinity by BthTX-I from *B. jararacussu*, with great contacts with His48, Lys49, and Gly30, residues also observed in the analogue interactions and in other phospholipases from some *Bothrops* sp venoms [15]. Although BPB interacted with essential amino acids to anchor with the analogue (Figure 3), it exhibits lower values of affinity when compared with Var. This inhibitor is known to decrease the enzymatic activity of phospholipase A2, binding covalently to His48 of PrTX-I [13,19]. This interaction leads to a distortion of the Ca^{2+}-binding loop plus a C-terminus rearrangement, decreasing the myotoxic activity of this protein. In this study, BPB also exhibits interactions with His48, Lys49, Leu5, Gly30, and Lys69 of BthTX-I and with the main residues of BthTX-II and CdtsPLA2 active site. In addition, earlier studies reveal that BPB also fits well in the hydrophobic channel with extensive hydrophobic interactions with the surrounding residues, especially Phe5, Cys45, and Gly30 from bovine pancreatic PLA_2, besides reducing edema induced by PrTX-I in rat and rabbit [22,23]. Moreover, it has already demonstrated the His48 chemical modification of the acid phospholipase A2 from *B. jararacussu* using BPB [24].

Cavity analysis has been used to identify potential binding sites on the protein surface besides ranking them based on ligandability and druggability scores [25]. In this study, this web server helped to confirm and guide the best sites of the protein to focus on molecular docking. To better understand the compounds' potential to interact with these toxins, it is essential to emphasize the residues involved in the anchoring between these three toxins with the phosphonate transition-state analogue. In Figure 2a, BthTX-I revealed that Lys20, Gly30, Lys49, Tyr22, His120, and some N-terminal residues are involved in anchoring the compound-toxin. BthTX-II exhibited contacts with Tyr28, Gly32,33, and Asp49, and CdtsPLA2 shows great interactions with Gly30, 32, His48, and Asp49. Similarly, Q shows great contact with some of these residues of BthTX-I, such as Lys20, Lys49, Gly30, and His48. Quercetin has already been shown to bind with the dimmer interface and active site of MTX-II, a Lys-49 PLA2 from the *Bothrops brazili* venom [26]. Furthermore, in this study, Q fits well in the active site of BthTX-II, and these data are supported by a previous study in which the compound inhibited the protein enzymatic activity [5]. Q with CdtsPLA2 exhibited similar interactions with BPB, such as Lys69, Gly30, and some residues of the N-terminal region. In addition, Q anchors in the interface between monomers B and D in both analyses, with a great value of affinity, higher than that observed with BthTX-II. This toxin in the solution presents dimeric or tetrameric oligomers [27]. Hence, it may be necessary to use a higher concentration of inhibitor compared to that was used in BthTX-II, once the protein is primarily in a monomeric form in its relaxed state with a fatty acid in its hydrophobic channel [28].

Rhm reveals a different region with high affinity with BthTX-I, the interface between chain A and B, with hydrogen bonds with Lys20 and Gly30, residues that match the phospholipid analogue interaction. Hydrophobic contacts can also be compared since the Tyr119 residue next to His120 presented in Figure 2b and Gly30 are included in Rhm with BthTX-II. Rhm has already revealed an antimyotoxic activity against BthTX-II in previous work [5] and could be a potential inhibitor of BthTX-I. The dimeric BthTX-II (tense-state) is necessary for myotoxic activity [8], hence, compounds that change this conformation could decrease its activity. This compound also shows contacts with Gly30 and other important residues, such as Trp31, Tyr120, and the N-terminal region, similar to the amino acids involved in the analogue and myotoxin, BthTX-I. There is no data concerning the inhibitory potential of this compound with CdtsPLA2; however, it has already been shown to decrease the inflammatory cytokines levels and oxidative stress in the mice aortic tissue, besides to inhibit enzymatic, edematogenic, and myotoxic effect of BthTX-II [5,29].

As observed in Rhm, Rhz shows similar interactions with CdtsPLA2. However, it is possible to notice more hydrophobic contact, and contacts with residues are not found in the active site. In addition, Rhz reveals an anchoring in the active site in BthTX-II; in fact, it has already been shown to inhibit the enzymatic activity of this protein [5]. Rhz and Rhm anchor between chains A and B of the BthTX-I and exhibits some important residues in common that were involved in this contact, such as His120, Lys69, His48, Val31, Gly30, and Lys20. Figure 3e,f exhibits that 3MQ also fits in the phospholipid channel, highlighting Lys49, N-terminal residues and His120 of BthTX-I. BthTX-II, besides its myotoxic activity, exhibits low enzymatic activity, and Figure 2b, exhibits the phospholipid analogue in the catalytic site and 3MQ shows to bind in this region. However, it reveals a lower affinity value and fewer interactions. Therefore, this compound has already been shown to poorly inhibit BthTX-II catalytic activity [5]. In a different manner, 3MQ exhibits great interactions in the active site of CdtsPLA2, with the same residues observed in the anchoring between the protein-analogue and protein-commercial inhibitors.

Considering that all these three toxins are from two different species of snake from the Viperidae family, and PLA2 is one of the most abundant components of both Cdt and Bj venoms, it is essential to investigate how distinct compounds can interact with different macromolecules [30,31]. These data indicate that Q, Rhm, 3MQ, and Rhz can anchor in different manners with each toxin. Studies concerning the structural interactions with in vitro analysis, besides the pharmacological assays, could help to better understand diverse mechanisms of inhibition of these compounds. Therefore, our results show that a great candidate to inhibit BthTX-I must interact with residues key-residues cited. BthTX-I docking analysis shows quercetin derivatives could potentially diminish the myotoxic activity by anchoring in the phospholipid hydrophobic channel or interacting with residues next to C-terminal region, which is an essential area to execute the biological effects of this toxin [10]. Nevertheless, it is necessary to perform in vitro and in vivo assays with BthTX-I and CdtsPLA2 to correlate and confirm all the results.

4. Materials and Methods

4.1. Evaluation of Compounds' Properties and Preparation

All compounds were analyzed using Molinspiration, SwissAdme and ChEMBL to better understand their characteristics and how they can influence their activities. Physical–chemical properties, the structure and the main biological targets were obtained in ChEMBL. TPSA, lipophilicity, and other features were found in SwissADME and Molinspiration. All molecules were prepared before the molecular anchoring analysis, adding polar hydrogen atoms, and aggregating the Kollman charges and converted to PDBQT.

4.2. Proteins Preparation

The PDB (Protein Data Bank—https://www.rcsb.org, accessed on 1 December 2022) was used in this analysis to find the 3D structure of all proteins. Information on the 3D structure of each compound was taken from the PubChem platform (https://puchem.ncbi.nlm.nih.gov, accessed on 1 December 2022). The crystallographic models chosen as the best model for the construction of the theoretical structural models to BthTX-I and BthTX-II from *Bothrops jararacussu* were 3hzd and 2oqd, respectively, and PLA2 from Cdt was 2qog and. First, a general analysis using Swissdock was made between each protein with all compounds to use as a guide for the best regions to interact. Chimera 1.14 program (Ucsf Chimera, 2004) was used to assemble the structural molecular model of the protein and to evaluate the general possibilities of the proteins binding with compounds. After these first steps, proteins were prepared using Autodock Tools, removing the water molecules, besides additional polar hydrogens and aggregating the Kollman charges. Then, the files were converted into PDBQT to perform the calculations of the energy maps (Grid Box) using Autodock Vina [32]. The size was chosen to enclose all amino acids from the catalytic sites or the C-terminal region. The results were obtained using the tools LIGPLOT+ and PyMOL v 2.4 to evaluate the binding energies and orientations of molecules in the microenvironment of the active site of PLA2.

5. Conclusions

The results with analogue and the inhibitors corroborate with the literature and indicate the key residues of each toxin to reduce their activity. Q, Rhm, 3MQ, and Rhz showed the highest values of affinity with CdtsPLA2; however, the analysis indicates that 3MQ could better inhibit the enzymatic activity. In BthTX-II, Q anchors in the active site and Rhm in the interface—in vitro and in vivo assays support these results. Rhz also shows great interaction with the active site, and 3MQ revealed less affinity with BthTX-II. Similar results were observed with the compounds and BthTX-I, including affinity values. These docking results emphasize the essential role of Gly30, His48, and Asp49 of BthTX-II and CdtsPLA2 in the interaction with the phospholipids and their hydrolysis. In the case of BthTX-I, the C-terminal region, plus Gly30, His48, Lys49, and His120, are essential to the interaction with the membrane phospholipids and their perturbation and the consequent inflammation induction.

Author Contributions: M.N.B., C.R.d.C.C., A.R., L.L.F.M., R.S., I.A. and M.H.T., were responsible for the conceptualization and methodology. Writing, data development, and analysis created by M.N.B. and M.H.T. M.A.d.O., S.F.S. and M.H.T. were responsible for the review and supervision of the development of this work. All authors have read and agreed to the published version of the manuscript.

Funding: This research was funded by FAPESP 2017/20291-0 and CNPQ productivity funding 2 process number: 309271/2022-3.

Institutional Review Board Statement: Not applicable.

Informed Consent Statement: Not applicable.

Data Availability Statement: Data are contained within the article.

Acknowledgments: We would like to thank FAPESP and CNPQ for the funding.

Conflicts of Interest: Authors declare no conflict of interest.

References

1. Liu, Y.; Li, Y.; Xu, L.; Shi, J.; Yu, X.; Wang, X.; Li, X.; Jiang, H.; Yang, T.; Yin, X.; et al. Quercetin Attenuates Podocyte Apoptosis of Diabetic Nephropathy Through Targeting EGFR Signaling. *Front. Pharmacol.* **2022**, *12*, 792777. [CrossRef] [PubMed]
2. Wen, L.; Jiang, Y.; Yang, J.; Zhao, Y.; Tian, M.; Yang, B. Structure, bioactivity, and synthesis of methylated flavonoids. *Ann. N. Y. Acad. Sci.* **2017**, *1398*, 120–129. [CrossRef]

3. Magar, R.T.; Sohng, J.K. A Review on Structure, Modifications and Structure-Activity Relation of Quercetin and Its Derivatives. *J. Microbiol. Biotechnol.* **2020**, *30*, 11–20. [CrossRef] [PubMed]
4. Karancsi, Z.; Kovács, D.; Pézsa, N.P.; Gálfi, P.; Jerzsele, Á.; Farkas, O. The Impact of Quercetin and Its Methylated Derivatives 3-o-Methylquercetin and Rhamnazin in Lipopolysaccharide-Induced Inflammation in Porcine Intestinal Cells. *Antioxidants* **2022**, *11*, 1265. [CrossRef] [PubMed]
5. Belchor, M.N.; Gaeta, H.H.; Rodrigues, C.F.B.; Costa, C.R.D.C.; Toyama, D.D.O.; Passero, L.F.D.; Laurenti, M.D.; Toyama, M.H. Evaluation of Rhamnetin as an Inhibitor of the Pharmacological Effect of Secretory Phospholipase A_2. *Molecules* **2017**, *22*, 1441. [CrossRef]
6. Devi, A.; Namsa, N.D.; Doley, R. In silico and in vitro neutralization of PLA2 activity of Daboxin P by butein, mimosine and bakuchiol. *Int. J. Biol. Macromol.* **2020**, *165*, 1066–1078. [CrossRef]
7. Batsika, C.S.; Gerogiannopoulou, A.-D.D.; Mantzourani, C.; Vasilakaki, S.; Kokotos, G. The design and discovery of phospholipase A_2 inhibitors for the treatment of inflammatory diseases. *Expert Opin. Drug Discov.* **2021**, *16*, 1287–1305. [CrossRef]
8. Fernandes, C.A.H.; Borges, R.J.; Lomonte, B.; Fontes, M.R.M. A structure-based proposal for a comprehensive myotoxic mechanism of phospholipase A_2-like proteins from viperid snake venoms. *Biochim. Biophys. Acta* **2014**, *1844*, 2265–2276. [CrossRef]
9. Azevedo, F.V.P.d.V.; Lopes, D.S.; Zóia, M.A.P.; Correia, L.I.V.; Saito, N.; Fonseca, B.B.; Polloni, L.; Teixeira, S.C.; Goulart, L.R.; Ávila, V.d.M.R. A New Approach to Inhibiting Triple-Negative Breast Cancer: In Vitro, Ex Vivo and In Vivo Antiangiogenic Effect of BthTx-II, a PLA2-Asp-49 from Bothrops jararacussu Venom. *Biomolecules* **2022**, *12*, 258.
10. Salvador, G.H.M.; Pinto, Ê.K.R.; Ortolani, P.L.; Fortes-Dias, C.L.; Cavalcante, W.L.G.; Soares, A.M.; Lomonte, B.; Lewin, M.R.; Fontes, M.R.M. Structural basis of the myotoxic inhibition of the *Bothrops pirajai* PrTX-I by the synthetic varespladib. *Biochimie* **2023**, *207*, 1–10. [CrossRef]
11. Faure, G.; Xu, H.; Saul, F.A. Crystal Structure of Crotoxin Reveals Key Residues Involved in the Stability and Toxicity of This Potent Heterodimeric β-Neurotoxin. *J. Mol. Biol.* **2011**, *412*, 176–191. [CrossRef]
12. Nemecz, D.; Ostrowski, M.; Ravatin, M.; Saul, F.; Faure, G. Crystal Structure of Isoform CBd of the Basic Phospholipase A_2 Subunit of Crotoxin: Description of the Structural Framework of CB for Interaction with Protein Targets. *Molecules* **2020**, *25*, 5290. [CrossRef]
13. Marchi-Salvador, D.P.; Fernandes, C.A.; Silveira, L.B.; Soares, A.M.; Fontes, M.R. Crystal structure of a phospholipase A2 homolog complexed with p-bromophenacyl bromide reveals important structural changes associated with the inhibition of myotoxic activity. *Biochim. Biophys. Acta* **2009**, *1794*, 1583–1590. [CrossRef] [PubMed]
14. Youngman, N.; Walker, A.; Naude, A.; Coster, K.; Sundman, E.; Fry, B.G. Varespladib (LY315920) neutralises phospholipase A2 mediated prothrombinase-inhibition induced by Bitis snake venoms. *Comp. Biochem. Physiol. Part C Toxicol. Pharmacol.* **2020**, *236*, 108818. [CrossRef]
15. Salvador, G.H.M.; Cardoso, F.F.; Gomes, A.A.; Cavalcante, W.L.G.; Gallacci, M.; Fontes, M.R.M. Search for efficient inhibitors of myotoxic activity induced by ophidian phospholipase A_2-like proteins using functional, structural and bioinformatics approaches. *Sci. Rep.* **2019**, *9*, 510. [CrossRef]
16. Souza, J.D.; Oliveira, I.C.F.; Yoshida, E.H.; Cantuaria, N.M.; Cogo, J.C.; Torres-Bonilla, K.A.; Hyslop, S.; Junior, N.J.S.; Floriano, R.S.; Gutiérrez, J.M.; et al. Effect of the phospholipase A_2 inhibitor Varespladib, and its synergism with crotalic antivenom, on the neuromuscular blockade induced by *Crotalus durissus* terrificus venom (with and without crotamine) in mouse neuromuscular preparations. *Toxicon* **2022**, *214*, 54–61. [CrossRef] [PubMed]
17. White, S.P.; Scott, D.L.; Otwinowski, Z.; Gelb, M.H.; Sigler, P.B. Crystal Structure of Cobra-Venom Phospholipase A_2 in a Complex with a Transition-State Analogue. *Science* **1990**, *250*, 1560–1563. [CrossRef]
18. Daina, A.; Michielin, O.; Zoete, V. SwissADME: A free web tool to evaluate pharmacokinetics, drug-likeness and medicinal chemistry friendliness of small molecules. *Sci. Rep.* **2017**, *7*, 42717. [CrossRef]
19. Sudharshan, S.; Dhananjaya, B.L. Antibacterial potential of a basic phospholipase A2 (VRV-PL-VIIIa) from *Daboia russelii* pulchella (Russell's viper) venom. *J. Venom. Anim. Toxins Incl. Trop. Dis.* **2015**, *21*, 17. [CrossRef]
20. Rifaioglu, A.S.; Atas, H.; Martin, M.J.; Cetin-Atalay, R.; Atalay, V.; Doğan, T. Recent applications of deep learning and machine intelligence on in silico drug discovery: Methods, tools and databases. *Brief. Bioinform.* **2019**, *20*, 1878–1912. [CrossRef] [PubMed]
21. Cotrim, C.A.; de Oliveira, S.C.B.; Filho, E.B.S.D.; Fonseca, F.V.; Baldissera, L., Jr.; Antunes, E.; Ximenes, R.M.; Monteiro, H.S.A.; Rabello, M.M.; Hernandes, M.Z.; et al. Quercetin as an inhibitor of snake venom secretory phospholipase A_2. Chemico-Biological Interactions. *Chem. Biol. Interact.* **2011**, *189*, 9–16. [CrossRef]
22. Zhao, H.; Tang, L.; Wang, X.; Zhou, Y.; Lin, Z. Structure of a snake venom phospholipase A_2 modified by p-bromo-phenacyl-bromide. *Toxicon* **1998**, *36*, 875–886. [CrossRef] [PubMed]
23. Landucci, E.C.; de Castro, R.C.; Toyama, M.; Giglio, J.R.; Marangoni, S.; De Nucci, G.; Antunes, E. Inflammatory oedema induced by the Lys-49 phospholipase A_2 homologue piratoxin-i in the rat and rabbit: Effect of polyanions and p-bromophenacyl bromide. *Biochem. Pharmacol.* **2000**, *59*, 1289–1294. [CrossRef]
24. Takeda, A.A.S.; Santos, J.I.D.; Marcussi, S.; Silveira, L.B.; Soares, A.M.; Fontes, M.R.M. Crystallization and preliminary X-ray diffraction analysis of an acidic phospholipase A_2 complexed with p-bromophenacyl bromide and alpha-tocopherol inhibitors at 1.9- and 1.45-A resolution. *Biochim. Biophys Acta* **2004**, *1699*, 281–284. [CrossRef]

25. Xu, Y.; Wang, S.; Hu, Q.; Gao, S.; Ma, X.; Zhang, W.; Shen, Y.; Chen, F.; Lai, L.; Pei, J. CavityPlus: A web server for protein cavity detection with pharmacophore modelling, allosteric site identification and covalent ligand binding ability prediction. *Nucleic Acids Res.* **2018**, *46*, W374–W379. [CrossRef]
26. Kumar, R.; Vijayalakshmi, S.; Nadanasabapathi, S. Health Benefits of Quercetin. *Def. Life Sci. J.* **2017**, *2*, 142–151. [CrossRef]
27. Wang, M.; Wu, Y.; Li, W. Rhamnetin ameliorates macrophage-mediated inflammation and pro-atherosclerosis pathways in apolipoprotein E-deficient mice. *J. Physiol. Pharmacol.* **2021**, *72*, 249–258.
28. Marchi-Salvador, D.P.; Corrêa, L.C.; Magro, A.J.; Oliveira, C.Z.; Soares, A.M.; Fontes, M.R.M. Insights into the role of oligomeric state on the biological activities of crotoxin: Crystal structure of a tetrameric phospholipase A_2 formed by two isoforms of crotoxin B from *Crotalus durissus* terrificus venom. *Proteins* **2008**, *72*, 883–891. [CrossRef]
29. Borges, R.J.; Salvador, G.H.M.; Campanelli, H.B.; Pimenta, D.C.; de Oliveira Neto, M.; Usón, I.; Fontes, M.R.M. BthTX-II from Bothrops jararacussu venom has variants with different oligomeric assemblies: An example of snake venom phospholipases A_2 versatility. *Int. J. Biol. Macromol.* **2021**, *191*, 255–266. [CrossRef]
30. Cedro, R.C.A.; Menaldo, D.L.; Costa, T.R.; Zoccal, K.F.; Sartim, M.A.; Santos-Filho, N.A.; Faccioli, L.H.; Sampaio, S.V. Cytotoxic and inflammatory potential of a phospholipase A_2 from *Bothrops jararaca* snake venom. *J. Venom. Anim. Toxins Incl. Trop. Dis.* **2018**, *24*, 33. [CrossRef] [PubMed]
31. Muller, S.P.; Silva, V.A.O.; Silvestrini, A.V.P.; de Macedo, L.H.; Caetano, G.F.; Reis, R.M.; Mazzi, M.V. Crotoxin from Crotalus durissus terrificus venom: In vitro cytotoxic activity of a heterodimeric phospholipase A_2 on human cancer-derived cell lines. *Toxicon* **2018**, *156*, 13–22. [CrossRef] [PubMed]
32. Trott, O.; Olson, A.J. AutoDock Vina: Improving the speed and accuracy of docking with a new scoring function, efficient optimization, and multithreading. *J. Comput. Chem.* **2010**, *31*, 455–461. [CrossRef] [PubMed]

pharmaceuticals

MDPI

Article

Accessing Lipophilicity and Biomimetic Chromatography Profile of Biologically Active Ingredients of Botanicals Used in the Treatment of Inflammatory Bowel Disease

Mario-Livio Jeličić [1], Daniela Amidžić Klarić [1], Jelena Kovačić [1], Donatella Verbanac [2] and Ana Mornar [1,*]

1 Department of Pharmaceutical Analysis, Faculty of Pharmacy and Biochemistry, University of Zagreb, A. Kovačića 1, 10000 Zagreb, Croatia
2 Department of Medical Biochemistry and Hematology, Faculty of Pharmacy and Biochemistry, University of Zagreb, A. Kovačića 1, 10000 Zagreb, Croatia
* Correspondence: amornar@pharma.hr

Abstract: In the present study, various procedures have been compared for the determination of lipophilicity, hydrophobicity, and plasma protein binding of curcuminoids, boswellic acids, andrographolides, and piperine as biologically active ingredients of botanicals used in IBD treatment. Our results have shown that IAM-HPLC assay is the most suitable one for lipophilicity determination of all analytes regardless of their class and botanical source. HSA-HPAC and AGP-HPAC assays revealed that all investigated compounds have a higher affinity for HSA which is the most abundant protein in human plasma. The high affinity of biologically active compounds to all biological structures (phospholipids and proteins) admonishes that their small portion is available for therapeutic effects in IBD patients. Our experimental research is complemented by various theoretical approaches based on different algorithms for pharmacokinetic properties prediction. The similarities between experimental and calculated values were evaluated using PCA and CA as a statistical tool. The statistical analysis implies that plasma protein binding is a complex process, and theoretical approaches still cannot fully replace experimental ones.

Keywords: inflammatory bowel diseases; botanicals; biomimetic chromatography; shake-flask method; immobilized artificial membrane; plasma protein binding

Citation: Jeličić, M.-L.; Amidžić Klarić, D.; Kovačić, J.; Verbanac, D.; Mornar, A. Accessing Lipophilicity and Biomimetic Chromatography Profile of Biologically Active Ingredients of Botanicals Used in the Treatment of Inflammatory Bowel Disease. *Pharmaceuticals* **2022**, *15*, 965. https://doi.org/10.3390/ph15080965

Academic Editor: Diana Roxana Pelinescu

Received: 11 July 2022
Accepted: 1 August 2022
Published: 4 August 2022

1. Introduction

Inflammatory bowel disease (IBD), in its definition, includes two primary conditions—ulcerative colitis (UC) and Crohn's disease (CD). These are chronic, relapsing gastrointestinal tract disorders that occur worldwide and affect people of all ages, with a high impact on their quality of life. More than 2 million Europeans and 1.5 million North Americans have IBD, and researchers report the rising incidence rates in newly industrialised countries in South America, Eastern Europe, Asia and Africa paralleled with the westernisation of diet and culture [1]. Proper IBD management requires early diagnosis, novel therapies, and management programs. Regardless recent improvements in IBD treatment, complementary and alternative therapeutic approaches for IBD have earned growing interest from patients, with a dynamic landscape of research in this area. According to Lin and Cheifetz [2], up to 60% of IBD patients use complementary and alternative medicine, with botanicals being the most popular. This growing popularity of botanicals is not surprising since patients perceive these products as safe and natural way of healing, free of side effects unlike conventional medicines. Amongst the botanicals, *Curcuma longa* L., *Zingiberaceae* (Turmeric), *Boswellia serrata Roxb. ex Colebr.*, *Burseraceae* (Indian frankincense) and *Andrographis paniculata (Burm. f.) Wall. ex Nees*, *Acanthaceae* (Green chiretta) are the most popular herbal remedies used in IBD treatment. Despite the results of the relevant clinical studies showing the role of these botanicals in preventing and alleviating the symptoms of

IBD, there is still considerable concern surrounding the bioavailability of their biologically active ingredients: curcuminoids as the most prominent compounds present in turmeric, boswellic acids as pentacyclic terpenoid molecules that make up 30% of frankincense resin and andrographolides as diterpene lactones (aglycone and glycosidic form) isolated from Green chiretta (Figure 1) [3–6]. Piperine is a piperidine alkaloid isolated from the fruits of *Piper nigrum* L., *Piperaceae* (Black pepper). Several studies unravel the therapeutic potential of piperine on amelioration of IBD as well as improvement in dissolution, stability of metabolic processes and membrane permeability of above mentioned biologically active compounds [7].

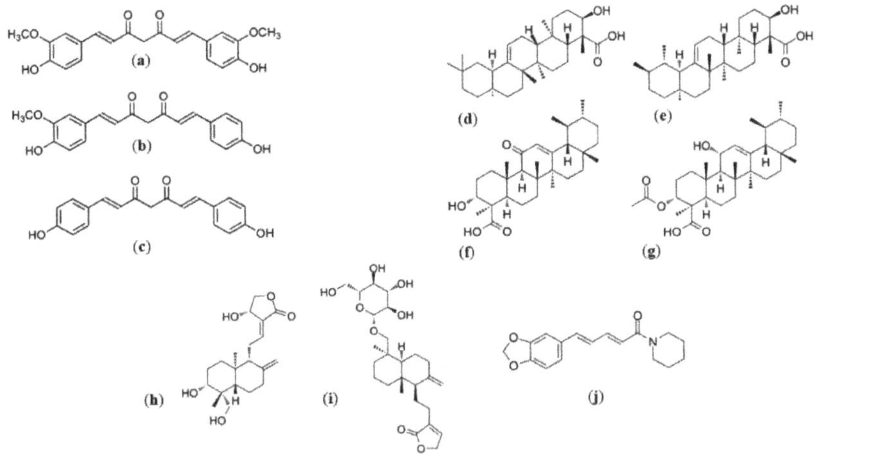

Figure 1. Chemical structures of curcuminoids (curcumin (**a**), demetoxycurcumin (**b**) and bisdemetoxycurcumin (**c**)), boswellic acids (α-boswellic acid (**d**), β-boswellic acid (**e**), 11-keto-β-boswellic acid (**f**) and 3-acetyl-11-keto-β-boswellic acid (**g**)), andrographolides (andrographolide (**h**) and neoandrographolide (**i**)) and piperine (**j**).

Numerous investigations have pointed out that research on herbal remedies, such as those mentioned above, has proved to be very difficult in all respects. Such a statement is attributed to their complex composition consisting of diverse chemical constituents each identical to a single active pharmaceutical ingredient of conventional medicine. However, a study providing deeper mechanistic insights into the bioavailability of biologically active ingredients of botanicals used in IBD treatment is essential to develop better treatment strategies in the future based on botanical products. The gold standard of bioavailability determination has long been considered a research study involving human volunteers. Albeit valuable, these studies are frequently demanding, unfeasible, subject to high overheads, and time consuming. In the light of these considerations, the scientific community's attention towards alternative methods is increasing. Chromatographic methods are undoubtedly the most prominent ones since these methods using immobilized artificial membrane (IAM) provide the introduction to the separation mode of biological structures playing an essential role in drugs' bioavailability, namely membrane phospholipids [8,9]. Likewise, high-performance affinity chromatography (HPAC) using human serum albumin (HSA) and α1-acid glycoprotein (AGP) as ligands within the chromatographic column has been successfully used to evaluate the interactions of drugs with serum proteins [10–12].

Unlike screening methodologies implemented on cell and animal models, these biomimetic chromatographic methods offer a superior reproducibility of the measurements, high-throughput data acquisition and environment-friendly, sustainable analytical approach [13]. Moreover, it is known that the biomimetic chromatography is based on physicochemical parameters, which allows elucidation of molecular mechanisms.

The main aim of this study was to assess the lipophilicity of biologically active ingredients of botanicals used in IBD treatment by miniaturized shake-flask procedure and evaluate their hydrophobicity, lipophilicity and affinity for plasma proteins using chromatographic techniques. Another goal of our research was to compare experimental indices with physicochemical parameters as well as adsorption and distribution properties estimated by artificial intelligence.

2. Results and Discussion

2.1. In Silico Calculation

Nowadays, computer-aided drug design is a crucial factor in the drug discovery and development process. As such, it was used as a starting point of our research to get some insight in physicochemical properties of biologically active ingredients of the most popular botanicals used in IBD treatment. The selection of parameters for in silico calculation was founded upon their relation to experimentally determined properties.

Lipophilicity expressed as the logarithm of the partition coefficient between *n*-octanol and aqueous phase (log *P*) is one of the most important physicochemical parameters that has been found to affect a number of pharmacokinetic parameters. Accordingly, various software packages based on different algorithms are available for prediction of these properties. The log *P* values of biologically active ingredients of botanicals used in IBD treatment estimated by means of 15 theoretical approaches are summarized in Table 1. The highest values were obtained for lipophilic boswellic acids (average log *P* values were in the range from 5.56 to 6.57) followed by piperine (average log *P* value was 3.12) and curcuminoids (average log *P* values were in the range from 3.05 to 3.13). Generally, the lowest values were found for both compounds from the andrographolides group (average log *P* values were 2.06 and 2.11). The highest difference between log *P* values obtained by different theoretical approaches (up to high 4.42 units) found for boswellic acid reflects the incoherence of in silico approaches for highly lipophilic compounds. The lowest values for boswellic acids (up to 2.58 units lower than average calculated value) were obtained by iLOGP prediction that uses a rather new approach relaying on free energies of solvation in *n*-octanol/water and solvent accessible surface area for lipophilicity calculation. On the other hand, the highest values for this group of biologically active compounds (up to 1.86 units higher than average calculated value) were obtained using XLOGP3 software based on the well-known atomic method including corrective factors and knowledge-based library. For other less lipophilic compounds (curcuminoids, andrographolides and piperine), differences between calculated values were rather reasonable. Silicos-IT LogP software using an hybrid theoretical approach relaying on fragments and topological descriptors slightly overestimated log *P* values for all three curcuminoids (differences between calculated and average values were less than 0.91 units) while miLogP values calculated using group contribution model were lowest ones for both bicyclic diterpenoid lactones andrographolides (differences between calculated and average values were less than 1.00).

The solubility of biologically active ingredients of botanicals used in IBD treatment estimated by means of five theoretical approaches is presented in Table 1 and expressed as the logarithm of molar concentration (log *S*). As expected, the high lipophilicity of investigated compounds resulted in reduced solubility. The discrepancy between predicted log *S* values was acceptable for all compounds ($\Delta \leq 1.74$ units) except lipophilic boswellic acids (Δ was in the range from 2.61 units to 3.70 units) which confirms that highly lipophilic compounds are challenging for in silico approaches.

After the log *P* and log *S* value evaluation was completed, we have moved to the prediction of pharmacokinetic properties of target compounds using three platforms. We have used well recognized preADMET platform as well as two novel approaches to the prediction of pharmacokinetic properties pkCSM which relies on graph-based signatures as well as admetSAR platform based on models trained by state-of-the-art machine learning methods such as support vector machine, k-nearest neighbors, neural network etc. (Table 1).

Table 1. Summary of the lipophilicity and pharmacokinetic properties obtained by theoretical approaches.

Parameter	Curcumin	Demetoxy-curcumin	Bisdemetoxy-curcumin	α-Boswellic Acid	β-Boswellic Acid	11-Keto-β-boswellic Acid	3-Acetyl-11-keto-β-boswellic Acid	Andro-grapholide	Neoandro-grapholide	Piperine
					Lipophilicity					
iLOGP	3.27	2.78	1.75	3.99	3.96	3.32	3.80	2.61	3.37	3.38
XLOGP3	3.20	3.32	3.26	8.41	8.26	7.20	7.22	2.16	2.63	3.46
WLOGP	3.15	3.14	3.13	7.23	7.09	6.27	6.84	1.96	1.85	2.51
MLOGP	1.47	1.80	2.13	5.82	5.82	4.87	5.14	1.98	1.26	2.39
Silicos-IT LogP	4.04	3.95	3.87	5.85	5.46	5.15	5.67	2.94	2.55	3.41
SwissADME LogP	3.03	3.00	2.83	6.26	6.12	5.36	5.74	2.33	2.33	3.03
ALOGPs	3.62	3.54	3.46	7.42	6.46	5.50	5.90	1.57	1.88	3.38
miLogP	2.30	2.48	2.67	6.72	6.79	5.69	6.39	1.05	1.17	3.33
Mcule logP	3.37	3.36	3.35	7.23	7.09	6.27	6.84	1.96	1.85	2.94
cLogP	2.95	3.02	3.09	6.06	6.00	5.29	5.78	1.88	1.74	3.60
MolLogP	2.83	2.88	2.93	5.86	5.68	4.66	5.17	2.32	2.51	3.47
ALOGP98	3.22	3.24	3.26	6.42	6.47	5.54	5.92	2.06	2.38	2.86
pkCSM logP	3.37	3.36	3.35	7.23	7.09	6.27	6.84	1.96	1.85	3.00
SKlog_P	3.72	3.49	3.25	6.78	6.61	5.72	6.02	2.10	2.42	3.03
ALOGP	3.37	3.36	3.35	7.23	7.09	6.27	6.84	1.96	1.85	3.00
					Solubility					
$\log S_{SILICOS\text{-}IT}$	−4.45	−4.34	−4.23	−6.12	−5.67	−5.55	−6.15	−2.69	−2.70	−3.00
$\log S_{ALOGPS}$	−4.81	−4.70	−4.46	−5.86	−5.97	−5.51	−6.06	−3.09	−3.37	−3.28
$\log S_{AquaSol}$	−3.92	−4.01	−4.06	−5.13	−5.20	−4.71	−4.87	−2.78	−3.42	−3.55
$\log S_{preADMET}$	−4.53	−4.05	−3.56	−7.35	−7.49	−6.92	−7.28	−3.12	−4.44	−3.71
$\log S_{admetSAR}$	−3.36	−3.09	−2.88	−3.85	−3.79	−3.45	−4.67	−2.85	−3.31	−3.40
					Absorption					
HIA_{pkCMS} (%)	82.190	91.393	91.159	94.363	97.466	98.826	99.510	95.357	62.263	94.444
$HIA_{preADMET}$ (%)	94.403	94.029	93.750	95.996	95.996	96.714	99.202	87.688	81.330	98.180
$HIA_{admetSAR}$ (%)	97.70	97.70	97.37	98.53	98.53	99.01	99.19	98.22	81.24	96.39
$HOB_{admetSAR}$ (%)	60.00	64.29	64.29	50.00	52.86	51.43	51.43	61.43	65.71	57.14
					Plasma protein binding					
$PPB_{preADMET}$ (%)	88.030	89.913	93.826	100.000	100.000	100.000	94.061	94.873	93.373	90.449
$PPB_{admetSAR}$ (%)	64.0	68.9	42.1	51.4	65.8	62.9	79.4	49.4	52.5	100.0

All prediction procedures labelled boswellic acids as biologically active compounds with the high human intestinal absorption (obtained values were higher than 94%). Despite the widespread use of Indian frankincense, only few preliminary pharmacokinetic studies were conducted [14]. AdmetSAR web service offered us prediction not only of boswellic acids absorption but also their human oral bioavailability. This term is used to indicate the fraction of an orally administered dose that reaches the systemic circulation as intact drug, taking into account absorption, gastrointestinal stability and local metabolic degradation. Although the highest absorption among investigated biologically active compounds was predicted for boswellic acids at the same time their lowest bioavailability (from 50% to 53%) might be attributed to gastrointestinal instability of boswellic acids, high accumulation within the enterocytes, intestinal metabolism, extensive phase I metabolism observed for non-acetylated boswellic acids in human liver microsomes as well as saturable kinetics [15,16]. The most intriguing high absorption (obtained values were in the range from 82% to 98%) and bioavailability (obtained values were in the range from 60% to 64%) by these theoretical approaches was outlined for curcuminoids. Despite their demonstrated biological effects, the potential health benefits of curcuminoids are limited by their poor solubility, intestinal instability at a pH lower than 3 and higher than 6 and extensive first-pass intestinal and hepatic metabolism. As a result, different innovative strategies have been pursued to improve the absorption of curcuminoids including nanocrystals, emulsions, liposomes, etc. [14]. Amongst the investigated compounds the lowest human intestinal absorption (obtained values were between 62% and 81%) was predicted for diterpene neoandrographolide which can be related both to its pyranose ring and α,β-unsaturated lactone.

Regarding the plasma protein binding data, we found a discrepancy between values obtained by two tested theoretical approaches (Table 1). The recent studies confirmed the stable binding of biologically active ingredients of botanicals used in IBD treatment with plasma proteins and their complex formation supporting the data obtained by preADMET system [14,17]. Because boswellic acids represent lipophilic acids, it appeared reasonable to speculate that high plasma protein binding will be revealed. The predicted plasma protein binding of 100% for three boswellic acids needs to be interpreted with attention as such tight albumin-binding raises doubts regarding the efficiency of Indian frankincense in vivo and questions its pharmacological relevance.

Although diverse in silico values were obtained for biologically active ingredients of botanicals used in the treatment of IBD, this approach was our starting point in experimental design for lipophilicity, hydrophobicity and biomimetic chromatography profile assessment.

2.2. *Shake-Flask Method*

So far, the different approaches for the log P evaluation have been developed and described in literature [18]. We have chosen the reference shake-flask method on account of the fact that the gained results are scalable and easily transferable to other experimental and theoretical procedures. In our previous work [19] we have developed a miniaturized shake-flask methodology to increase the experimental throughput and to reduce the experimental effort and costs. Principles of green analytical chemistry were implemented in all analytical processes from sample preparation to their analysis. From the results presented in Table 1 it is evident that boswellic acids are highly lipophilic and such as that these compounds are not good candidate for lipophilicity determination by shake-flask methodology. Our preliminary investigation showed that shake-flask procedure failed for those compounds due to solubility issues even with the use of dimethylsulfoxide as a modifier. Therefore, at this point our investigation was directed to validation of procedure for curcuminoids, andrographolides and piperine according to ICH quidlines [20]. Results presented in Table 2 reveal linear (correlation coefficients (r) were higher than 0.999), sensitive (Limits of Quantitation (LOQ) were lower than 10 μg/mL), accurate (recoveries were within ±5%) and precise (Relative Standard Deviations (RSD) were lower than 3.61%) analytical methodology.

The obtained log $D_{7.4}$ values were in the range from 1.51 to 1.94 supporting the well-known applicability of the method for moderately lipophilic compounds (Table 3). Curcumin is a symmetric compound whose structure comprises three chemical units consisting of 2 aromatic ring systems containing *O*-methoxy phenolic groups linked through an α,β-unsaturated β-diketone moiety. Lower log $D_{7.4}$ values obtained for demetoxycurcumin ($\Delta = 0.42$ units) and bisdemetoxycurcumin ($\Delta = 0.22$ units) are related with loss of one or both *O*-methoxy groups, respectively. On the other hand, pyranose ring in the structure of neoandrographolide slightly increased lipophilicity ($\Delta = 0.19$ units) of this Green chiretta active ingredient compared to andrographolide. Moderate lipophilicity of curcumin and its novel synthetic structural analogues is observed in previous studies [21,22].

Table 2. Shake-flask method validation data.

Analyte	Linearity Range (μg/mL)	Regression Equation *	*r*	LOD (μg/mL) **	LOQ (μg/mL) ***
Curcumin	1–375	y = 987.3 *x* + 0.935	0.9999	0.3	1
Demetoxycurcumin	1–375	y = 421.0 *x* + 0.610	0.9998	0.3	1
Bisdemetoxycurcumin	10–375	y = 134.4 *x* + 0.200	0.9999	3	10
Andrographolide	5–300	y = 15,039 *x* + 35.28	0.9996	2	5
Neoandrographolide	5–250	y = 1723 *x* + 3.188	0.9995	2	5
Piperine	5–300	y = 10,601 *x* + 10.15	0.9999	2	5
Analyte	**Precision (RSD, %) ********		**Accuracy (Recovery and RSD, %) ***********		
	Repeatability (*n* = 6)	**Intermediate Precision (*n* = 9)**	**Low (*n* = 3)**	**Medium (*n* = 3)**	**High (*n* = 3)**
Curcumin	0.42	1.59	102.5 ± 3.89	95.0 ± 3.55	97.5 ± 2.95
Demetoxycurcumin	0.38	0.80	102.9 ± 3.81	95.4 ± 3.41	98.8 ± 3.01
Bisdemetoxycurcumin	0.98	0.27	103.0 ± 3.40	96.2 ± 3.31	99.4 ± 2.55
Andrographolide	1.61	3.61	97.2 ± 3.66	99.1 ± 1.98	98.4 ± 1.59
Neoandrographolide	1.73	3.39	98.1 ± 2.90	102.1 ± 2.50	96.5 ± 2.34
Piperine	0.73	0.59	102.5 ± 0.92	104.3 ± 0.94	103.0 ± 0.91

* Linearity was examined on at least five concentration levels in three individual standard solution preparations from which a single regression line was constructed. ** Limit of Detection (LOD) was determined using signal-to-noise value 3. *** Limit of Quantitation (LOQ) was determined using signal-to-noise value 10. **** Repeatability was assessed analysing six individual samples on the same day, while intermediate precision was examined on three individual samples over three days. Results are expressed as Relative Standard Deviations (RSD). ***** Accuracy of the method was examined by analysis of standard solutions in triplicate on three concentration levels. Results are expressed both as recoveries and Relative Standard Deviations (RSD).

Table 3. Log $D_{7.4}$ values * of biologically active ingredients of botanicals used in IBD treatment determined by applied shake-flask procedure.

Analyte	log $D_{7.4}$ ± RSD ** (%) (*n* = 3)
Curcumin	1.94 ± 2.59
Demetoxycurcumin	1.52 ± 3.78
Bisdemetoxycurcumin	1.72 ± 4.21
Andrographolide	1.32 ± 3.25
Neoandrographolide	1.51 ± 2.55
Piperine	1.67 ± 3.55

* Log $D_{7.4}$—the logarithm of the distribution coefficient between *n*-octanol and buffer phase; describes the distribution of all forms of the compound at a specific pH. ** RSD—Relative Standard Deviation.

2.3. *Chromatographic Methods*

2.3.1. Hydrophobicity Evaluation

In the course of the research, various stationary phases were used, differing in chemical structure and physicochemical properties. First, we have used C18 stationary phase which is alkyl-bonded phase modification of silica gel to which octadecyl carbon chains are attached.

This stationary phase offers simple hydrophobic interaction with analyte and is available as thin-layer chromatography (TLC) plate and high performance liquid chromatography (HPLC) column. To get insight into hydrophobicity of investigated biologically active compounds, both solutions were used in this research.

In TLC measurements R_F value depends on the chemical structure of analyte and its interaction with stationary and mobile phase. All investigated compounds were strongly retained by hydrophobic stationary phase and for this reason mixtures of buffer and organic solvent were used as mobile phase. Methanol appeared to be the most suitable organic modifier for this purpose because it does not disturb the hydrogen-bonding network of water. As it was expected, with increase in the percentage of organic modifier in the mobile phase the increase in R_F values can be seen with accompanied decrease in R_M values expressed as $R_M = \log(1/R_F - 1)$. Measured R_F values were in the range from 0.1 to 0.7 obtained in a wide range of mobile phase organic modifier concentrations. Due to strong interaction with the stationary phase for hydrophobic analytes (11-keto-β-boswellic acid and 3-acetyl-11-keto-β-boswellic acid) this range was somewhat narrower, and mobile phases rich in an organic modifier had to be used. For the most hydrophobic analytes (α- and β-boswellic acids) no spot migration was observed regardless the percentage of the organic modifier in the mobile phase. The TLC hydrophobicity index, R_{M0}, was determined by extrapolation of the organic modifier in the mobile phase to the zero concentration (Table 4) and regression coefficient for all analytes were higher than 0.979 with small values of standard error of the regression model (lower than 0.1041), which proves the high significance for hydrophobicity determination. Increasing concentration of the organic modifier in the mobile phase caused a slower decrease in the R_M values for less hydrophobic analytes than for more hydrophobic ones. Based on these results, generally, bicyclic diterpenoid lactones andrographolides show the least hydrophobic character followed by piperine and curcuminoids while extremely high was found for boswellic acids. Increase in the hydrophobicity within curcuminoids and andrographolides is related to increase in lipophilicity determined by shake-flask method (Table 3).

Functional groups present in the structure of these compounds increase their hydrophobicity and lipophilicity in the same manner. The presence of keto group in the structure of boswellic acids was crucial for applicability of TLC procedure in determination of hydrophobicity of these compounds. As it was expected, the acetyl group present in the structure of 3-acetyl-11-keto-β-boswellic acid increased hydrophobicity of this boswellic acid compared to 11-keto-β-boswellic acid ($\Delta = 1.53$ units).

Hydrophobicity of analytes was also investigated using high-performance liquid chromatography and C18 stationary phase. The analytical approach to both indirect chromatographic determination of hydrophobicity was coherent, still each technique provided advantages and limitations. The adaptability of TLC allowed us the analysis of all investigated compounds simultaneously, while the most important practical advantage of HPLC was process automatization and online detection.

The principles of hydrophobicity determination by HPLC are characterized by the logarithm of retention factor (k), defined as $\log k = \log{(t_R - t_0)}/t_0$, where t_R is retention time of analyte and t_0 is the dead time. The direct measurement of $\log k$ in buffer of investigated biologically active compounds was impossible due to their high hydrophobicity which led to very long retention time and at the same time to extensive broadening of the peaks. The percentage of methanol in the mobile phase was optimized so that the analytes had retention time between dead time and a maximum of 20 min keeping the analysis time as short as possible, and the peak shape acceptable (peak asymmetry factor were between 0.8 and 1.2). By evaluating the profiles of $\log k$ values for all methanol fractions, the regular changes in retention with increasing methanol ratios were observed. The HPLC hydrophobicity index, $\log k_0$, was determined by extrapolation of the organic modifier in the mobile phase to its zero concentration. The results of linear regression are listed in Table 4. For all analytes, high values of regression coefficients (higher than 0.970) and small values of standard error of the regression model (lower than 0.0581) were achieved. The interaction of analytes with stationary phase was consistent within both C18 chromatographic systems. Still, except in the case of neoandrographolide the obtained HPLC hydrophobicity indices were slightly lower than those obtained by C18-TLC assay (from 0.18 to 1.07 units). Although a good chromatographical separation of curcuminoids and piperine in different mobile phases was observed, this method failed in discriminating logk_0 values of curcuminoids and piperine (Δ = 0.01 units). As it can be seen in the case of the highly hydrophobic substances boswellic acids, these compounds are beyond the reach of C18-HPLC assay. Their retention time using mobile phase with high 80% of methanol was over 17 min. The polarization of the stationary phase in the presence of increasing content of methanol as well as increased hydrophobic interactions of analytes with stationary phase explains the disruption of the linearity of the regression and unsuitability of C18-HPLC system for such highly hydrophobic compounds.

Table 4. Hydrophobicity and lipophilicity parameters determined by chromatographic assays.

Analyte	Linear Equation *	*r*	Standard Error
	C18-TLC Assay		
Curcumin	y = −0.0540 x + 4.4855	0.9916	0.0835
Demetoxycurcumin	y = −0.0504 x + 4.1781	0.9932	0.0699
Bisdemetoxycurcumin	y = −0.0503 x + 4.1751	0.9937	0.0675
11-keto-β-boswellic acid	y = −0.0564 x + 5.3756	0.9950	0.0399
3-acetyl-11-keto-β-boswellic acid	y = −0.0722 x + 6.9059	0.9798	0.1041
Andrographolide	y = −0.0302 x + 2.0908	0.9970	0.0277
Neoandrographolide	y = −0.0334 x + 2.1672	0.9964	0.0266
Piperine	y = −0.0526 x + 4.4044	0.9789	0.0913
	C18-HPLC Assay		
Curcumin	y = −0.0442 x + 3.4187	0.9918	0.0368
Demetoxycurcumin	y = −0.0448 x + 3.4349	0.9827	0.0542
Bisdemetoxycurcumin	y = −0.0447 x + 3.4174	0.9818	0.0556
Andrographolide	y = −0.0271 x + 1.9065	0.9695	0.0581
Neoandrographolide	y = −0.0359 x + 2.8595	0.9782	0.0489
Piperine	y = −0.0394 x + 3.4218	0.9796	0.0519
	IAM-HPLC Assay		
Curcumin	y = −0.0565 x + 4.3463	0.9969	0.0530
Demetoxycurcumin	y = −0.0555 x + 4.4229	0.9941	0.0554
Bisdemetoxycurcumin	y = −0.0568 x + 4.6386	0.9940	0.0569
α-boswellic acid	y = −0.0788 x + 6.3109	0.9956	0.0526
β-boswellic acid	y = −0.0816 x + 6.5157	0.9982	0.0346
11-keto-β-boswellic acid	y = −0.0570 x + 4.9194	0.9874	0.0645
3-acetyl-11-keto-β-boswellic acid	y = −0.0703 x + 5.1157	0.9944	0.0526
Andrographolide	y = −0.0464 x + 2.2763	0.9796	0.0753
Neoandrographolide	y = −0.0620 x + 3.4572	0.9854	0.0755
Piperine	y = −0.0512 x + 3.2026	0.9830	0.0872

* Depending on the applied methodology, the intercept of linear regression represents hydrophobicity/lipophilicity indices R_{M0}, logk_0 and log$k_{0\ IAM}$, respectively. The indices were derived by extrapolation using the following equations: $R_M = R_{M0}$—S x φ for C18-TLC assay, log$k$ = logk_0—S x φ for C18-HPLC assay, and logk_{IAM} = log$k_{0\ IAM}$—S x φ for IAM-HPLC assay. φ represents the volume fraction of the organic modifier in the mobile phase and S a constant derived by linear regression analysis.

2.3.2. Lipophilicity Evaluation

Both C18-TLC and C18-HPLC assays gave us insight into hydrophobicity of investigated compounds. Although these results are valuable, to evaluate bioavailability of biologically active compounds more data should be collected and discussed. IAM columns provide potential to simulate membrane permeability since the amphiphilic character of phospholipid functional groups plays an important role in IAM retention especially when charged molecules are analysed. In order to clarify similarities and dissimilarities between chromatographic approaches the analytical procedure used for determination of HPLC lipophilicity index, log$k_{0\ IAM}$, was comparable with above-described procedure for C18-HPLC assay.

Our findings would seem to imply that IAM-HPLC assay is the most suitable one for determination of lipophilicity of all analytes regardless of structure and botanical source (Table 4). In addition to the wide applicability of the method, the high values of correlation coefficients (higher than 0.980) and the small values of standard error of the regression model (lower than 0.0872) support the idea that IAM-HPLC assay would lend itself well for use in membrane permeability studies. The all investigated compounds retained less on IAM stationary phase than on C18 one, still $\log k_{0\,IAM}$ values obtained for all analytes except piperine (Δ = 0.01 units) are generally higher than $\log k_0$ values (differences were in the range between 0.37 to 1.22 units). In the case of boswellic acids, the position of two methyl groups on C-19/C-20 (α-type: germinal groups on C-20; β-type: vicinal groups on C-19/C-20) had an influence on interaction with phospholipids. The oleanane (β) type interacted strongly with phospholipids compared to ursane (α) type. The pharmacologically interesting 11-keto-β-boswellic acid had a lower affinity compared to β-boswellic acid (Δ = 2.60 units), most likely due to carbonyl group at C-11 of pentacyclic triterpene. Conversely, the acetyl function at C-3 position increased $\log k_{0\,IAM}$ for more than one unit. The evidence from this study points toward the idea that IAM stationary phase may be useful for the evaluation of lipophilicity of highly lipophilic compounds, such as boswellic acids, which distinguish this experimental approach among the others. On the other hand, low solubility and high affinity of boswellic acids for phospholipids leads to question of their bioavailability and pharmacological relevance. Consequently, future work in the field of pharmaceutical development needs to be performed to get the most of Boswellia serrata extract as it was assigned orphan drug status in 2002 by the European Medicine Agency for treatment of peritumoral edema [23]. All analytes from curcuminoid group compared to boswellic acids showed to some extent lower affinity to phospholipid. The less retained curcuminoid on IAM stationary phase was curcumin whose structure comprises of the two phenolic-methoxy groups in the opposite sides of curcumin backbone. The IAM assay revealed the increase in $\log k_{0\,IAM}$ values for each methoxy function attached to phenolic backbone from 0.1 to 0.2 units. Still, as in the case of Boswellia serrata extract an innovative improvement in pharmaceutical development of Curcuma extract is considered a very reasonable approach for improving their bioavailability [24]. Moderate affinity of andrographolides to IAM stationary phase among investigated compounds is in accordance with their lower retention in C18 stationary phase. Still, the presence of sugar residue in the structure of labdane glucoside diterpene neoandrographolide considerably affected its affinity to phospholipids (difference in $\log k_{0\,IAM}$ values was 1.18 units) than alkyl chains (difference in $\log k_0$ values was 0.95 units and R_{M0} values 0.08 units). Before interpreting results for piperine, it should be stated that the IAM stationary phase surface is mainly zwitterionic at pH 7.4. The positively charged choline moieties are located in the outer part of the IAM layer. By way of contrast, the negatively charged phosphate groups are present in the phase's inner part. This would appear to indicate that retention of piperidine alkaloid piperine in IAM stationary phase is mostly due to interaction of its positively charged nitrogen atom and phosphate groups more in-depth of stationary phase.

2.3.3. Plasma Protein Affinity Evaluation

In the consecutive set of studies, plasma protein binding of biologically active compounds was evaluated by bioaffinity chromatography using both assays HSA-HPAC and AGP-HPAC, respectively. The selection of HSA-HPAC assay for protein binding assessment purposes was substantiated by the fact that HSA is the most abundant protein found in human plasma (around 60%), and accordingly the most emerged versatile carrier for therapeutic agents. Albeit the AGP concentration in the plasma ranges up to only 3%, its concentration depends on the disease state and can increase significantly in patients with IBD. Therefore, our study has gone some way towards enhancing our understanding of plasma protein binding of selected biologically active compounds particularly in patients with IBD [25].

The gradient retention times obtained by HSA-HPAC and AGP-HPAC assays were standardized using a calibration set of compounds with known percentage of plasma protein binding data and procedure described in our previously published study [12]. Although the utmost care was put in preserving the performance of both biomimetic columns over the time verification of protein stationary phase was essential and was ensured by analysis of racemic mixture of warfarin during six consecutive days that revealed the repeatable separation of enantiomers (average resolution factors between enantiomers were 2.27 with RSD value 1.93% for HSA while 1.63 with RSD value 1.04% for AGP) (Table S1). This verification assured that the protein affinity indices obtained by HPAC assays depict not only unspecific generally lipophilicity-driven interactions, but also highly specific recognition forces responsible of enantioselectivity.

Table 5 highlights that HSA strongly attracted biologically active ingredients of botanicals used in IBD treatment. Beside andrographolides, all investigated compounds have exceptional affinity to HSA (more than 95%). The high affinity for protein molecule possibly relies on the hydrophobic regions of compounds (aliphatic and aromatic rings) that improved their ability to penetrate the hydrophobic cavity. Hydrogen bonds might also be involved in the sharing of hydroxylic or phenolic protons with numerous amide carbonyl moieties of HSA [17,26].

Table 5. Plasma protein binding data (%) determined by HSA-HPAC and AGP-HPAC assays.

Analyte	HSA Binding ± RSD * (%) (n = 3)	AGP Binding ± RSD (%) (n = 3)
Curcumin	97.52 ± 0.20	83.07 ± 1.10
Demetoxycurcumin	97.97 ± 0.35	82.92 ± 1.15
Bisdemetoxycurcumin	98.37 ± 0.10	83.14 ± 1.15
α-boswellic acid	98.94 ± 0.35	89.22 ± 0.30
β-boswellic acid	99.79 ± 0.30	89.15 ± 2.04
11-keto-β-boswellic acid	98.09 ± 0.02	79.01 ± 0.20
3-acetyl-11-keto-β-boswellic acid	99.83 ± 0.47	83.10 ± 0.37
Andrographolide	80.07 ± 0.74	37.13 ± 1.11
Neoandrographolide	87.71 ± 0.41	64.95 ± 0.41
Piperine	95.48 ± 0.41	71.98 ± 1.33

* RSD—Relative Standard Deviation.

We have found that our analytes had generally lower affinity for AGP than HSA (AGP binding proportions were from 10% to 42% lower than values obtained for HSA) (Table 5). The differences were specific for each botanical group. The lowest deviations were obtained for highly lipophilic boswellic acids (differences were around 10%), a difference of 15% was obtained for all curcuminoids, while the least lipophilic andrographolides and piperine had the highest differences between HSA and AGP binding proportions from 23% to 43% (neoandrographolide, the only compound in glycosidic form).

High affinity to phospholipids demonstrated by IAM-HPLC assay supported with reversed but still noteworthy inactivation by protein binding presented by both assays reinforces the claim that a small portion of biologically active ingredients of botanicals used in IBD treatment are available for therapeutic effects [14]. Moreover, the most lipophilic compounds and consequently ones with the most compromising bioavailability have the highest affinity for both plasma proteins which makes them therapeutically inactive.

2.4. *Comparison of Computational and Experimentally Observed Values Using Statistical Methods*

To evaluate the similarities between calculated and experimentally observed values, and their mutual correlation, statistical tools such as Principal Component Analysis (PCA) and Cluster Analysis (CA) were used. Obtained results of PCA showed that first two principal components describe 85.49% of data variability, where first component (PC1) and second (PC2) describe 76.90% and 8.59% of data variability, respectively (Figure 2).

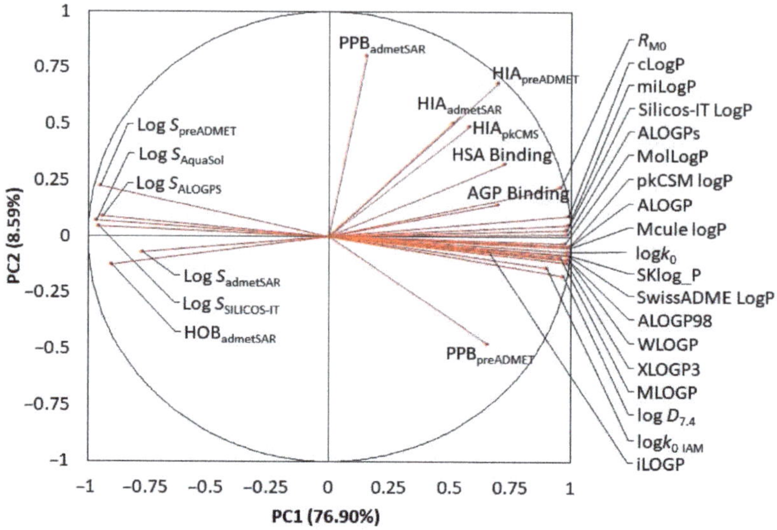

Figure 2. Results of PCA analysis reflecting correlation between predicted and experimentally observed values.

Looking at the bi-plot we can see that most of the data is densely placed far right on the PC1 axis, mostly comprised of predicted log *P* values, as well as experimentally observed log $D_{7.4}$ values and hydrophobicity and lipophilicity indices obtained using TLC and HPLC techniques. Similar orientation of these vectors, with small angles between them and similar PC1 values mean that predicted values are relatively close to those experimentally observed, which implies that used calculation tools have good prediction capability.

The second cluster formed is mostly comprised of predicted log *S* values representing the predicted solubility/hydrophilicity of examined biologically active compounds. Seeing that they are placed opposite of predicted and experimentally observed lipophilicity values says that there is a negative correlation between them, which is in line with our expectations. The only one pharmacokinetic parameter that correlates with predicted log *S* values is human oral bioavailability observed by the admetSAR software package. As mentioned above, the bioavailability of xenobiotics is a complex process. According to our results the solubility of investigated biologically active ingredients of herbals used in IBD treatment plays a crucial role in their oral bioavailability.

The third cluster is formed from human intestinal absorption values obtained by different prediction procedures. These values are closed to various hydrophobicity and lipophilicity parameters indicating acceptable relationship.

The last cluster, rather dispersed, is comprised of predicted plasma protein binding values accompanied by experimentally observed affinity to AGP and HSA protein. However, predicted plasma protein binding values are poorly corelated with experimentally observed ones which implies that prediction of affinity of our biologically active compounds to both plasma proteins is rather complex.

On the other hand, Figure 3 highlights the results of CA that reflect the similarity of observed values. As expected, the formation of two main clusters which are dissimilar one to another, one comprised of data related to lipophilicity and solubility and other representing the data related to plasma protein binding, human intestinal absorption and human oral bioavailability of biologically active ingredients of botanicals is evident (Figure 3a).

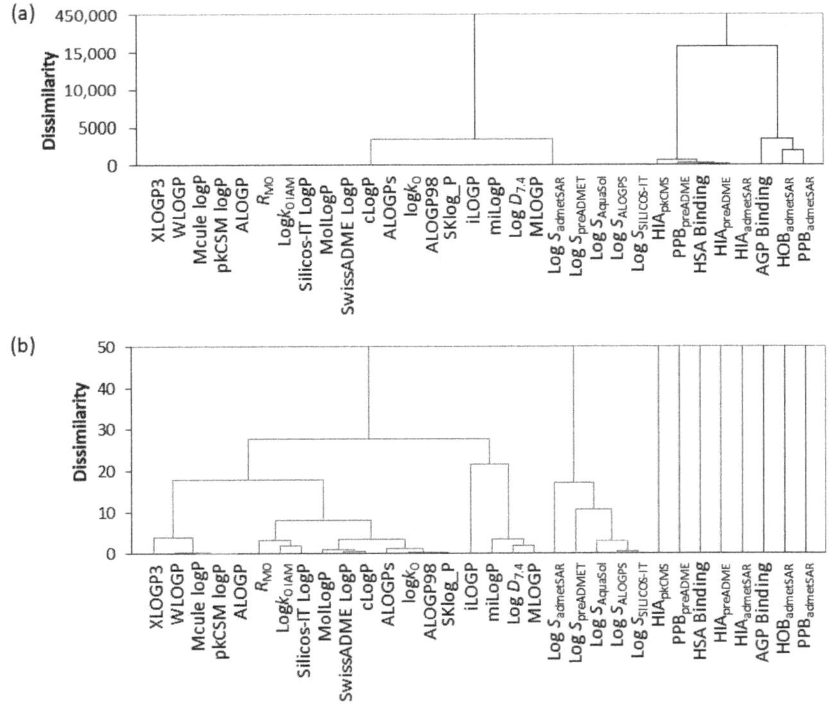

Figure 3. Results of CA showing (**a**) observed dendrogram in full scale from 0–450,000 with brake after 15,000, and (**b**) dendrogram showing only 0–50 range of dissimilarity values.

For the better visualisation of CA results, due to the high dissimilarity of those two clusters, Figure 3b shows a fraction of the main dendrogram. We can see that the green cluster is comprised of two subclusters. The first subcluster shows results related to lipophilicity. Experimentally obtained log $D_{7.4}$ values are closest to values predicted by the MLOGP procedure, followed by procedure of Molinspiration Cheminformatics Group. Log P values predicted by ALOGP98 and SKlog_P are closest to those experimentally observed by C18-HPLC assay, followed by predicted ALOGPs values. With best accuracy Silicos-IT LogP values predict $\log k_{0\ IAM}$ ones collected using IAM-HPLC assay. Following the dendrogram we can notice which model predicted log P values with the best accuracy, leaving iLOGP with the highest dissimilarity implying that it showed poor performance. Second subcluster is comprised of log S values showing the order of similarity between used prediction tools.

The second main cluster, related to plasma protein bindings, shows that the predicted values are not as accurate as ones observed experimentally, which was also observed in PCA.

3. Materials and Methods

3.1. Chemicals

Analytical grade standards of biologically active ingredients of botanicals (curcumin, demetoxycurcumin, bisdemetoxycurcumin, α-boswellic acid, β-boswellic acid, 11-keto-β-boswellic acid, 3-acetyl-11-keto-β-boswellic acid, neoandrographolide and piperine) were purchased from Sigma-Aldrich (St. Louis, MO, USA), while andrographolide (≥98.0%) was obtained from TCI (Tokyo, Japan). Warfarin (PESTANAL®, analytical standard) was obtained by Sigma-Aldrich. Buffer solutions were prepared using phosphate buffer saline (PBS) tablets (Sigma Aldrich), di-sodium hydrogen phosphate dihidrate (buffer substance for chromatography) and sodium dihydrogen phosphate dihydrate (EMSURE® reagent Ph. Eur.) both by Merck KGaA, Darmstadt, Germany. Organic solvents *n*-octanol (gradient grade for liquid chromatography, ≥99%) and methanol (gradient grade for liquid chromatography LiChrosolv® Reag. Ph Eur.) as well as formic acid (for LC-MS, LiChropur®, 97.5–98.5%) and dimethyl sulfoxide (suitable for HPLC, ≥99.7%) were obtained by Merck KGaA. Dead volume of HPLC system was evaluated using sodium nitrate (reagent for USP/NF monographs) by J.T. Baker, Gliwice, Poland. Ultrapure water was produced using an Ultra Clear UV water purifying system (SG Water, Barsbuttel, Germany); resistivity > 18 MΩ/cm at 25 °C and total organic carbon < 5 ppb.

3.2. Methods

3.2.1. In Silico Calculation

Several software packages were used for lipophilicity calculations, whereas each is based on different algorithms. Six different log *P* values (iLOGP, XLOGP3, WLOGP, MLOGP, Silicos-IT LogP, SwissADME LogP) were attained using a web service developed by the Molecular Modeling Group of the Swiss Institute of Bioinformatic (available online: http://www.swissadme.ch, accessed on 1 June 2022). ALOGPs values were calculated using a software developed by Virtual Computational Chemistry Laboratory (available online: http://www.vcclab.org, accessed on 1 June 2022). Free web property calculation service developed and maintained by Molinspiration Cheminformatics Group was used for calculation of miLogP values (available online: https://www.molinspiration.com, accessed on 1 June 2022), while Mcule logP values were derived using an online drug discovery platform developed by Mcule, Inc. (available online: https://mcule.com/apps/property-calculator/, accessed on 1 June 2022). OSIRIS Property Explorer as Integral part of Actelion's system (available online: https://www.organic-chemistry.org/prog/peo/, accessed on 1 June 2022) calculated cLogP values. Free on-line software tool (available online: https://molsoft.com/mprop/, accessed on 1 June 2022) from Molsoft L.L.C. was used for acquiring MolLogP values. In the same vein, the web-based application for drug-likeness prediction preADMET QSARhub available online: https://preadmet.qsarhub.com/druglikeness/ (accessed on 1 June 2022) was used for calculation of ALOGP98 values.

Software packages by the Molecular Modeling Group of the Swiss Institute of Bioinformatic and Virtual Computational Chemistry Laboratory enabled solubility prediction (log $S_{SILICOS\text{-}IT}$ and log S_{ALOGPS}). This parameter as AquaSol was also available at ChemDB Portal available at https://re.edugen.wiley.com/, accessed on 1 June 2022.

The pkCSM integrated freely available web server platforms for predicting small molecule pharmacokinetic properties from the University of Melbourne and University of Cambridge available at: http://biosig.unimelb.edu.au/pkcsm/, accessed on 1 June 2022 was used for calculation of log *P* values and pharmacokinetic property (Human Intestinal Absorption). Web service preADMET, available online at https://preadmet.webservice.bmdrc.org/adme/, accessed on 1 June 2022 from Yonsei University, was used to predict lipophilicity (SKlog_P), solubility (log $S_{preADMET}$) and biological properties (Human Intestinal Absorption and Plasma Protein Binding). Finally, an upgraded version of the comprehensive source and free tool for evaluating ALOGP values, solubility (log $S_{admetSAR}$) and similar chemical pharmacokinetic properties (Human Intestinal Absorption, Human Oral Bioavailability and Plasma Protein Binding) admetSAR is available online: http://lmmd.ecust.edu.cn/admetsar2/, accessed on 1 June 2022.

3.2.2. Shake-Flask Method with Chromatographic Analysis

Sample Preparation

Stock solution of biologically active ingredients of botanicals used in IBD treatment were prepared using dimethylsulfoxide in concentration of 2 mg/mL. Working solutions were prepared daily by dilution of the stock solutions with PBS (20 mM, pH 7.4) pre-saturated with *n*-octanol. Before the injection into the HPLC system all solutions were sonicated for 15 min at room temperature in Elmasonic XtraTT (Elma Schmidbauer,

Singen, Germany) and filtered through 0.2 μm polyethersulfone filters (Obrnuta faza, Pazin, Croatia).

The pre-saturated solutions of *n*-octanol and PBS (20 mM, pH 7.4) were used for determination of logarithm of distribution coefficient (log $D_{7.4}$). The *n*-octanol and aqueous phase (mixture 1:1, *v*/*v*) were mutually saturated for 24 h in an orbital shaker-incubator ES-20/60 (Biosan, Riga, Latvia) at 100 oscillations per min at 25.0 ± 0.1 °C. The prepared mixture was left resting for at least 12 h in the thermostat at 25.0 ± 0.1 °C to ensure complete separation of the two phases. Until analysis, separated saturated solutions were kept in refrigerator at 4 °C.

Shake-flask procedure was conducted in amber 2 mL HPLC vials. 300 μL of the working solution was added to 1500 μL of *n*-octanol pre-saturated with PBS. To prevent material loss due to volatilisation, the two-phase system nearly filled the entire volume of the test vessels. All *n*-octanol–buffer mixtures were first vortexed for 1 min and then shaken for 1 h (150 oscillations per min) at room temperature (25.0 ± 0.1 °C) to reach equilibrium and phase distribution. Afterward, the samples were centrifuged using a Z 326 K tabletop centrifuge (HERMLE Labortechnik, Wehingen, Germany) for 30 min at 4000× *g* and 25 °C to ensure that possible emulsions were removed.

Sample Analysis

After equilibration and phase separation, the aqueous phase of all samples was analysed on an Agilent HPLC 1100 (Agilent Technologies, Waldbronn, Germany) with diode array detection. Analysis was conducted with a Zorbax SB C18 column (length 150 mm, i.d. 4.6 mm, particle size 5 μm) by Agilent Technologies. 10 min chromatographic runs were performed at 25.0 ± 0.1 °C in isocratic mode using a methanol/water mixture (60/40, *v*/*v*) with addition of 0.1% of formic acid as mobile phase modifier at a flow rate of 1.0 mL/min. The mobile phase was filtered through a membrane filter, no. 66, diameter 47 mm, pore size 0.45 μm (Supelco, Bellefonte, PA, USA). The vials were placed in the rack on the autosampler at 25 °C. The injection volume was 10 μL, with a needle offset of 0.5 mm. To avoid any carryover of the analyte-containing *n*-octanol phase, the outside of the syringe was subsequently washed in methanol before injection. The absorbance of the analytes during a chromatographic run was collected in the spectral range 200–400 nm, and the detection wavelength for each analyte was the one providing the maximum peak height.

3.2.3. Chromatographic Methods

Sample Preparation

Stock solutions of analytes and calibration standards were prepared by dissolving 10 mg of each solute in 10 mL of methanol and kept at 4 °C. Working solutions were freshly prepared at the beginning of each day by dilution of the stock solutions to 100 μg/mL with methanol for TLC measurements or mobile phase for HPLC measurements. Before use, the samples were sonicated (Elmasonic XtraTT, Elma Schmidbauer) for 10 min at 25 °C and all working solutions were filtered through 0.2 μm polyethersulfone filters (Obrnuta faza). To protect analytes from photodegradation, the amber glass HPLC vials were used.

C18-TLC Assay

Analyses were carried out using a commercially available 10 cm x 20 cm RP-18 TLC plates with fluorescent indicator (Merck KGaA). 5 μL of each working solution (100 μL/mL) was applied to the plates as 5 mm bends, 10 mm from the lower edge and 15 mm from sides of plates. The mobile phases were prepared by mixing the respective amounts of PBS (20 mM, pH 7.4) and methanol from 50% to 80% of organic solvent in 5% increment. The vertical flat-bottom chamber (Camag, Muttenz, Switzerland) with a stainless-steel lid was saturated with mobile phase for 30 min. The plates were developed to a distance of 9 cm at room temperature, dried in the open air for 5 min and visualized under λ = 254 nm UV light (UV Cabinet with Dual Wavelength UV lamp by Camag).

C18- and IAM-HPLC Assays

C18-HPLC assay was carried out at Agilent HPLC 1100 and Zorbax SB C18 chromatographic column (length 250 mm.6 mm, particle size 5 μm) by Agilent Technologies, while IAM-HPLC assay was conducted using IAM P.C.DD2 chromatographic column (length 50 mm, i.d. 3.0 mm, particle size 300 Å) by Regis Technologies (Morton Grove, IL, USA). Retention data were collected at 25.0 ± 0.1 °C using isocratic method working with mobile phases, mixing the respective amounts of PBS (10 mM, pH 7.4) and methanol from 35% to 80% of organic solvent in 5% increment. Each mobile phase was shaken vigorously and filtrated through membrane filter, no. 66, diameter 47 mm, pore size 0.45 μm by Supelco and degassed by sonication 5 min before use. The injection volume was set at 10 μL, and the measurements were carried out at flow rate 1.0 mL/min. Natrium nitrate solution (0.1 mg/mL in mobile phase) was used as the marker of dead-time. The absorbance of the analytes during a chromatographic run was collected in the spectral range of 200–400 nm, and the detection wavelength for each analyte was the one providing the maximum peak height.

HSA- and AGP-HPAC Assays

The binding of biologically active compounds to the plasma proteins was investigated using affinity chromatographic columns by ChromTech (Cedex, France) containing immobilized human serum albumin (HSA) (Chiralpak-HSA, length 50 mm, i.d. 4.6 mm, particle size 5 μm) and α1-acid glycoprotein (AGP) (Chiralpak-AGP, length 50 mm, i.d. 4.6 mm, particle size 5 μm) and Agilent HPLC 1100 instrument. The 20 mM potassium phosphate buffer with the pH adjusted to 7.4 and iso-propanol were used as mobile phase components A and B. Retention data were collected at 25.0 ± 0.1 °C using gradient program as follows: 0 min/0% B, 6 min/30% B, 15 min/30% B and 20 min/0% B. Injection volume was set at 10 μL and the measurements were carried out at 25.0 ± 0.1 °C with flow rate of 1.5 mL/min. The absorbance of the analytes during a chromatographic run was collected in the spectral range 200–400 nm, and the detection wavelength for each analyte was the one providing the maximum peak height.

3.2.4. Statistical Analysis

Statistical analyses for shake-flask method and biomimetic chromatography were conducted using Microsoft Excel v16.0.14026.20270 (Microsoft Corporation, Redmond, WA, USA).

PCA and CA analysis was performed using XLSTAT by Addinsoft (Paris, France) with the aim of correlation and better visualization of experimental and calculated data. A matrix 30 × 10 was created where number of rows represented data related to each analyte and columns represented data obtained from each prediction tool. Statistical data processing was performed using the principal component analysis with Pearson's correlation and cluster analysis using Euclidean distance measures and Ward's agglomerative clustering.

4. Conclusions

This research describes the shake-flask and chromatographic investigations used to understand the interactions of bioavailability challenged biologically active ingredients of botanicals used in IBD treatment with cell membrane and plasma protein models.

Our findings imply that IAM-HPLC assay is recommended as the most suitable method for determining the lipophilicity of these classes of biologically active compounds, regardless of their structure and botanical source. Moreover, the excellent linearity of the regression models supports the idea that IAM-HPLC assay is useful in membrane permeability studies. The standardized HSA- and AGP-HPAC assays were successfully applied to evaluate of plasma protein binding of selected biologically active compounds. Our results revealed that investigated compounds have a higher affinity for HSA than AGP. A high affinity to phospholipids supported with meaningful inactivation by plasma proteins presented by both assays underpins the claim that a small portion of biologically active ingredients of botanicals used in IBD treatment are available for therapeutic effects.

Our research was empowered by 16 theoretical approaches innovatively applied, based on different algorithms to solve pharmacokinetics matters. The similarities between experimental and calculated values were evaluated using PCA and CA as a statistical tool. The outcomes from statistical analysis imply that among investigated parameters, plasma protein binding is the most complex, and further work in the improvement of in silico approaches in this area is wanted.

Supplementary Materials: The following supporting information can be downloaded at: https://www.mdpi.com/article/10.3390/ph15080965/s1, Table S1. Verification data of HSA-HPAC and AGP-HPAC assays.

Author Contributions: Conceptualization, A.M. and D.A.K.; methodology, A.M. and M.-L.J.; software, M.-L.J. and J.K.; formal analysis, A.M. and M.-L.J.; resources, A.M. and D.V.; data curation, D.A.K. and J.K.; writing—original draft preparation, A.M. and M.-L.J.; writing—review and editing, D.A.K., J.K. and D.V.; visualization, J.K.; supervision, A.M.; project administration, A.M. and D.V.; funding acquisition, A.M. and D.V. All authors have read and agreed to the published version of the manuscript.

Funding: This research was funded by CROATIAN SCIENCE FOUNDATION, grant number HRZZ-UIP-2017-05-3949 and HAMAG-BICRO_IRI, grant number KK.01.2.1.02.0142.

Institutional Review Board Statement: Not applicable.

Informed Consent Statement: Not applicable.

Data Availability Statement: Data are contained within the article and supplementary materials.

Conflicts of Interest: The authors declare no conflict of interest.

References

1. Jairath, V.; Feagan, B.G. Global burden of inflammatory bowel disease. *Lancet Gastroenterol. Hepatol.* **2020**, *5*, 2–3. [CrossRef]
2. Lin, S.C.; Cheifetz, A.S. The use of complementary and alternative medicine in patients with inflammatory bowel disease. *Gastroenterol. Hepatol.* **2018**, *14*, 415–425.
3. Governa, P.; Marchi, M.; Cocetta, V.; De Leo, B.; Saunders, P.T.K.; Catanzaro, D.; Miraldi, E.; Montopoli, M.; Biagi, M. Effects of Boswellia Serrata Roxb. and Curcuma longa L. in an in vitro intestinal inflammation model using immune cells and Caco-2. *Pharmaceuticals* **2018**, *11*, 126. [CrossRef] [PubMed]
4. Mishra, A.; Shaik, H.A.; Sinha, R.K.; Shah, B.R. Andrographolide: A herbal-chemosynthetic approach for enhancing immunity, combating viral infections, and its implication on human health. *Molecules* **2021**, *26*, 7036. [CrossRef]
5. Coelho, M.R.; Romi, M.D.; Ferreira, D.M.T.P.; Zaltman, C.; Soares-Mota, M. The use of curcumin as a complementary therapy in ulcerative colitis: A systematic review of randomized controlled clinical trials. *Nutrients* **2020**, *12*, 2296. [CrossRef] [PubMed]
6. Algieri, F.; Rodriguez-Nogales, A.; Rodriguez-Cabezas, M.E.; Risco, S.; Angeles Ocete, M.; Galvez, J. Botanical drugs as an emerging strategy in inflammatory bowel disease: A review. *Mediat. Inflamm.* **2015**, *2015*, 179616. [CrossRef]
7. Li, Q.; Zhai, W.; Jiang, Q.; Huang, R.; Liu, L.; Dai, J.; Gong, W.; Du, S.; Wu, Q. Curcumin-piperine mixtures in self-microemulsifying drug delivery system for ulcerative colitis therapy. *Int. J. Pharm.* **2015**, *490*, 22–31. [CrossRef]
8. Ermondi, G.; Vallaro, M.; Caron, G. Learning how to use IAM chromatography for predicting permeability. *Eur. J. Pharm. Sci.* **2018**, *114*, 385–390. [CrossRef]
9. Carrasco-Correa, E.J.; Ruiz-Allica, J.; Rodríguez-Fernández, J.F.; Miró, M. Human artificial membranes in (bio)analytical science: Potential for in vitro prediction of intestinal absorption-A review, *TrAC-Trend. Anal. Chem.* **2021**, *145*, 116446. [CrossRef]
10. Obradović, D.; Radan, M.; Đikić, T.; Popović Nikolić, M.; Oljačić, S.; Nikolić, K. The evaluation of drug-plasma protein binding interaction on immobilized human serum albumin stationary phase, aided by different computational approaches. *J. Pharmaceut. Biomed. Anal.* **2022**, *211*, 114593. [CrossRef]
11. Jeličić, M.-L.; Brusač, E.; Klarić, D.A.; Nigović, B.; Turk, N.; Mornar, A. A chromatographic approach to development of 5-aminosalicylate/folic acid fixed-dose combinations for treatment of Crohn's disease and ulcerative colitis. *Sci. Rep.* **2020**, *10*, 20838. [CrossRef] [PubMed]
12. Brusač, E.; Jeličić, M.-L.; Amidžić Klarić, D.; Nigović, B.; Turk, N.; Klarić, I.; Mornar, A. Pharmacokinetic profiling and simultaneous determination of thiopurine immunosuppressants and folic acid by chromatographic methods. *Molecules* **2019**, *24*, 3469. [CrossRef] [PubMed]

13. Russo, G.; Grumetto, L.; Baert, M.; Lynen, F. Comprehensive two-dimensional liquid chromatography as a biomimetic screening platform for pharmacokinetic profiling of compound libraries in early drug development. *Anal. Chim. Acta* **2021**, *1142*, 157–168. [CrossRef] [PubMed]
14. Abdel-Tawab, M. Considerations to be taken when carrying out medicinal plant research—what we learn from an insight into the IC50 values, bioavailability and clinical efficacy of exemplary anti-inflammatory herbal components. *Pharmaceuticals* **2021**, *14*, 437. [CrossRef]
15. Sharma, T.; Jana, S. Investigation of molecular properties that influence the permeability and oral bioavailability of major β-boswellic acids. *Eur. J. Drug Metab. Pharmacokinet.* **2020**, *45*, 243–255. [CrossRef]
16. Krüger, P.; Daneshfar, R.; Eckert, G.P.; Klein, J.; Volmer, D.A.; Bahr, U.; Müller, W.E.; Karas, M.; Schubert-Zsilavecz, M.; Abdel-Tawab, M. Metabolism of boswellic acids in vitro and in vivo. *Drug Metab. Dispos.* **2008**, *36*, 1135–1142. [CrossRef]
17. Godugu, D.; Rupula, K.; Sashidhar, R.B. Binding studies of andrographolide with human serum albumin: Molecular docking, chromatographic and spectroscopic studies. *Protein Pept. Lett.* **2018**, *25*, 330–338. [CrossRef]
18. Kempińska, D.; Chmiel, T.; Kot-Wasik, A.; Mróz, A.; Mazerska, Z.; Namieśnik, J. State of the art and prospects of methods for determination of lipophilicity of chemical compounds. *TrAC-Trend. Anal. Chem.* **2019**, *113*, 54–73. [CrossRef]
19. Brusač, E.; Jeličić, M.-L.; Amidžić Klarić, D.; Mornar, A. Miniaturized shake-flask HPLC method for determination of distribution coefficient of drugs used in inflammatory bowel diseases. *Acta Pharm.* **2019**, *69*, 649–660. [CrossRef]
20. International Conference on Harmonization of Technical Requirements for Registration of Pharmaceuticals for Human Use, ICH Harmonised Tripartite Guideline. Validation of Analytical Procedures: Text and Methodology Q2(R1), Current Step 4 Version. November 2005. Available online: https://www.ich.org/fileadmin/Public_Web_Site/ICH_Products/Guidelines/Quality/Q2_R1/Step4/Q2_R1__Guideline.pdf (accessed on 20 June 2022).
21. Kotagale, N.R.; Charde, P.B.; Helonde, A.; Gupta, K.R.; Umekar, M.J.; Raut, N.S. Studies on bioavailability enhancement of curcumin. *Int. J. Pharm. Pharm. Sci.* **2020**, *12*, 20–25. [CrossRef]
22. Jithavech, P.; Suwattananuruk, P.; Muangnoi, C.; Thitikornpong, W.; Towiwat, P.; Vajragupta, O.; Rojsitthisak, P. Physico-chemical investigation of a novel curcumin diethyl γ-aminobutyrate, a carbamate ester prodrug of curcumin with enhanced antineuroinflammatory activity. *PLoS ONE* **2022**, *17*, e0265689. [CrossRef] [PubMed]
23. Roy, N.K.; Parama, D.; Banik, K.; Bordoloi, D.; Devi, A.K.; Thakur, K.K.; Padmavathi, G.; Shakibaei, M.; Fan, L.; Sethi, G.; et al. An Update on pharmacological potential of boswellic acids against chronic diseases. *Int. J. Mol. Sci.* **2019**, *20*, 4101. [CrossRef] [PubMed]
24. Dei Cas, M.; Ghidoni, R. Dietary curcumin: Correlation between bioavailability and health potential. *Nutrients* **2019**, *11*, 2147. [CrossRef] [PubMed]
25. Boyle, C.A.; Coatney, R.W.; Wickham, A.; Mukherjee, S.K.; Meunier, L.D. Alpha-1 acid glycoprotein as a biomarker for subclinical illness and altered drug binding in rats. *Comp. Med.* **2021**, *71*, 123–132. [CrossRef]
26. Dezhampanah, H.; Shabanzade, Z. Investigation of binding interaction between human serum albumin with zirconium complex of curcumin and curcumin. *J. Biomol. Struct. Dyn.* **2022**, *40*, 722–732. [CrossRef]

pharmaceuticals

Review

Effects of Natural Product-Derived Compounds on Inflammatory Pain via Regulation of Microglial Activation

Joon Park [1,2,3], Changho Lee [1] and Yun Tai Kim [1,2,*]

1 Division of Functional Food Research, Korea Food Research Institute, Wanju 55365, Republic of Korea; bioscienceрark@gmail.com (J.P.); chang@kfri.re.kr (C.L.)
2 Department of Food Biotechnology, Korea University of Science and Technology, Daejeon 34113, Republic of Korea
3 Department of Anesthesiology, College of Medicine, The University of Arizona, Tucson, AZ 85724, USA
* Correspondence: ytkim@kfri.re.kr; Tel.: +82-63-219-9295; Fax: +82-63-219-9876

Abstract: Inflammatory pain is a type of pain caused by tissue damage associated with inflammation and is characterized by hypersensitivity to pain and neuroinflammation in the spinal cord. Neuroinflammation is significantly increased by various neurotransmitters and cytokines that are expressed in activated primary afferent neurons, and it plays a pivotal role in the development of inflammatory pain. The activation of microglia and elevated levels of pro-inflammatory cytokines are the hallmark features of neuroinflammation. During the development of neuroinflammation, various intracellular signaling pathways are activated or inhibited in microglia, leading to the regulation of inflammatory proteins and cytokines. Numerous attempts have been conducted to alleviate inflammatory pain by inhibiting microglial activation. Natural products and their compounds have gained attention as potential candidates for suppressing inflammatory pain due to verified safety through centuries of use. Many studies have also shown that natural product-derived compounds have the potential to suppress microglial activation and alleviate inflammatory pain. Herein, we review the literature on inflammatory mediators and intracellular signaling involved in microglial activation in inflammatory pain, as well as natural product-derived compounds that have been found to suppress microglial activation. This review suggests that natural product-derived compounds have the potential to alleviate inflammatory pain through the suppression of microglial activation.

Keywords: natural products; inflammatory pain; microglial activation; microglia; neuroinflammation

Citation: Park, J.; Lee, C.; Kim, Y.T. Effects of Natural Product-Derived Compounds on Inflammatory Pain via Regulation of Microglial Activation. *Pharmaceuticals* **2023**, *16*, 941. https://doi.org/10.3390/ph16070941

Academic Editor: Diana Roxana Pelinescu

Received: 8 June 2023
Revised: 22 June 2023
Accepted: 26 June 2023
Published: 29 June 2023

1. Introduction

The International Association for the Study of Pain (IASP) describes pain as an unpleasant sensory and emotional experience associated with, or resembling that associated with, actual or potential tissue damage. If pain persists for more than 3 months, which is considered the tissue healing period, it is diagnosed as chronic pain. In 2019, approximately 20% of adults in the US were diagnosed with chronic pain, and the number of patients with chronic pain is increasing with the increase in the aging population [1]. According to a National Institute of Health (NIH) report, the cost of treating chronic pain exceeds that of a few major diseases related to the highest morbidity and mortality, such as cardiovascular diseases (USD 309 billion), cancers (USD 243 billion), and injuries (USD 205 billion) [2]. Patients with chronic pain have a poor quality of life for reasons such as difficulty engaging in daily activities and prolonged treatment. Patients with chronic pain also suffer from mental disorders such as depression, anxiety disorders, and sleep disturbances [3]. As the average lifespan continues to increase, the importance of pain management is also increasing.

Pain perception is a complex and highly orchestrated process involving a series of sequential events [4]. Following inflammation-related tissue damage, nociceptors, which are sensory neurons responsible for detecting harmful stimuli, initiate a response. Nociceptors convert stimuli into electrical signals, which are then transmitted to the central nervous system (CNS). Subsequently, these electrical signals are transmitted to secondary afferent neurons located in the dorsal horn of the spinal cord. With the repetitive transmission of pain signals, neuroinflammation is strongly induced within the spinal cord. Neuroinflammation profoundly influences synaptic transmission, thereby contributing to the persistence of pain.

The causes of chronic pain include nerve injury, cancer, muscle injury, and inflammation [5]. Among the various causes, pain caused by inflammation is called inflammatory pain. Currently, non-steroidal anti-inflammatory drugs (NSAIDs) are the most commonly used drugs to treat inflammatory pain [6]. The main mechanism underlying the analgesic and anti-inflammatory effects of NSAIDs is inhibition of the cyclooxygenase (COX2) enzyme, which produces prostaglandins. Prostaglandin is a representative inflammatory mediator that induces fever, inflammation, and pain. Aspirin, naproxen, and ibuprofen are some of the commonly used NSAIDs. However, since NSAIDs have adverse effects, such as indigestion, stomach ulcers, headaches, drowsiness, and dizziness, interest in natural products as agents for alleviating inflammatory pain has recently increased. Natural products have been used for centuries to treat various diseases related with inflammation, without causing side effects [7]. Therefore, natural products have the potential to be developed into new drugs against inflammatory pain.

In this review, we aim to summarize the current understanding of inflammatory factors and intracellular signaling involved in the development of inflammatory pain as well as to highlight the potential of natural products in treating inflammatory pain.

2. Mechanism Underlying the Development of Inflammatory Pain

Inflammatory pain is characterized by a heightened sensitivity to pain due to tissue damage resulting from an inflammatory or immune response. The two typical symptoms of pain are hyperalgesia and allodynia. Hyperalgesia is characterized by an abnormally increased sensitivity to pain and an extreme response to pain. Allodynia is recognized as pain for common stimuli that do not normally cause pain. Chronic inflammatory diseases and infections, such as arthritis, shingles, and tissue injury are representative causes of inflammatory pain [8]. Inflammatory mediators are released locally by immune cells at the site of inflammation and can directly activate sensory neurons in peripheral tissues. Activated sensory neurons then release neuropeptides, such as substance P, calcitonin gene-related peptide, and prostanoids, into the dorsal horn of the spinal cord. Repetitive and persistent stimulation of sensory neurons can lead to the over-release of neuropeptides, resulting in neuroinflammation of the spinal cord.

Spinal neuroinflammation, caused by peripheral inflammation, is characterized by the activation of microglia and increased expression of inflammatory mediators in the spinal cord [9]. Activated microglia are major sources of pro-inflammatory cytokines and inflammation-related proteins that are regulated by various intracellular signaling (Figure 1). Along with spinal neuroinflammation, microglial activation is significantly induced, resulting in pain hypersensitivity through central sensitization. Central sensitization, a leading cause of chronic pain, represents the reinforcement of the function between pre- and post-synaptic neurons in the nociceptive pathway caused by increasing excitatory transmission and strengthening of synapses in response to inflammation and nerve injury. The increased levels of pro-inflammatory cytokines can enhance synaptic transmission by increasing excitatory synaptic transmission and decreasing inhibitory synaptic transmission in the dorsal horn of spinal cord [10,11]. Pro-inflammatory cytokines may induce gene expression by activating cAMP response element-binding protein (CREB) transcription factors, leading to long-term potentiation [12]. These results comprehensively contribute to the persistence and hypersensitivity of pain through central sensitization.

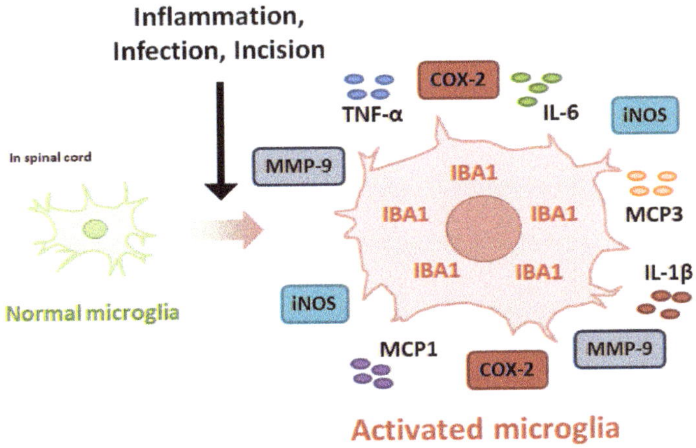

Figure 1. Inflammatory mediators associated with the development of microglial activation-mediated inflammatory pain. Tissue damage associated with inflammation lead to microglial activation by releasing diverse signaling molecules from sensory neurons. Activated microglia induce neuroinflammation through increasing the expression of inflammatory mediators, such as excessive NO, iNOS, COX-2, TNF-α, IL-1β, IL-6, MCP1, and MCP3. These mediators lead to central sensitization, resulting in increased sensitivity to pain. Activated microglia can trigger inflammatory pain via inflammatory mediator-related signaling.

Microglia, the resident macrophage-like cells located in the CNS, play an important role in the development of chronic pain associated with neuroinflammation [13,14]. As the first immune cell in the CNS, microglia are essential for brain maintenance and homeostasis as they are involved in removing cell debris, infectious agents, and other unnecessary elements. Although the primary function of microglia is to protect the CNS, they can have destructive effects on neurons. Various signaling molecules are released from sensory neurons damaged by inflammation and eventually activate microglia [13]. Microglial activation is characterized by the increased production of inflammatory mediators, such as iNOS, COX-2, MMP-9, TNF-α, IL-1β, IL-6, MCP1, and MCP3. The expression of these inflammatory mediators is regulated by intracellular signaling pathways, including NF-B, MAPK, JAK2-STAT3, Nrf2, and autophagy. The expression and secretion of inflammatory mediators are induced by intracellular signaling in activated microglia, resulting in an increase in neuroinflammation [15]. Inflammatory mediators contribute to increased neuroinflammation to damage to cells. Furthermore, these mediators and cytokines are involved in the induction and maintenance of central sensitization by upregulating the NMDA receptor in excitatory synaptic neurons [16]. In agreement with these findings, pro-inflammatory cytokines, such as TNF-α, IL-1β, or IL-6, modulate the function of receptors associated with central sensitization in the spinal cord [12]. Experimental studies using rodent models have shown that intrathecal injection of pro-inflammatory cytokines induces pain hypersensitivity [10,17,18]. A previous study showed that the specific deletion of microglia in the spinal cord had an inhibitory effect on formalin-induced inflammatory pain via the modulation of central sensitization [19]. Inhibition of microglial activation using chemogenetic approaches, specifically DREADD, alleviate neuroinflammation and chronic pain following nerve injury [20]. Inhibition of microglial activation attenuated nerve injury-induced pain hypersensitivity in the early phase but not the late phase. Therefore, understanding and targeting the processes and factors involved in microglial activation-induced neuroinflammation may offer an effective approach to prevent the early phase of inflammatory pain, which has the potential to become chronic pain.

3. Expression of Inflammatory Mediators in Activated Microglia

3.1. Inducible Nitric Oxide Synthase

Nitric oxide synthase (NOS) is an enzyme that catalyzes the production of nitric oxide (NO) from L-arginine. Among the different isoforms of NOS, inducible nitric oxide synthase (iNOS) plays a significant role in the development of inflammatory pain. iNOS is induced in various cells and tissues by cytokines and other molecules. Although previous studies have reported the role of iNOS in various inflammatory diseases except for the CNS, it has been recently confirmed that iNOS contributes to the development of chronic pain [21,22]. iNOS continuously produces large amounts of NO until it is degraded. High amounts of NO result in the production of high levels of reactive nitrogen oxide species (RNOS), causing damage to the surrounding tissue and cells. Among various cells in the CNS, microglia are the major cellular sources of iNOS [23]. Many previous studies have shown that activated microglia remarkably increase the expression of iNOS, leading to the excessive production of NO [24–26]. Additionally, iNOS expression and neuroinflammation in the dorsal horn of the spinal cord were found to have significantly increased in CFA-injected mice [27]. Osborne et al. reported that carrageenan-induced thermal hyperalgesia was significantly alleviated by the intrathecal injection of the nonselective NOS inhibitor L-NAME in rats. Furthermore, a selective iNOS inhibitor suppressed thermal hypersensitivity during carrageenan-induced inflammatory pain [28]. Formalin-induced pain behavior was attenuated in iNOS knockout mice. Moreover, nerve injury-induced pain hypersensitivity and microglial activation in the spinal cord were suppressed in the iNOS knockout mice compared with in wild-type mice [26]. These results demonstrated that iNOS expression in activated microglia can exacerbate neuroinflammation, resulting in increased pain sensitivity.

3.2. Cyclooxygenase-2

Cyclooxygenase-2 (COX-2) is a primary target for reducing inflammation and pain. COX-2 converts arachidonic acid to prostaglandin E2 (PGE2), which is associated with inflammation and pain. COX-2 is induced in response to inflammatory stimulation and is primarily expressed by monocytes, macrophages, fibroblasts, neurons, and microglia [29,30]. COX-2 inhibition is expected to reduce inflammation and pain without causing side effects [31]. Many studies have shown that lipopolysaccharide (LPS) treatment increased COX-2 expression in microglia [32–35]. These studies suggest that neuroinflammation can be suppressed by inhibition of COX-2 expression in microglia. In addition, various COX-2 inhibitors suppressed neuroinflammation by inhibiting the release of inflammatory mediators by microglia [36]. Naproxen is a representative oral NSAID; it attenuated CFA-induced pain hypersensitivity by inhibiting COX-2 expression in the spinal cord [37]. Therefore, microglia-specific inhibition of COX-2 expression has the potential to alleviate inflammatory pain.

3.3. Matrix Metalloproteinases-9

Matrix metalloproteinases-9 (MMP-9) is a member of the zinc metalloproteinase family involved in the degradation of the extracellular matrix and is strongly implicated in the development of various neuroinflammation-related diseases [38]. Microglia are a major source of MMP-9. Nerve injury has been found to increase the expression of MMP-9 rapidly and temporarily in the dorsal root ganglion (DRG), leading to the induction of neuropathic pain [39,40]. Microglial activation is also increased by nerve injury-induced MMP-9 expression in the spinal cord. Intrathecal injection of siMMP-9 significantly attenuated nerve injury-induced pain hypersensitivity and inhibited microglial activation in the spinal cord. A previous study found that activated microglia significantly increased the expression of MMP-9 in LPS-treated microglial cells [41]. Additionally, in one study, MMP-9 expression was notably upregulated in the DRG and spinal cord in a CFA-induced inflammatory pain model [42]. This study showed that inhibition of MMP-9 had inhibitory effects on

CFA-induced pain hypersensitivity in rats. These studies indicated that MMP-9 is involved in the development of microglial activation-mediated inflammatory pain.

3.4. Pro-Inflammatory Cytokines

Cytokines are secreted mainly by the immune and glial cells of the CNS. These cytokines act as intercellular mediators that control the function and differentiation of other cells [43]. In response to peripheral inflammation and tissue injury, microglia can be activated to secrete pro-inflammatory cytokines, such as TNF-α, interleukin-1β (IL-1β), IL-6, monocyte chemotactic protein-1 (MCP1), and MCP3 [24,33,34,37]. These pro-inflammatory cytokines are strongly involved in neuroinflammation and mainly contribute to the exacerbation of chronic pain.

3.4.1. TNF-α

Tumor necrosis factor-α (TNF-α) is a cytokine that causes inflammation-related diseases and might be a potential therapeutic target. Moreover, TNF-α is proposed to be a pro-inflammatory cytokine that plays a critical role in the development of chronic pain [43]. Microglia express and secrete TNF-α in response to stimuli and can also be activated by TNF-α via TNF receptors (TNFRs) [44]. In one study, TNF-α induced microglial activation, as evidenced by the increased expression of iNOS, IL-1β, and IL-6 in primary cells [45]. This study showed that LPS-induced microglial activation was partially blocked by treatment with TNFR type 1 (TNFR1) antibody. Moreover, intrathecal injection of TNF-α significantly induced pain hypersensitivity in mice. In addition, TNFR1 knockout mice exhibited better inhibition of pain hypersensitivity than TNFR2 knockout mice in CFA and formalin-induced inflammatory pain models [10]. Thus, TNF-α can activate microglia via binding to TNFR1 to increase pain sensitivity.

3.4.2. Interleukin-1β

Interleukin-1β (IL-1β) is one of the important mediators of the inflammatory response and is implicated in microglial activation-mediated inflammatory pain. The IL-1β precursor is cleaved by cytosolic caspase 1 and activated to mediate the inflammatory response. Previous studies showed that IL-1β could activate microglia, as revealed by the increased expression of IL-6, MCP1, and CXCL10 in human microglia cells [46,47]. IL-1β is a major mediator that increases the expression of COX-2 in the spinal cord, resulting in the development of CFA-induced inflammatory pain [48]. In one study of a mouse model, inhibition of IL-1β in the spinal cord decreased sensitivity to pain by decreasing the expression of COX2. Another study reported that intravenous injection of IL-1β significantly increased pain hypersensitivity and microglial activation in dorsal horn of the spinal cord [49].

3.4.3. Interelukin-6

Interelukin-6 (IL-6) is also a well-known inflammatory cytokine along with TNF-α and IL-1β. IL-6 is an important mediator of fever and pathogenesis of chronic pain. In one study, exposure to LPS remarkably induced the expression of IL-6 in microglia [50]. Furthermore, formalin- and CFA-injected mice showed notably increased microglial activation and IL-6 expression in the spinal cord [51,52]. In addition, nerve injury-induced IL-6 expression was decreased by microglia inhibitor in serum and spinal cord [17,53]. This study showed that intrathecal injection of IL-6 significantly increased microglial activation in the spinal cord. These data indicate the presence of a positive feedback loop between IL-6 and microglial activation, resulting in the pathogenesis of microglial activation-mediated inflammatory pain.

3.4.4. Monocyte Chemoattractant Protein-1

MCP1 is one of the key chemokines that control the migration and infiltration of monocyte. MCP1 plays an important role in the development of chronic pain. Although MCP1 is known to interact with several receptors, C-C chemokine receptor type 2 (CCR2) is its preferred receptor [54]. Direct injection of MCP1 into the spinal cord induced pain hypersensitivity, whereas co-treatment with an MCP1 inhibitor reduces sensitivity to pain by blocking central sensitization [11]. This study showed that CFA-induced pain hypersensitivity in mice was significantly attenuated by intrathecal injection of a CCR2 inhibitor. Additionally, the intrathecal injection of MCP1 induced microglial activation in the spinal cord [55]; spinal microglial activation was also markedly decreased by an antibody against MCP1. A previous study reported that MCP1 was strongly increased in DRG neurons due to peripheral inflammation and was transported into the spinal cord [56]. In addition, neuron-derived MCP1 notably induced microglial activation, as evidenced by the upregulated expression of iNOS, COX2, IL-1β, and IL-6 [57]. Several in vitro experiments showed that activated microglia could secrete MCP1 [24,58]. These results suggest that peripheral inflammation induces MCP1 expression in the neurons. MCP1 is transported into the spinal cord, followed by autocrine activation of microglia by MCP1. Therefore, peripheral inflammation-induced MCP1 expression may increase pain sensitivity by activation of microglia through a positive feedback loop.

3.4.5. Monocyte Chemoattractant Protein-3

MCP3 is a small cytokine that is closely related to MCP1. MCP3 has been found to play a role in the development of chronic pain. In one study, nerve injury remarkably induced MCP3 expression in the spinal cords of mice [18]. Moreover, nerve injury-induced microglial activation was decreased in the spinal cord of CCR2 knockout mice compared to in normal mice. Moreover, intrathecal injection of MCP3 induced pain hypersensitivity in a dose-dependent manner, while intrathecal injection of a CCR2 inhibitor or antibody against MCP3 reduced sensitivity to pain in mice. This study suggests that MCP3 is primarily expressed in astrocytes. Astrocyte-derived MCP3 plays a key role in the development of neuropathic pain. However, many studies have reported that MCP3 is also expressed in activated microglia [24,59]. Furthermore, CFA-induced pain hypersensitivity was attenuated by inhibition of microglial activation and MCP3 expression in the spinal cord. Microglial activation was inhibited by MCP3 knockdown. Therefore, inhibition of MCP3 expression may alleviate inflammation-induced pain hypersensitivity by regulating microglial activation in the spinal cord.

4. Intracellular Signaling in Activated Microglia

Activated microglia release inflammatory mediators that may contribute to hypersensitivity to inflammatory pain. Many studies have demonstrated the importance of intracellular signaling pathways that are strongly involved in microglial activation-mediated inflammatory pain. Activation of microglia leads to the induction of cascades of numerous intracellular signaling pathways. These signaling pathways may contribute to changes in the function of microglia, and gene expression resulting from these signaling pathways may influence the functions and structures of nearby cells and tissues, resulting in exacerbated inflammatory pain. Understanding how intracellular signaling pathways work in activated microglia may help identify new therapeutic targets for inflammatory pain (Figure 2).

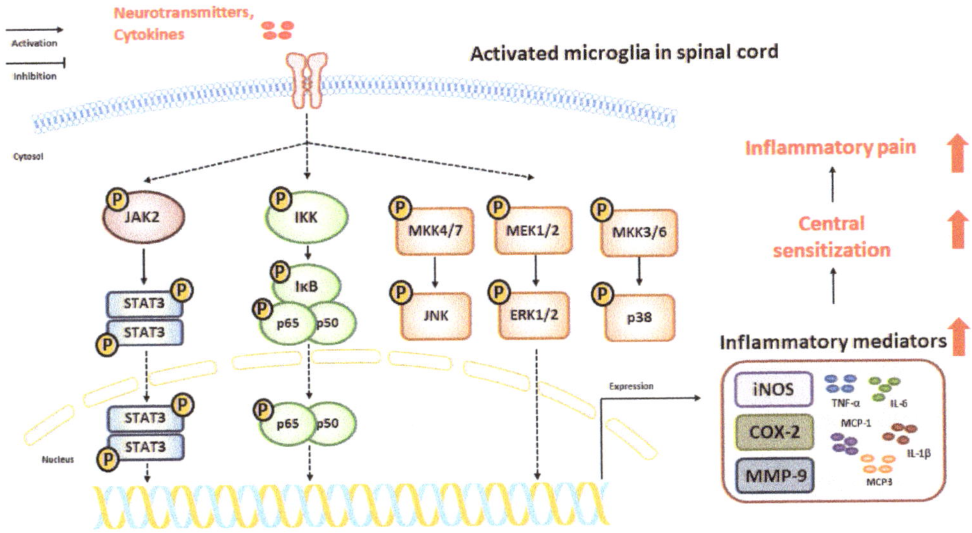

Figure 2. Intracellular signaling that can induce inflammatory mediators in activated microglia. NF-κB, MAPK, and JAK2/STAT3 signaling can be activated in microglia. And this signaling can induce expression of inflammatory mediators in activated microglia, leading to exacerbation of inflammatory pain.

4.1. Nuclear Factor-κB

Nuclear factor-κB (NF-κB) is a representative of the family of transcript factors associated with the inflammatory response. The activation of NF-κB signaling involves two major pathways: canonical and noncanonical. Although both pathways are important for regulating the inflammatory response, the noncanonical pathway is particularly involved in regulating specific functions of the adaptive immune system [60]. This chapter focuses on canonical NF-κB signaling, which is associated with microglial activation. Canonical NF-κB signaling is triggered by various stimuli [61]. The first step in the canonical NF-κB signaling pathway is the activation of the IκB kinase (IKK) complex, which comprises IKKα, IKKβ, and IKKγ subunits [62]. The IKK complex phosphorylates IκBα, leading to ubiquitylation and proteasomal degradation. This results in the phosphorylation and nuclear translocation of the NF-κB dimer (p65 and p50). The translocated NF-κB dimer binds to a specific DNA sequence and promotes the transcription of target genes. NF-κB signaling is strongly associated with the development of microglial activation, as evidenced by increased expression of inflammatory mediators and cytokines [63]. Previous studies have shown that microglial activation was significantly reduced by treatment with NF-κB inhibitors [64,65]. Moreover, LPS-induced expression of inflammatory cytokines and mediators was reduced by suppression of NF-κB activity through IKK-specific deletion in microglia [66]. Inhibition of NF-κB in the spinal cord showed an alleviative effect on CFA-induced pain hypersensitivity and microglial activation [67]. Furthermore, nerve injury-induced upregulation of pain sensitivity was alleviated by the suppression of NF-κB signaling in microglial activation [68]. These data demonstrate that activation of NF-κB signaling in microglia can lead to increased pain sensitivity in inflammatory pain.

4.2. Mitogen-Activated Protein Kinase

The mitogen-activated protein kinase (MAPK) signaling pathway, which comprises the c-Jun N-terminal kinase (JNK), extracellular signal-regulated kinase (ERK), and p38 mitogen-activated protein (p38) kinase, plays a crucial role in regulating various cellular functions such as proliferation, differentiation, development, and migration [69]. MAPK is involved in protein kinase cascades, in which they are activated in a sequential manner by the upstream signals such as MAPKK and MAPKKK.

In the nervous system, JNK is implicated in the pathogenesis of various neuro inflammation-related diseases [70,71]. JNK is activated by its upstream signals, MKK4 and MKK7, leading to the phosphorylation of the downstream signal, c-Jun. The activated JNK could induce the production of inflammatory mediators and cytokines in the CNS. Thus, the inhibition of JNK has been considered a therapeutic target for the treatment of neurodegenerative diseases. In addition, the inhibition of JNK in microglia has been suggested to attenuate inflammatory pain. Previously, LPS-induced microglial activation was remarkably suppressed by treatment with a JNK inhibitor, as revealed by the reduced expression of inflammatory mediators and cytokines [72]. Additionally, CFA-induced pain hypersensitivity was attenuated by intrathecal injection of a JNK inhibitor in rats [73].

Among the MAPK family members, ERK1/2 activation by MEK1/2 is generally considered to regulate cell survival, proliferation, and differentiation. The activation of ERK in microglia leads to neuroinflammation by increasing the expression of pro-inflammatory cytokines and inflammatory mediators [74]. These effects have become the cornerstone in the development of neurodegenerative diseases. A previous study identified that ERK activation in microglia was significantly increased in the early stages of nerve injury-induced chronic pain [75]. In in vitro experiments, LPS-induced expression of iNOS, COX-2, and pro-inflammatory cytokines was notably reduced by treatment with ERK inhibitor in microglia [76,77]. In addition, direct injection of ERK inhibitor into the spinal cord showed inhibitory effects on CFA-induced pain hypersensitivity in mice [78,79]; these effects were accompanied by a reduction in COX-2 expression in the spinal cord.

p38 is more strongly involved in the development of chronic pain related to microglial activation than other MAPK family members. p38 is activated by the upstream kinases MKK3 and MKK6 and plays an important role in the inflammatory response. p38 inhibitors have been found to alleviate inflammatory diseases [80]. During the development of chronic pain, p38 activation is notably increased in microglia compared to that in other cells [81]. In addition, the activation of p38 in microglia was upregulated in CFA-induced inflammatory pain. Moreover, intrathecal injection of p38 inhibitor attenuated CFA- or formalin-induced pain-like behaviors and significantly suppressed expression of IL-1β and IL-6 in the spinal cord [82–84]. In in vitro studies, LPS-induced NO overproduction and expression of iNOS and COX2 were reduced by treatment with a p38 inhibitor in a dose-dependent manner [37]. Taken together, these results demonstrate that MAPK signaling is critical for microglial activation and leads to the pathogenesis of inflammatory pain.

4.3. *Janus Kinase 2 (JAK2)/Signal Transducer and Activator of Transcription 3*

The Janus kinase 2 (JAK2)/signal transducer and activator of transcription 3 (STAT3) pathway is an intracellular signaling pathway activated by cytokines. The JAK2/STAT3 pathway is involved in immune cell division, development, recruitment, and activation. Many studies have indicated that microglial activation is dependent on the phosphorylation of the JAK2/STA3 signaling pathway caused by various stimuli [85–87]. These data revealed that inhibition of JAK2/STAT3 signaling suppressed microglial activation, as shown by the decreased expression of inflammatory cytokines and mediators. Additionally, CFA-induced pain hypersensitivity and spinal microglial activation were significantly reduced by inhibition of the JAK2/STAT3 signaling pathway in rodents [24,88,89]. These results indicate that JAK2/STAT3 signaling plays an important role in microglial activation and development of inflammatory pain.

4.4. *Nuclear Factor-Erythroid 2-Related Factor 2*

Nuclear factor-erythroid 2-related factor 2 (Nrf2) is a transcription factor that regulates antioxidant enzymes to protect against damage caused by oxidative stress. Oxidative stress is a hallmark of neuroinflammation and neurodegeneration that leads to disease progression [90]. Accumulating evidence has shown that microglial activation was significantly increased in the CNS of Nrf2-deficient mice, contributing to the exacerbation of neurodegenerative diseases (Figure 3) [91,92]. Previous studies have shown that microglial activation was increased in Nrf2 knockout mice, as revealed by the upregulated expression of pro-inflammatory cytokines and inflammatory mediators [93,94]. Nrf2 induces gene expression via interaction with an antioxidant response element (ARE) that is known to encode antioxidant enzymes. Nrf2-dependent gene expression exerts a protective effect against oxidative stress in microglia, resulting in the suppression of neuroinflammation. Moreover, neuroinflammation-mediated chronic pain is regulated by the activation of Nrf2 in microglia. Among Nrf2-dependent genes, the heme oxygenase-1 (*HO-1*) gene is a representative gene with strong antioxidant effects. Previously, Nrf2-dependent HO-1 expression in microglia showed inhibitory effects against CFA-induced pain hypersensitivity and microglial activation in mice [75]. Additionally, the administration of an HO-1 inducer significantly attenuated formalin-induced pain-like behavior in mice. However, the attenuative effect of the HO-1 inducer was reversed in Nrf2-knockout mice [95]. These results indicated that Nrf2 inhibits inflammation-induced microglial activation and pain hypersensitivity via HO-1 induction.

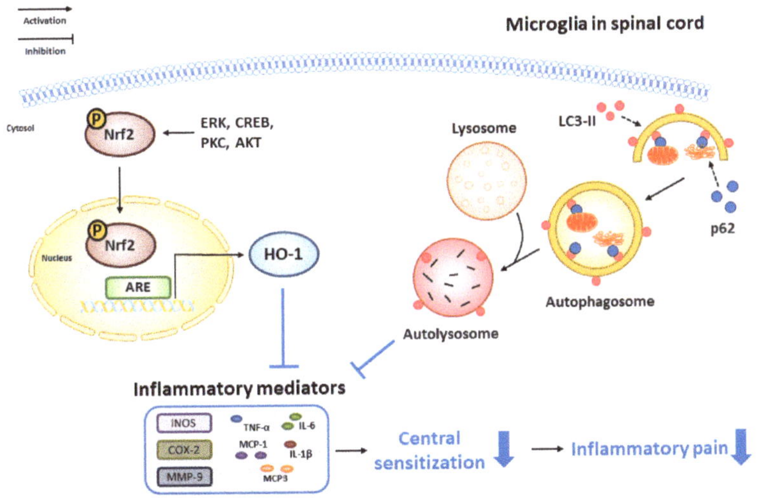

Figure 3. Intracellular signaling that can regulate inflammatory mediators in microglia. Nrf2 signaling and autophagy have a protective effect on microglial activation. Inflammatory mediators expressed in activated microglia are suppressed by activation of Nrf2 signaling and autophagy.

4.5. Autophagy

Autophagy is a lysosomal degradation pathway responsible for the removal and recycling of unnecessary or dysfunctional molecules maintain cellular homeostasis [96]. The process begins by the marking of unwanted or damaged molecules for removal, followed by the formation of an autophagosome that envelopes the unwanted molecules. This autophagosome then combines with a lysosome to degrade the cargo, after which unwanted molecules are removed and recycled. Autophagy has a protective effect, as demonstrated by its ability to remove amyloid-β, a hallmark of Alzheimer's disease, and prevent neurodegeneration in mice [97]. Additionally, microglial autophagy has been found to play a role in regulating neuroinflammation. LPS-treated microglia showed inhibited autophagic activity, leading to increased neuroinflammation. However, treatment with the autophagy inducer, rapamycin, significantly reduced LPS-induced neuroinflammation in microglia [98]. CFA-induced pain hypersensitivity was attenuated by the induction of autophagy in the spinal cord [24,99,100]. In addition, CFA-induced expression of pro-inflammatory cytokines and microglial activation were decreased by enhanced autophagy in the spinal cord. These results suggested that autophagy activation may have a protective effect against microglial activation and inflammatory pain.

5. Natural Product-Derived Compounds against Microglial Activation-Mediated Inflammatory Pain

Multiple studies have demonstrated that microglial activation contributes significantly to the development of inflammatory pain. Therefore, targeting microglial activation through the regulation of inflammatory mediators has been proposed as a therapeutic strategy for the treatment of inflammatory pain. Many natural products and their compounds have been found to exert protective effects against inflammation [101]. These studies suggest that natural product-derived compounds with anti-inflammatory effects inhibit inflammatory pain by suppressing microglial activation. Table 1 and Figure 4 present a summary of natural product-derived compounds that have been found to have the potential to alleviate microglial activation-mediated inflammatory pain.

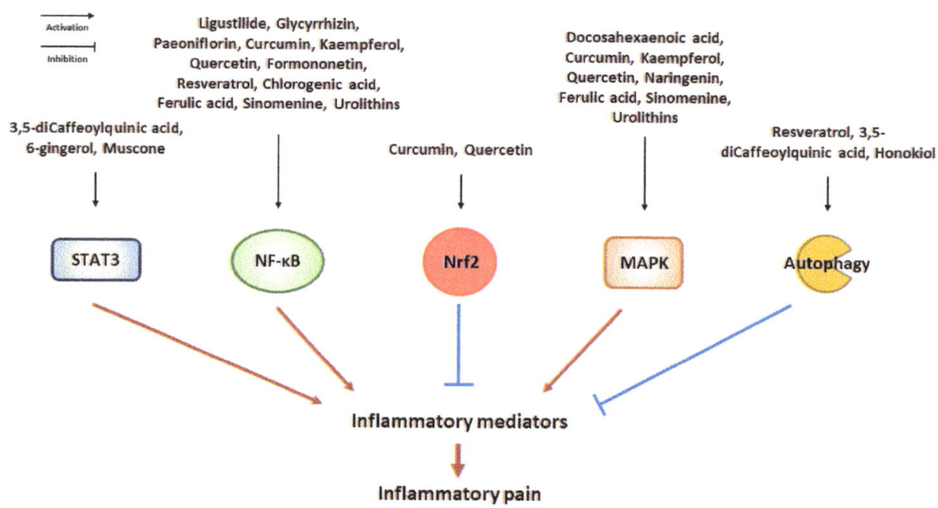

Figure 4. Effect of natural product-derived compounds on inflammatory pain via the suppression of microglial activation. Natural products and phytochemicals may have suppressive effects on microglial activation. Each compound may inhibit microglial activation by regulating the intracellular signaling pathways. The suppression of microglial activation by the modulation of intracellular signaling via natural product-derived compounds shows potential for attenuating inflammatory pain.

5.1. 3,5-Dicaffeoylquinic Acid

3,5-Dicaffeoylquinic acid (3,5-DCQA) is a phenolic nutraceutical present in *Arctium lappa* and *Aster yomena*; it has shown inhibitory effects on the LPS-induced expression of iNOS and COX-2 and secretion of TNF-α, IL-1β, IL-6, MCP1, and MCP3 in BV2 microglial cells [24]. Suppression of MCP3 expression by 3,5-DCQA enhanced autophagy by suppressing LPS-induced activation of JAK2-STAT3, resulting in the reduction of microglial activation. Furthermore, CFA-induced pain hypersensitivity was attenuated by the administration of 3,5-DCQA. Additionally, the administration of 3,5-DCQA suppressed microglial activation in the spinal cord of CFA-injected mice.

5.2. Chlorogenic Acid

Chlorogenic acid, an ester of caffeic and quinic acids, is a natural phenolic compound found in plants. One study showed that chlorogenic acid significantly inhibited LPS-induced NO production and expression of iNOS and TNF-α in primary microglia [102]. In addition, LPS-induced phosphorylation of NF-κB signaling was suppressed by treatment with chlorogenic acid. Carrageenan-induced foot swelling and formalin-induced pain-like behavior were significantly reduced by oral administration of chlorogenic acid in mice [103]. In a clinical study, plasma antioxidant capacity was significantly increased in the group that consumed chlorogenic acid-rich coffee/day. Participants had consumed a maximum of 480 mg/day chlorogenic acid for 8 weeks and did not experience any adverse effects [104]. Furthermore, a clinical study identified that the consumption of chlorogenic acid resulted in improved neuronal function [105].

5.3. Ferulic Acid

Ferulic acid, a well-known phenolic compound, is a bioactive compound found in medicinal herbs, including *Ferula asafoetida*. One study showed that LPS-induced expression of iNOS and TNF-α was reduced after treatment with ferulic acid in BV2 microglial cells in a dose-dependent manner [106]. Additionally, LPS-induced phosphorylation of JNK and NF-κB was reduced by ferulic acid. Moreover, formalin-induced pain-like behavior in mice was alleviated by intraperitoneal injection of ferulic acid [107]. In a clinical study, participants were administered 1000 mg/day ferulic acid for 6 weeks. No toxicity associated with this dose of ferulic acid was observed. The oxidative stress marker was significantly reduced in the group supplemented with ferulic acid. Moreover, TNF-α was remarkably reduced in blood samples [108]. Another clinical study also showed clinical positive effective of ferulic acid on neuronal functioning [109].

5.4. 6-Gingerol

6-gingerol, present in *Zingiber officinale*, is a bioactive phenolic compound, which is known to have a neuroprotective effect. Previously, LPS-induced NO production and expression of iNOS, IL-1β, and IL-6 were dose-dependently suppressed by treatment with 6-gingerol, leading to inhibition of microglial activation [110]. LPS-induced phosphorylation of STAT3 in microglia was significantly reduced by 6-gingerol treatment. Furthermore, intraperitoneal injection of 6-gingerol was found to attenuate acetic acid- and formalin-induced pain-like behaviors, such as writhing and licking, in mice [111]. Carrageenan-induced paw swelling was also suppressed by the administration of 6-gingerol. No clinical trials have specifically investigated analgesic effects of 6-gingerol alone. However, numerous studies have reported alleviative effects of ginger on inflammation-related pain in humans [112]. Given that 6-gingerol had no adverse effects at a concentration of 10 mg twice a day for 12 weeks, it is necessary to evaluate the analgesic effects of 6-gingerol using this dosage [113].

5.5. Curcumin

Curcumin is a bright yellow bioactive component found in *Curcuma longa*. LPS-induced expression of iNOS, TNF-α, and IL-1β was decreased after treating BV2 microglial cells with curcumin [114]. Additionally, lipoteichoic acid (LTA) treatment increased the production of NO and PGE3, and expression of iNOS, COX-2, and TNF-α, reversed by treatment with curcumin in BV2 microglial cells. In one study, curcumin showed a suppressive effect on microglial activation via inhibition of NF-κB and MAPK signaling and induction of Nrf2 in BV2 microglial cells [115]. Moreover, in another study CFA-induced hyperalgesia was attenuated via suppression of TNF-α, IL-1β, and IL-6 in the spinal cord by administration of curcumin [116]. In a clinical study identifying anti-inflammatory effects of curcumin, the reduction in pain sensitivity and inflammation at the surgical site was evaluated in a group of patients who receive 400 mg of curcumin three times a day for 6 days [117].

5.6. Kaempferol

Kaempferol, one of the most common flavonoids found in numerous medicinal herbs, is known to have antioxidant and anti-inflammatory effects. In one study, LPS-induced microglial activation was suppressed by treatment with kaempferol, as revealed by decreased production of NO and PGE2 and decreased expression of iNOS, COX-2, MMP9, TNF-α, and IL-1β in microglia [118]. The underlying inhibitory mechanism of kaempferol is the inhibition of NF-κB and MAPK signaling pathways in microglia. Moreover, formalin-induced pain hypersensitivity was alleviated by intrathecal injection of kaempferol in mice [119]. Administration of kaempferol showed an inhibitory effect on the formalin-induced expression of TNF-α, IL-1β, and IL-6 in the spinal cord. A variety of clinical studies had provided evidence for the preventive effects of kaempferol on diseases associated with inflammation [120,121]. The consumption of 50 mg a day of kaempferol for 4 weeks is safe in adults [122].

5.7. *Quercetin*

Quercetin is considered an antioxidant, anti-inflammatory, and anti-nociceptive compound. Studies have shown that quercetin inhibits LPS-induced NO production and iNOS expression in BV2 microglial cells. Furthermore, LPS-induced NF-κB activation was reduced by quercetin treatment [123]. Additionally, in one study, quercetin led to the activation of Nrf/HO-1 signaling, resulting in the inhibition of NO production in microglia. Moreover, CFA-induced chronic inflammatory hyperalgesia was attenuated by the inhibition of ERK1/2 and NF-κB in the spinal cord after the administration of quercetin [124]. CFA-induced TNF-α expression was also decreased in the spinal cord by quercetin administration. A previous study investigated the effect of quercetin supplementation on inflammation and pain in women diagnosed with rheumatoid arthritis [125]. The patients were given 500 mg of quercetin a day for 8 weeks. The results revealed significant reductions in plasma levels of TNF-α and improvements in symptoms related to swelling and pain in patients following quercetin supplementation. Notably, no side effects were observed in the patients.

5.8. *Formononetin*

Formononetin is a bioactive isoflavone found in various plants including *Trifolium pratense* L. In one study, LPS-induced microglial activation was reduced by treatment of formononetin, as revealed by a decrease in the expression of TNF-α, IL-1β, and IL-6 [126]. Additionally, LPS-induced expression of iNOS and COX-2 was suppressed in BV2 microglial cells. Moreover, formononetin showed an inhibitory effect on LPS-induced activation of NF-κB signaling. In the CFA-induced inflammatory pain model, the administration of formononetin alleviated mechanical allodynia and thermal hyperalgesia in mice [127]. Formononetin has been studied in preclinical tests for other diseases, but clinical studies for the use of formononetin alone have yet to be performed. A previous study showed that extracts containing rich-formononetin (50 mg/day for at least 1 year) exhibited beneficial effects on the bone, with no significantly adverse effects [128]. However, further clinical research is required to determine the safety and effects of formononetin specifically on inflammatory pain.

5.9. *Naringenin*

Naringenin is a flavonoid with antioxidant, anti-inflammatory, and anti-cancer properties. One study revealed that naringenin blocked transformation into LPS-induced activation, as evidenced by expression of iNOS, TNF-α, and IL-1β in BV2 microglia cells [129]. In addition, the LPS-induced phosphorylation of MAPK members, including JNK, ERK, and p38, was notably inhibited by naringenin treatment. In mice with inflammatory pain, carrageenan-, capsaicin-, CFA-, and PGE2-induced mechanical hyperalgesia was significantly alleviated by the oral administration of naringenin without gastric or hepatic toxicity [130]. A study conducted on healthy adults to evaluate the safety and pharmacokinetics of naringenin reported that the half-life was 3 h and almost disappeared from the serum after 24 h of ingestion [131]. No adverse events were reported up to 900 mg of naringenin. Clinical trials using orange juice, which is known to contain naringenin, showed anti-inflammatory effects as evidenced by increasing pro-inflammatory cytokines [132].

These findings suggest that naringenin has the potential to attenuate pain through the regulation of inflammation.

5.10. Resveratrol

Resveratrol is a bioactive component produced in grapes and is a representative inducer of autophagy. One study reported that LPS/interferon γ (IFNγ)-induced expression of iNOS, TNF-α, and IL-1β was suppressed by resveratrol treatment in N9 microglial cells [133]. In the case of LPS/IFNγ-induced microglial activation, activation of NF-κB was inhibited by resveratrol treatment. Moreover, in one study, CFA-induced temporomandibular disorders, resveratrol dose-dependently attenuated pain-like behavior in mice [134]. TNF-α in activated microglia of spinal trigeminal nucleus caudalis is also inhibited by resveratrol treatment. In a clinical study, resveratrol was evaluated for its effects on inflammation and pain in patients with knee osteoarthritis [135]. A total of 110 patients were treated with 500 mg/day resveratrol for 90 days. In the group that received oral administration of resveratrol, pain sensitivity and pro-inflammatory cytokines in serum were significantly decreased compared to the control group.

5.11. Honokiol

Honokiol, a natural polyphenolic compound, is extracted from the bark and seeds of *Magnolia officinalis*. Honokiol is an autophagy inducer that suppresses skin cancer [136]. LPS-induced NO production and expression of iNOS, IL-1β, and IL-6 in primary microglia were suppressed after treatment with honokiol [137]. In addition, in one study, carrageenan- and CFA-induced mechanical hyperalgesia, allodynia, and thermal hyperalgesia were alleviated by intraperitoneal injection of honokiol in mice [138]. In a clinical study aimed at evaluating safety, 50 mg per kg of honokiol was intravenously injected into cancer patients [139]. There were no serious adverse effects and a positive clinical response was achieved in patients. Therefore, an evaluation is needed of the anti-inflammatory and analgesic effects of honokiol at the same concentration in humans.

5.12. Ligustilide

Ligustilide, a major compound found in the roots of *Angelica sinensis*, has a protective effect against inflammation in microglia. Studies have shown that LPS-induced NO production and expression of iNOS and COX2 were remarkably reduced in microglia after ligustilide treatment. LPS-induced production of TNF-α, IL-1β, IL-6, and MCP1 was also suppressed by ligustilide treatment in previous studies [140,141]. Furthermore, in one study CFA-induced pain hypersensitivity and microglial activation in the spinal cord were significantly reduced after ligustilide treatment; ligustilide treatment alleviated acetic acid- and formalin-induced pain in mice [142]. Following a safety evaluation of ligustilide in rats, oral administration of 90 mg/kg ligustilide had good health status, without any histopathological change [143]. Moreover, tissue analysis indicated that ligustilide could penetrate the blood–brain barrier. Based on these results, it is suggested that ligustilide needs to be evaluated for safety and effects on inflammatory pain in clinical settings.

5.13. Glycyrrhizin

Glycyrrhizin, a triterpene saponin present in *Glycyrrhiza glabra*, has shown inhibitory effects on inflammatory pain by suppressing microglial activation [144]. In one study, LPS-induced microglial activation was significantly reduced by treatment with glycyrrhizin, as evidenced by decreased NO production and expression of pro-inflammatory cytokines. Glycyrrhizin inhibited LPS-induced HMGB1/TLR4/NF-κB signaling in microglia, leading to reducing microglial activation. Additionally, CFA-induced pain hypersensitivity was attenuated by the administration of glycyrrhizin in mice. Glycyrrhizin suppresses CFA-induced expression of pro-inflammatory cytokines and activation of NF-κB in the spinal cord. In a clinical study, glycyrrhizin was administered to patients with the selective serotonin reuptake inhibitor (SSRI) in order to evaluate the effects on depression and inflammation [145]. The results showed serum levels of TNF-α, and IL-1β were significantly reduced and depressive symptoms were improved. No patients experienced severe adverse events with 150 mg/3 times a day glycyrrhizin in combination with a 10 mg/day SSRI for 4 weeks.

5.14. Docosahexaenoic Acid

Docosahexaenoic acid (DHA) is the major bioactive omega-3 polyunsaturated fatty acid. DHA is known to regulate the inflammatory responses in neurodegenerative diseases. In one study, carrageenan-induced inflammatory pain and microglial activation were inhibited by the administration of DHA [146]. Carrageenan-induced mechanical allodynia was inhibited by intrathecal injection of DHA in mice. After DHA treatment, carrageenan-induced microglial activation was suppressed by p38 inhibition in spinal microglia. Further, LPS-induced expression of TNF-α, IL-1β, IL-6, MCP1, CCL3, and CXCL10 was significantly suppressed by DHA treatment of BV2 microglial cells. The consumption of omega-3 fatty acid showed improvements in pain and function in patients with osteoarthritis [147]. The patients who consumed 0.45 omega-3 fatty acids/day for 24 months experienced beneficial effects on osteoarthritis without any adverse events. Thus, it is necessary to evaluate the analgesic effects and safety specifically using DHA alone in clinical settings.

5.15. Paeoniflorin

Paeoniflorin, the main active ingredient of *Paeonia lactiflora*, is known to reduce CFA-induced pain hypersensitivity and mRNA expression of TNF-α, IL-1β, and IL-6 in the spinal cord [148]. Furthermore, in one study, CFA-induced microglial activation in the dorsal horn was inhibited by paeoniflorin. In in vitro experiments, LPS-induced pro-inflammatory cytokines were reduced by treatment with paeoniflorin via inhibiting AKT- NF-κB in microglia. In a clinical study, intravenous injection of powders containing 35.8 mg/day paeoniflorin for 7 days showed no adverse events in healthy adults [149].

5.16. Sinomenine

Sinomenine is found in *Sinomenium acutum* and is known to have various pharmacological effects such as anti-cancer, anti-inflammation, and antioxidant effects. One study found that treatment with sinomenine suppressed amyloid-β-induced microglial activation, as evidenced by the reduction in NO production and expression of TNF-α, IL-1β, and MCP1 in BV2 cells [150]. Moreover, CFA-induced pain hypersensitivity was inhibited by intraperitoneal injection of sinomenine in mice [151]. In addition, CFA-induced expression of TNF-α, IL-1β, IL-6, and COX-2 and PGE2 production was inhibited by the activation of p38 and NF-κB in the spinal cord. In a clinical study, patients with osteoarthritis were orally administered 20 mg/2 times a day sinomenine for 3 months [152]. The results indicated that disease symptoms were attenuated and plasma levels of pro-inflammatory cytokines were significantly reduced by sinomenine in patients. No adverse events were observed during the study. Based on these findings, further evaluation is necessary to assess the attenuative effects of sinomenine on inflammatory pain.

5.17. Muscone

Muscone is found in musk, which is a glandular secretion of musk deer; it is a pharmacologically bioactive compound that has been used in medicine for centuries. One study revealed that muscone had an inhibitory effect on LPS-induced NO production and expression of iNOS, IL-1β, and IL-6 in BV2 microglial cells [89]. In addition, LPS-induced activation of JAK2/STAT3 signaling was significantly suppressed by muscone treatment of BV2 cells. In an inflammatory pain model, CFA-induced pain hypersensitivity was attenuated by the intraperitoneal injection of muscone in mice. Moreover, muscone administration suppressed CFA-induced expression of pro-inflammatory cytokines and phosphorylation of JAK2/STAT3 signaling in the spinal cord of mice. Muscone exhibited liver toxicity in Kunming mice at doses exceeding 50 mg/kg [153]. As a result, additional studies in pro-clinical and clinical trials are required to further investigate the safety and effects of muscone on inflammatory pain.

5.18. Urolithins

Urolithins are secondary metabolites formed by gut microbiome from ellagic acid and ellagitannins found in foods like pomegranate. Urolithins suppressed LPS-induced production of NO and mRNA expression of TNF-α, IL-1β, IL-6, iNOS, and COX-2 in BV2 microglial cells. Additionally, LPS-induced activation of ERK, p38, and NF-κB signaling were significantly reduced by urolithins in BV2 cells [154]. Another group induced experimental osteoarthritis to study the effects of urolithins on inflammatory pain. The meniscotibial and medial collateral ligaments were transected in the knees of mice to induce osteoarthritis as inflammatory pain model. Mice were given a diet containing urolithins to identify the effects on pain. Results showed urolithins reduced pain hypersensitivity and slowed down disease progression in mice [155]. In clinical trials of older adults, supplementation with 1000 mg/day urolithin for 4 months showed no adverse events [156]. Additionally, plasma levels of inflammatory biomarkers were significantly reduced in a group that consumed urolithin. However, further experiments are needed to evaluate the effects of urolithins on pain relief.

Table 1. Natural product-derived compounds attenuating microglial activation-mediated inflammatory pain.

Class of Phytochemicals	Subclass	Major Compound	Source	Targeting Inflammatory Mediators	Targeting Intracellular Signaling	Inducer in Animal Model	Safety Dosage in Clinical Study	Effects in Clinical Study	Reference
Phenolics	Phenolic acid	3,5-Dicaffeoylquinic acid	*Arctium lappa, aster yomena*	TNF-α, IL-1β, IL-6, MCP1, MCP3, iNOS, COX2	JAK2/STAT3, Autophagy	CFA			[24]
Phenolics	Phenolic acid	Chlorogenic acid		NO, iNOS, TNF-α	NF-κB	Carrageenan, Formalin	480 mg/day for 8 weeks	Improvements in neuronal function	[102–105]
Phenolics	Phenolic acid	Ferulic acid	*ferula asafetida*	TNF-α, iNOS	NF-κB, JNK	Formalin	1000 mg/day for 6 weeks	Anti-oxidant, anti-inflammation	[106–109].
Phenolics	Phenolic acid	6-gingerol	*zingiber officinale*	NO, iNOS, IL-1β, IL-6	STAT3	Acetic acid, Formalin, Carrageenan	20 mg/day for 12 weeks		[110,111, 113]
Phenolics	flavonoids	Curcumin	*Curcuma longa*	NO, PGE2, iNOS, COX2, TNF-α, IL-1β, IL-6	NF-κB, MAPK, Nrf2	CFA	1200 mg/day for 6 days	Analgesic effects, anti-inflammation	[114–117]
Phenolics	flavonoids	Kaempferol	Tea, broccoli	NO, PGE2, iNOS, COX2, MMP-9, TNF-α, IL-1β, IL-6	NF-κB, JNK, ERK, p38	Formalin	50 mg/day for 4 weeks	Anti-inflammation	[118–122]
Phenolics	flavonoids	Quercetin		NO, iNOS, TNF-α	NF-κB, Nrf2, ERK	CFA	500 mg/day for 8 weeks	Analgesic effects, anti-inflammation	[123–125]
Phenolics	flavonoids	Formononetin	*Trifolium pretense* L.	TNF-α, IL-1β, IL-6, iNOS, COX2	NF-κB	CFA			[126,127]
Phenolics	flavonoids	Naringenin		TNF-α, IL-1β, iNOS	MAPK	Carrageenan, Capsaicin, CFA, PGE2	900 mg for a day		[129–131]
Phenolics	stilbenes	Resveratrol	grape	TNF-α, IL-1β, iNOS	NF-κB, Autophagy	CFA	500 mg/day for 90 days	Analgesic effects, anti-inflammation	[133–135]
Phenolics	lignan	Honokiol	*Magnolia officinlis*	NO, iNOS, TNF-α, IL-1β, IL-6	Autophagy	Carrageenan, CFA	50 mg/kg for a week		[136–139]
Non-phenolics	phthalide	Ligustilide	the roof of *Angelica sinensis*	NO, iNOS, COX-2, TNF-α, IL-1β, IL-6, MCP1	NF-κB	CFA, Acetic acid, Formalin			[140–142]
Non-phenolics	saponin	Glycyrrhizin	*Glycyrrhiza glabra*	NO, TNF-α, IL-1β, IL-6	NF-κB	CFA	450 mg/day for 4 weeks	Anti-inflammation	[144,145]
Non-phenolics	omega-3 fatty acid	Docosahexaenoic acid	Omega-3 polyunsaturated fatty acid	TNF-α, IL-1β, IL-6, MCP1, CCL3, CXCL10	p38	Carrageenan			[146]
Non-phenolics	monoterpene	Paeoniflorin	*Paeonia lactiflora*	TNF-α, IL-1β, IL-6	NF-κB	CFA	35.8 mg/day for 7 days		[148,149]
Non-phenolics	alkaloid	Sinomenine	*Sinomenium acutum*	NO, TNF-α, IL-1β, IL-6, MCP1	NF-κB, p38	CFA	40 mg/day for 3 months	Anti-inflammation	[150–152]
		Muscone	Musk	NO, TNF-α, IL-1β, IL-6	JAK2/STAT3	CFA			[89]
		Urolithins	Secondary metabolite	TNF-α, IL-1β, IL-6, iNOS, and COX-2	ERK, p38, and NF-κB	Surgery	1000 mg/day for 4 months	Anti-inflammation	[154–156]
		Muscone	Musk	NO, TNF-α, IL-1β, IL-6	JAK2/STAT3	CFA			[89]
		Urolithins	Secondary metabolite	TNF-α, IL-1β, IL-6, iNOS, and COX-2	ERK, p38, and NF-κB	Surgery	1000 mg/day for 4 months	Anti-inflammation	[154–156]

6. Methods

Reference lists were searched for articles published until 19–20 June 2023, using the keywords "Microglial activation", "Neuroinflammation", "Inflammatory biomarkers", "Inflammatory pain", "Chronic pain", "Intracellular signaling", "Natural products", "Pharmaceuticals".

In our search for this review, we applied no limits for country of origin or study design. Articles published in a language other than English were excluded.

7. Conclusions

Inflammation in peripheral tissues can lead to the activation of microglia in the dorsal horn of the spinal cord, which is a significant contributor to neuroinflammation and inflammatory pain. Several studies have shown that natural products and their compounds have the ability to regulate microglial activation. Additionally, many studies have proposed that microglial activation-mediated inflammatory pain can be modulated using natural product-derived compounds. However, it is important to consider the possibility of drug–drug interactions (DDIs) when multiple drugs are consumed in combination, as these interactions can potentially affect the pharmacological effects of each drug [157]. In the case of muscone, a previous study reported that it reduced the hypnotic and analgesic effects of ketamine, which is a widely used anesthetic [158]. Taken together, although toxicological and pharmacological studies are required to determine their safety in humans, natural product-derived compounds are potential therapeutic candidates for the treatment of inflammatory pain.

Author Contributions: J.P.: Writing—Original Draft, Conceptualization, C.L.: Funding acquisition, Y.T.K.: Writing—Review and Editing, Supervision. All authors have read and agreed to the published version of the manuscript.

Funding: This work was supported by the Main Research Program of the Korea Food Research Institute (KFRI), funded by the Korean Ministry of Science and ICT, grant number E0210201-03.

Institutional Review Board Statement: Not applicable.

Informed Consent Statement: Not applicable.

Data Availability Statement: Not applicable.

Conflicts of Interest: The authors declare no conflict of interest.

References

1. Zelaya, C.E.; Dahlhamer, J.M.; Lucas, J.W.; Connor, E.M. Chronic Pain and High-Impact Chronic Pain among US Adults, 2019. 2020. Available online: https://stacks.cdc.gov/view/cdc/97308 (accessed on 7 June 2023).
2. Lee, J.; Jotwani, R.; Robert, S.W. The economic cost of racial disparities in chronic pain. *J. Comp. Eff. Res.* **2020**, *9*, 903–906. [CrossRef] [PubMed]
3. McWilliams, L.A.; Cox, B.J.; Enns, M.W. Mood and anxiety disorders associated with chronic pain: An examination in a nationally representative sample. *Pain* **2003**, *106*, 127–133. [CrossRef]
4. Reddi, D.; Curran, N.; Stephens, R. An introduction to pain pathways and mechanisms. *Br. J. Hosp. Med.* **2013**, *74* (Suppl. S12), C188–C191. [CrossRef] [PubMed]
5. Treede, R.D.; Rief, W.; Barke, A.; Aziz, Q.; Bennett, M.I.; Benoliel, R.; Cohen, M.; Evers, S.; Finnerup, N.B.; First, M.B.; et al. Chronic pain as a symptom or a disease: The IASP Classification of Chronic Pain for the International Classification of Diseases (ICD-11). *Pain* **2019**, *160*, 19–27. [CrossRef] [PubMed]
6. Ghlichloo, I.; Gerriets, V. *Nonsteroidal Anti-Inflammatory Drugs (NSAIDs)*; StatPearls: Treasure Island, FL, USA, 2022.
7. Harvey, A.L.; Edrada-Ebel, R.; Quinn, R.J. The re-emergence of natural products for drug discovery in the genomics era. *Nat. Rev. Drug Discov.* **2015**, *14*, 111–129. [CrossRef] [PubMed]
8. Ji, R.R.; Xu, Z.Z.; Gao, Y.J. Emerging targets in neuroinflammation-driven chronic pain. *Nat. Rev. Drug Discov.* **2014**, *13*, 533–548. [CrossRef]
9. Matsuda, M.; Huh, Y.; Ji, R.R. Roles of inflammation, neurogenic inflammation, and neuroinflammation in pain. *J. Anesth.* **2019**, *33*, 131–139. [CrossRef]
10. Zhang, L.; Berta, T.; Xu, Z.Z.; Liu, T.; Park, J.Y.; Ji, R.R. TNF-α contributes to spinal cord synaptic plasticity and inflammatory pain: Distinct role of TNF receptor subtypes 1 and 2. *Pain* **2011**, *152*, 419–427. [CrossRef]

11. Xie, R.G.; Gao, Y.J.; Park, C.K.; Lu, N.; Luo, C.; Wang, W.T.; Wu, S.X.; Ji, R.R. Spinal CCL2 Promotes Central Sensitization, Long-Term Potentiation, and Inflammatory Pain via CCR2: Further Insights into Molecular, Synaptic, and Cellular Mechanisms. *Neurosci. Bull.* **2018**, *34*, 13–21. [CrossRef]
12. Kawasaki, Y.; Zhang, L.; Cheng, J.K.; Ji, R.R. Cytokine mechanisms of central sensitization: Distinct and overlapping role of interleukin-1β, interleukin-6, and tumor necrosis factor-α in regulating synaptic and neuronal activity in the superficial spinal cord. *J. Neurosci.* **2008**, *28*, 5189–5194. [CrossRef]
13. Chen, G.; Zhang, Y.Q.; Qadri, Y.J.; Serhan, C.N.; Ji, R.R. Microglia in Pain: Detrimental and Protective Roles in Pathogenesis and Resolution of Pain. *Neuron* **2018**, *100*, 1292–1311. [CrossRef]
14. Vergne-Salle, P.; Bertin, P. Chronic pain and neuroinflammation. *Jt. Bone Spine* **2021**, *88*, 105222. [CrossRef]
15. Lim, E.Y.; Kim, Y.T. Food-Derived Natural Compounds for Pain Relief in Neuropathic Pain. *BioMed Res. Int.* **2016**, *2016*, 7917528. [CrossRef] [PubMed]
16. Ji, R.R.; Nackley, A.; Huh, Y.; Terrando, N.; Maixner, W. Neuroinflammation and Central Sensitization in Chronic and Widespread Pain. *Anesthesiology* **2018**, *129*, 343–366. [CrossRef] [PubMed]
17. Latremoliere, A.; Mauborgne, A.; Masson, J.; Bourgoin, S.; Kayser, V.; Hamon, M.; Pohl, M. Differential implication of proinflammatory cytokine interleukin-6 in the development of cephalic versus extracephalic neuropathic pain in rats. *J. Neurosci.* **2008**, *28*, 8489–8501. [CrossRef]
18. Imai, S.; Ikegami, D.; Yamashita, A.; Shimizu, T.; Narita, M.; Niikura, K.; Furuya, M.; Kobayashi, Y.; Miyashita, K.; Okutsu, D.; et al. Epigenetic transcriptional activation of monocyte chemotactic protein 3 contributes to long-lasting neuropathic pain. *Brain* **2013**, *136*, 828–843. [CrossRef] [PubMed]
19. Gu, N.; Yi, M.H.; Murugan, M.; Xie, M.; Parusel, S.; Peng, J.; Eyo, U.B.; Hunt, C.L.; Dong, H.; Wu, L.J. Spinal microglia contribute to sustained inflammatory pain via amplifying neuronal activity. *Mol. Brain* **2022**, *15*, 86. [CrossRef]
20. Yi, M.H.; Liu, Y.U.; Liu, K.; Chen, T.; Bosco, D.B.; Zheng, J.; Xie, M.; Zhou, L.; Qu, W.; Wu, L.J. Chemogenetic manipulation of microglia inhibits neuroinflammation and neuropathic pain in mice. *Brain Behav. Immun.* **2021**, *92*, 78–89. [CrossRef]
21. Zamora, R.; Vodovotz, Y.; Billiar, T.R. Inducible nitric oxide synthase and inflammatory diseases. *Mol. Med.* **2000**, *6*, 347–373. [CrossRef]
22. Rocha, P.A.; Ferreira, A.F.B.; Da Silva, J.T.; Alves, A.S.; Martins, D.O.; Britto, L.R.G.; Chacur, M. Effects of selective inhibition of nNOS and iNOS on neuropathic pain in rats. *Mol. Cell Neurosci.* **2020**, *105*, 103497. [CrossRef]
23. Lull, M.E.; Block, M.L. Microglial activation and chronic neurodegeneration. *Neurotherapeutics* **2010**, *7*, 354–365. [CrossRef] [PubMed]
24. Park, J.; Kim, Y.; Lee, C.; Kim, Y.T. 3,5-Dicaffeoylquinic acid attenuates microglial activation-mediated inflammatory pain by enhancing autophagy through the suppression of MCP3/JAK2/STAT3 signaling. *Biomed. Pharmacother.* **2022**, *153*, 113549. [CrossRef] [PubMed]
25. Zhang, L.Q.; Gao, S.J.; Sun, J.; Li, D.Y.; Wu, J.Y.; Song, F.H.; Liu, D.Q.; Zhou, Y.Q.; Mei, W. DKK3 ameliorates neuropathic pain via inhibiting ASK-1/JNK/p-38-mediated microglia polarization and neuroinflammation. *J. Neuroinflamm.* **2022**, *19*, 129. [CrossRef] [PubMed]
26. Kuboyama, K.; Tsuda, M.; Tsutsui, M.; Toyohara, Y.; Tozaki-Saitoh, H.; Shimokawa, H.; Yanagihara, N.; Inoue, K. Reduced spinal microglial activation and neuropathic pain after nerve injury in mice lacking all three nitric oxide synthases. *Mol. Pain* **2011**, *7*, 50. [CrossRef]
27. Chen, Y.; Zhou, Y.; Li, X.C.; Ma, X.; Mi, W.L.; Chu, Y.X.; Wang, Y.Q.; Mao-Ying, Q.L. Neuronal GRK2 regulates microglial activation and contributes to electroacupuncture analgesia on inflammatory pain in mice. *Biol. Res.* **2022**, *55*, 5. [CrossRef]
28. Osborne, M.G.; Coderre, T.J. Effects of intrathecal administration of nitric oxide synthase inhibitors on carrageenan-induced thermal hyperalgesia. *Br. J. Pharmacol.* **1999**, *126*, 1840–1846. [CrossRef]
29. Jiang, M.; Deng, H.; Chen, X.; Lin, Y.; Xie, X.; Bo, Z. The efficacy and safety of selective COX-2 inhibitors for postoperative pain management in patients after total knee/hip arthroplasty: A meta-analysis. *J. Orthop. Surg. Res.* **2020**, *15*, 39. [CrossRef]
30. Hoozemans, J.J.; Rozemuller, A.J.; Janssen, I.; De Groot, C.J.; Veerhuis, R.; Eikelenboom, P. Cyclooxygenase expression in microglia and neurons in Alzheimer's disease and control brain. *Acta Neuropathol.* **2001**, *101*, 2–8. [CrossRef]
31. Sinatra, R. Role of COX-2 inhibitors in the evolution of acute pain management. *J. Pain Symptom Manag.* **2002**, *24*, S18–S27. [CrossRef]
32. Cho, N.; Moon, E.H.; Kim, H.W.; Hong, J.; Beutler, J.A.; Sung, S.H. Inhibition of Nitric Oxide Production in BV2 Microglial Cells by Triterpenes from Tetrapanax papyriferus. *Molecules* **2016**, *21*, 459. [CrossRef]
33. Sun, Z.; Li, G.; Tong, T.; Chen, J. Micheliolide suppresses LPS-induced neuroinflammatory responses. *PLoS ONE* **2017**, *12*, e0186592. [CrossRef]
34. Kim, S.Y.; Jin, C.Y.; Kim, C.H.; Yoo, Y.H.; Choi, S.H.; Kim, G.Y.; Yoon, H.M.; Park, H.T.; Choi, Y.H. Isorhamnetin alleviates lipopolysaccharide-induced inflammatory responses in BV2 microglia by inactivating NF-κB, blocking the TLR4 pathway and reducing ROS generation. *Int. J. Mol. Med.* **2019**, *43*, 682–692. [CrossRef]
35. Park, J.; Lim, E.Y.; Kim, Y.T. The inhibitory effects of Aster yomena extract on microglial activation-mediated inflammatory response and pain by modulation of the NF-κB and MAPK signaling pathways. *J. Funct. Foods* **2021**, *85*, 104659. [CrossRef]

36. Dhapola, R.; Hota, S.S.; Sarma, P.; Bhattacharyya, A.; Medhi, B.; Reddy, D.H. Recent advances in molecular pathways and therapeutic implications targeting neuroinflammation for Alzheimer's disease. *Inflammopharmacology* **2021**, *29*, 1669–1681. [CrossRef]
37. Park, J.; Kim, Y.T. Erythronium japonicum Alleviates Inflammatory Pain by Inhibiting MAPK Activation and by Suppressing NF-κB Activation via ERK/Nrf2/HO-1 Signaling Pathway. *Antioxidants* **2020**, *9*, 626. [CrossRef] [PubMed]
38. Konnecke, H.; Bechmann, I. The role of microglia and matrix metalloproteinases involvement in neuroinflammation and gliomas. *Clin. Dev. Immunol.* **2013**, *2013*, 914104. [CrossRef] [PubMed]
39. Ji, R.R.; Xu, Z.Z.; Wang, X.; Lo, E.H. Matrix metalloprotease regulation of neuropathic pain. *Trends Pharmacol. Sci.* **2009**, *30*, 336–340. [CrossRef]
40. Kawasaki, Y.; Xu, Z.Z.; Wang, X.; Park, J.Y.; Zhuang, Z.Y.; Tan, P.H.; Gao, Y.J.; Roy, K.; Corfas, G.; Lo, E.H.; et al. Distinct roles of matrix metalloproteases in the early- and late-phase development of neuropathic pain. *Nat. Med.* **2008**, *14*, 331–336. [CrossRef]
41. Lee, E.J.; Kim, H.S. Inhibitory mechanism of MMP-9 gene expression by ethyl pyruvate in lipopolysaccharide-stimulated BV2 microglial cells. *Neurosci. Lett.* **2011**, *493*, 38–43. [CrossRef]
42. Kular, L.; Rivat, C.; Lelongt, B.; Calmel, C.; Laurent, M.; Pohl, M.; Kitabgi, P.; Melik-Parsadaniantz, S.; Martinerie, C. NOV/CCN3 attenuates inflammatory pain through regulation of matrix metalloproteinases-2 and -9. *J. Neuroinflamm.* **2012**, *9*, 36. [CrossRef]
43. Leung, L.; Cahill, C.M. TNF-α and neuropathic pain—A review. *J. Neuroinflamm.* **2010**, *7*, 27. [CrossRef]
44. Duan, Y.W.; Chen, S.X.; Li, Q.Y.; Zang, Y. Neuroimmune Mechanisms Underlying Neuropathic Pain: The Potential Role of TNF-α-Necroptosis Pathway. *Int. J. Mol. Sci.* **2022**, *23*, 7191. [CrossRef] [PubMed]
45. Kuno, R.; Wang, J.; Kawanokuchi, J.; Takeuchi, H.; Mizuno, T.; Suzumura, A. Autocrine activation of microglia by tumor necrosis factor-α. *J. Neuroimmunol.* **2005**, *162*, 89–96. [CrossRef] [PubMed]
46. Hankittichai, P.; Lou, H.J.; Wikan, N.; Smith, D.R.; Potikanond, S.; Nimlamool, W. Oxyresveratrol Inhibits IL-1β-Induced Inflammation via Suppressing AKT and ERK1/2 Activation in Human Microglia, HMC3. *Int. J. Mol. Sci.* **2020**, *21*, 6054. [CrossRef]
47. Davis, R.L.; Buck, D.J.; McCracken, K.; Cox, G.W.; Das, S. Interleukin-1β-induced inflammatory signaling in C20 human microglial cells. *Neuroimmunol. Neuroinflamm.* **2018**, *5*, 50. [CrossRef]
48. Samad, T.A.; Moore, K.A.; Sapirstein, A.; Billet, S.; Allchorne, A.; Poole, S.; Bonventre, J.V.; Woolf, C.J. Interleukin-1β-mediated induction of Cox-2 in the CNS contributes to inflammatory pain hypersensitivity. *Nature* **2001**, *410*, 471–475. [CrossRef] [PubMed]
49. Gui, W.S.; Wei, X.; Mai, C.L.; Murugan, M.; Wu, L.J.; Xin, W.J.; Zhou, L.J.; Liu, X.G. Interleukin-1β overproduction is a common cause for neuropathic pain, memory deficit, and depression following peripheral nerve injury in rodents. *Mol. Pain* **2016**, *12*, 1744806916646784. [CrossRef]
50. Jang, S.; Kelley, K.W.; Johnson, R.W. Luteolin reduces IL-6 production in microglia by inhibiting JNK phosphorylation and activation of AP-1. *Proc. Natl. Acad. Sci. USA* **2008**, *105*, 7534–7539. [CrossRef]
51. Lin, T.; Li, K.; Zhang, F.Y.; Zhang, Z.K.; Light, A.R.; Fu, K.Y. Dissociation of spinal microglia morphological activation and peripheral inflammation in inflammatory pain models. *J. Neuroimmunol.* **2007**, *192*, 40–48. [CrossRef]
52. Yang, Y.; Zhou, W.; Xu, X.; Ge, X.; Wang, F.; Zhang, G.Q.; Miao, L.; Deng, X. Aprepitant Inhibits JNK and p38/MAPK to Attenuate Inflammation and Suppresses Inflammatory Pain. *Front. Pharmacol.* **2021**, *12*, 811584. [CrossRef]
53. Zanjani, T.M.; Sabetkasaei, M.; Mosaffa, N.; Manaheji, H.; Labibi, F.; Farokhi, B. Suppression of interleukin-6 by minocycline in a rat model of neuropathic pain. *Eur. J. Pharmacol.* **2006**, *538*, 66–72. [CrossRef]
54. Bose, S.; Cho, J. Role of chemokine CCL2 and its receptor CCR2 in neurodegenerative diseases. *Arch. Pharm. Res.* **2013**, *36*, 1039–1050. [CrossRef]
55. Thacker, M.A.; Clark, A.K.; Bishop, T.; Grist, J.; Yip, P.K.; Moon, L.D.; Thompson, S.W.; Marchand, F.; McMahon, S.B. CCL2 is a key mediator of microglia activation in neuropathic pain states. *Eur. J. Pain* **2009**, *13*, 263–272. [CrossRef] [PubMed]
56. Dansereau, M.A.; Midavaine, E.; Begin-Lavallee, V.; Belkouch, M.; Beaudet, N.; Longpre, J.M.; Melik-Parsadaniantz, S.; Sarret, P. Mechanistic insights into the role of the chemokine CCL2/CCR2 axis in dorsal root ganglia to peripheral inflammation and pain hypersensitivity. *J. Neuroinflamm.* **2021**, *18*, 79. [CrossRef]
57. Zhang, L.; Tan, J.; Jiang, X.; Qian, W.; Yang, T.; Sun, X.; Chen, Z.; Zhu, Q. Neuron-derived CCL2 contributes to microglia activation and neurological decline in hepatic encephalopathy. *Biol. Res.* **2017**, *50*, 26. [CrossRef] [PubMed]
58. Cai, Q.; Li, Y.; Pei, G. Polysaccharides from Ganoderma lucidum attenuate microglia-mediated neuroinflammation and modulate microglial phagocytosis and behavioural response. *J. Neuroinflamm.* **2017**, *14*, 63. [CrossRef] [PubMed]
59. Kwiatkowski, K.; Popiolek-Barczyk, K.; Piotrowska, A.; Rojewska, E.; Ciapala, K.; Makuch, W.; Mika, J. Chemokines CCL2 and CCL7, but not CCL12, play a significant role in the development of pain-related behavior and opioid-induced analgesia. *Cytokine* **2019**, *119*, 202–213. [CrossRef] [PubMed]
60. Sun, S.C.; Liu, Z.G. A special issue on NF-κB signaling and function. *Cell Res.* **2011**, *21*, 1–2. [CrossRef]
61. Zhang, H.; Sun, S.C. NF-κB in inflammation and renal diseases. *Cell Biosci.* **2015**, *5*, 63. [CrossRef]
62. Hayden, M.S.; Ghosh, S. Shared principles in NF-κB signaling. *Cell* **2008**, *132*, 344–362. [CrossRef]
63. Brás, J.P.; Bravo, J.; Freitas, J.; Barbosa, M.A.; Santos, S.G.; Summavielle, T.; Almeida, M.I. TNF-α-induced microglia activation requires miR-342: Impact on NF-kB signaling and neurotoxicity. *Cell Death Dis.* **2020**, *11*, 415. [CrossRef] [PubMed]
64. Wen, X.; Xiao, L.; Zhong, Z.; Wang, L.; Li, Z.; Pan, X.; Liu, Z. Astaxanthin acts via LRP-1 to inhibit inflammation and reverse lipopolysaccharide-induced M1/M2 polarization of microglial cells. *Oncotarget* **2017**, *8*, 69370–69385. [CrossRef] [PubMed]

65. Gibson, C.J.; Hossain, M.M.; Richardson, J.R.; Aleksunes, L.M. Inflammatory regulation of ATP binding cassette efflux transporter expression and function in microglia. *J. Pharmacol. Exp. Ther.* **2012**, *343*, 650–660. [CrossRef] [PubMed]
66. Cho, I.H.; Hong, J.; Suh, E.C.; Kim, J.H.; Lee, H.; Lee, J.E.; Lee, S.; Kim, C.H.; Kim, D.W.; Jo, E.K.; et al. Role of microglial IKKbeta in kainic acid-induced hippocampal neuronal cell death. *Brain* **2008**, *131*, 3019–3033. [CrossRef]
67. Zhang, J.; Deng, X. Bupivacaine effectively relieves inflammation-induced pain by suppressing activation of the NF-κB signalling pathway and inhibiting the activation of spinal microglia and astrocytes. *Exp. Ther. Med.* **2017**, *13*, 1074–1080. [CrossRef]
68. Li, X.; Wu, G.; Li, M.; Zhang, Z. Oleanolic acid administration alleviates neuropathic pain after a peripheral nerve injury by regulating microglia polarization-mediated neuroinflammation. *RSC Adv.* **2020**, *10*, 12920–12928. [CrossRef]
69. Zhang, W.; Liu, H.T. MAPK signal pathways in the regulation of cell proliferation in mammalian cells. *Cell Res.* **2002**, *12*, 9–18. [CrossRef]
70. Yarza, R.; Vela, S.; Solas, M.; Ramirez, M.J. c-Jun N-terminal Kinase (JNK) Signaling as a Therapeutic Target for Alzheimer's Disease. *Front. Pharmacol.* **2015**, *6*, 321. [CrossRef]
71. Peng, J.; Andersen, J.K. The role of c-Jun N-terminal kinase (JNK) in Parkinson's disease. *IUBMB Life* **2003**, *55*, 267–271. [CrossRef]
72. Waetzig, V.; Czeloth, K.; Hidding, U.; Mielke, K.; Kanzow, M.; Brecht, S.; Goetz, M.; Lucius, R.; Herdegen, T.; Hanisch, U.K. c-Jun N-terminal kinases (JNKs) mediate pro-inflammatory actions of microglia. *Glia* **2005**, *50*, 235–246. [CrossRef]
73. Wang, Y.R.; Xu, H.; Tao, M.; Xu, L.H.; Fu, X.C. Ligustilide Relieves Complete Freund's Adjuvant-induced Mechanical Hyperalgesia through Inhibiting the Activation of Spinal c-Jun N-terminal Kinase/c-Jun Pathway in Rats. *Pharmacogn. Mag.* **2017**, *13*, 634–638. [CrossRef]
74. Gasco, H.A.; Ros-Bernal, F.; Castillo-Gómez, E.; Olucha-Bordonau, F. MAPK/ERK Dysfunction in Neurodegenerative Diseases. Available online: https://pdfs.semanticscholar.org/25e1/3caa0ba99099e466e319cf166cfe22d4a2a9.pdf (accessed on 7 June 2023).
75. Zhuang, Z.Y.; Gerner, P.; Woolf, C.J.; Ji, R.R. ERK is sequentially activated in neurons, microglia, and astrocytes by spinal nerve ligation and contributes to mechanical allodynia in this neuropathic pain model. *Pain* **2005**, *114*, 149–159. [CrossRef] [PubMed]
76. Kang, C.-H.; Jayasooriya, R.G.P.T.; Dilshara, M.G.; Choi, Y.H.; Jeong, Y.-K.; Kim, N.D.; Kim, G.-Y. Caffeine suppresses lipopolysaccharide-stimulated BV2 microglial cells by suppressing Akt-mediated NF-κB activation and ERK phosphorylation. *Food Chem. Toxicol.* **2012**, *50*, 4270–4276. [CrossRef]
77. Ryu, K.-Y.; Lee, H.-J.; Woo, H.; Kang, R.-J.; Han, K.-M.; Park, H.; Lee, S.M.; Lee, J.-Y.; Jeong, Y.J.; Nam, H.-W. Dasatinib regulates LPS-induced microglial and astrocytic neuroinflammatory responses by inhibiting AKT/STAT3 signaling. *J. Neuroinflamm.* **2019**, *16*, 190. [CrossRef]
78. Zhang, H.; Ma, S.B.; Gao, Y.J.; Xing, J.L.; Xian, H.; Li, Z.Z.; Shen, S.N.; Wu, S.X.; Luo, C.; Xie, R.G. Spinal CCL2 Promotes Pain Sensitization by Rapid Enhancement of NMDA-Induced Currents Through the ERK-GluN2B Pathway in Mouse Lamina II Neurons. *Neurosci. Bull.* **2020**, *36*, 1344–1354. [CrossRef] [PubMed]
79. Cao, D.L.; Zhang, Z.J.; Xie, R.G.; Jiang, B.C.; Ji, R.R.; Gao, Y.J. Chemokine CXCL1 enhances inflammatory pain and increases NMDA receptor activity and COX-2 expression in spinal cord neurons via activation of CXCR2. *Exp. Neurol.* **2014**, *261*, 328–336. [CrossRef]
80. Kumar, S.; Boehm, J.; Lee, J.C. p38 MAP kinases: Key signalling molecules as therapeutic targets for inflammatory diseases. *Nat. Rev. Drug Discov.* **2003**, *2*, 717–726. [CrossRef] [PubMed]
81. Ji, R.R.; Suter, M.R. p38 MAPK, microglial signaling, and neuropathic pain. *Mol. Pain* **2007**, *3*, 33. [CrossRef] [PubMed]
82. Ji, R.R.; Samad, T.A.; Jin, S.X.; Schmoll, R.; Woolf, C.J. p38 MAPK activation by NGF in primary sensory neurons after inflammation increases TRPV1 levels and maintains heat hyperalgesia. *Neuron* **2002**, *36*, 57–68. [CrossRef]
83. Qi, J.; Chen, C.; Meng, Q.X.; Wu, Y.; Wu, H.; Zhao, T.B. Crosstalk between Activated Microglia and Neurons in the Spinal Dorsal Horn Contributes to Stress-induced Hyperalgesia. *Sci. Rep.* **2016**, *6*, 39442. [CrossRef]
84. Taves, S.; Berta, T.; Liu, D.L.; Gan, S.; Chen, G.; Kim, Y.H.; Van de Ven, T.; Laufer, S.; Ji, R.R. Spinal inhibition of p38 MAP kinase reduces inflammatory and neuropathic pain in male but not female mice: Sex-dependent microglial signaling in the spinal cord. *Brain Behav. Immun.* **2016**, *55*, 70–81. [CrossRef] [PubMed]
85. Fan, Z.; Zhang, W.; Cao, Q.; Zou, L.; Fan, X.; Qi, C.; Yan, Y.; Song, B.; Wu, B. JAK2/STAT3 pathway regulates microglia polarization involved in hippocampal inflammatory damage due to acute paraquat exposure. *Ecotoxicol. Environ. Saf.* **2022**, *234*, 113372. [CrossRef] [PubMed]
86. Xiong, J.; Wang, C.; Chen, H.; Hu, Y.; Tian, L.; Pan, J.; Geng, M. Aβ-induced microglial cell activation is inhibited by baicalin through the JAK2/STAT3 signaling pathway. *Int. J. Neurosci.* **2014**, *124*, 609–620. [CrossRef] [PubMed]
87. Porro, C.; Cianciulli, A.; Trotta, T.; Lofrumento, D.D.; Panaro, M.A. Curcumin Regulates Anti-Inflammatory Responses by JAK/STAT/SOCS Signaling Pathway in BV-2 Microglial Cells. *Biology* **2019**, *8*, 51. [CrossRef]
88. Liu, J.; Xie, X.; Qin, K.; Xu, L.; Peng, J.; Li, X.; Li, X.; Liu, Z. Dexamethasone and potassium canrenoate alleviate hyperalgesia by competitively regulating IL-6/JAK2/STAT3 signaling pathway during inflammatory pain in vivo and in vitro. *Immun. Inflamm. Dis.* **2022**, *10*, e721. [CrossRef]
89. Yu, S.; Zhao, G.; Han, F.; Liang, W.; Jiao, Y.; Li, Z.; Li, L. Muscone relieves inflammatory pain by inhibiting microglial activation-mediated inflammatory response via abrogation of the NOX4/JAK2-STAT3 pathway and NLRP3 inflammasome. *Int. Immunopharmacol.* **2020**, *82*, 106355. [CrossRef]
90. Simpson, D.S.A.; Oliver, P.L. ROS Generation in Microglia: Understanding Oxidative Stress and Inflammation in Neurodegenerative Disease. *Antioxidants* **2020**, *9*, 743. [CrossRef]

91. Ren, P.; Chen, J.; Li, B.; Zhang, M.; Yang, B.; Guo, X.; Chen, Z.; Cheng, H.; Wang, P.; Wang, S.; et al. Nrf2 Ablation Promotes Alzheimer's Disease-Like Pathology in APP/PS1 Transgenic Mice: The Role of Neuroinflammation and Oxidative Stress. *Oxid. Med. Cell Longev.* **2020**, *2020*, 3050971. [CrossRef]
92. Rojo, A.I.; Pajares, M.; Garcia-Yague, A.J.; Buendia, I.; Van Leuven, F.; Yamamoto, M.; Lopez, M.G.; Cuadrado, A. Deficiency in the transcription factor NRF2 worsens inflammatory parameters in a mouse model with combined tauopathy and amyloidopathy. *Redox Biol.* **2018**, *18*, 173–180. [CrossRef]
93. Velagapudi, R.; El-Bakoush, A.; Olajide, O.A. Activation of Nrf2 Pathway Contributes to Neuroprotection by the Dietary Flavonoid Tiliroside. *Mol. Neurobiol.* **2018**, *55*, 8103–8123. [CrossRef]
94. Li, Y.; Lv, O.; Zhou, F.; Li, Q.; Wu, Z.; Zheng, Y. Linalool Inhibits LPS-Induced Inflammation in BV2 Microglia Cells by Activating Nrf2. *Neurochem. Res.* **2015**, *40*, 1520–1525. [CrossRef] [PubMed]
95. Rosa, A.O.; Egea, J.; Lorrio, S.; Rojo, A.I.; Cuadrado, A.; Lopez, M.G. Nrf2-mediated haeme oxygenase-1 up-regulation induced by cobalt protoporphyrin has antinociceptive effects against inflammatory pain in the formalin test in mice. *Pain* **2008**, *137*, 332–339. [CrossRef] [PubMed]
96. Aman, Y.; Schmauck-Medina, T.; Hansen, M.; Morimoto, R.I.; Simon, A.K.; Bjedov, I.; Palikaras, K.; Simonsen, A.; Johansen, T.; Tavernarakis, N.; et al. Autophagy in healthy aging and disease. *Nat. Aging* **2021**, *1*, 634–650. [CrossRef] [PubMed]
97. Fang, E.F.; Hou, Y.; Palikaras, K.; Adriaanse, B.A.; Kerr, J.S.; Yang, B.; Lautrup, S.; Hasan-Olive, M.M.; Caponio, D.; Dan, X.; et al. Mitophagy inhibits amyloid-β and tau pathology and reverses cognitive deficits in models of Alzheimer's disease. *Nat. Neurosci.* **2019**, *22*, 401–412. [CrossRef] [PubMed]
98. Ye, X.; Zhu, M.; Che, X.; Wang, H.; Liang, X.J.; Wu, C.; Xue, X.; Yang, J. Lipopolysaccharide induces neuroinflammation in microglia by activating the MTOR pathway and downregulating Vps34 to inhibit autophagosome formation. *J. Neuroinflamm.* **2020**, *17*, 18. [CrossRef]
99. Liu, C.; Zheng, X.; Liu, L.; Hu, Y.; Zhu, Q.; Zhang, J.; Wang, H.; Gu, E.W.; Yang, Z.; Xu, G. Caloric Restriction Alleviates CFA-Induced Inflammatory Pain via Elevating β-Hydroxybutyric Acid Expression and Restoring Autophagic Flux in the Spinal Cord. *Front. Neurosci.* **2022**, *16*, 828278. [CrossRef] [PubMed]
100. Liang, L.; Tao, B.; Fan, L.; Yaster, M.; Zhang, Y.; Tao, Y.X. mTOR and its downstream pathway are activated in the dorsal root ganglion and spinal cord after peripheral inflammation, but not after nerve injury. *Brain Res.* **2013**, *1513*, 17–25. [CrossRef]
101. Azab, A.; Nassar, A.; Azab, A.N. Anti-Inflammatory Activity of Natural Products. *Molecules* **2016**, *21*, 1321. [CrossRef]
102. Shen, W.; Qi, R.; Zhang, J.; Wang, Z.; Wang, H.; Hu, C.; Zhao, Y.; Bie, M.; Wang, Y.; Fu, Y.; et al. Chlorogenic acid inhibits LPS-induced microglial activation and improves survival of dopaminergic neurons. *Brain Res. Bull.* **2012**, *88*, 487–494. [CrossRef]
103. dos Santos, M.D.; Almeida, M.C.; Lopes, N.P.; de Souza, G.E. Evaluation of the anti-inflammatory, analgesic and antipyretic activities of the natural polyphenol chlorogenic acid. *Biol. Pharm. Bull.* **2006**, *29*, 2236–2240. [CrossRef]
104. Agudelo-Ochoa, G.M.; Pulgarin-Zapata, I.C.; Velasquez-Rodriguez, C.M.; Duque-Ramirez, M.; Naranjo-Cano, M.; Quintero-Ortiz, M.M.; Lara-Guzman, O.J.; Munoz-Durango, K. Coffee Consumption Increases the Antioxidant Capacity of Plasma and Has No Effect on the Lipid Profile or Vascular Function in Healthy Adults in a Randomized Controlled Trial. *J. Nutr.* **2016**, *146*, 524–531. [CrossRef] [PubMed]
105. Tajik, N.; Tajik, M.; Mack, I.; Enck, P. The potential effects of chlorogenic acid, the main phenolic components in coffee, on health: A comprehensive review of the literature. *Eur. J. Nutr.* **2017**, *56*, 2215–2244. [CrossRef] [PubMed]
106. Rehman, S.U.; Ali, T.; Alam, S.I.; Ullah, R.; Zeb, A.; Lee, K.W.; Rutten, B.P.F.; Kim, M.O. Ferulic Acid Rescues LPS-Induced Neurotoxicity via Modulation of the TLR4 Receptor in the Mouse Hippocampus. *Mol. Neurobiol.* **2019**, *56*, 2774–2790. [CrossRef]
107. Priebe, A.; Hunke, M.; Tonello, R.; Sonawane, Y.; Berta, T.; Natarajan, A.; Bhuvanesh, N.; Pattabiraman, M.; Chandra, S. Ferulic acid dimer as a non-opioid therapeutic for acute pain. *J. Pain Res.* **2018**, *11*, 1075. [CrossRef]
108. Bumrungpert, A.; Lilitchan, S.; Tuntipopipat, S.; Tirawanchai, N.; Komindr, S. Ferulic Acid Supplementation Improves Lipid Profiles, Oxidative Stress, and Inflammatory Status in Hyperlipidemic Subjects: A Randomized, Double-Blind, Placebo-Controlled Clinical Trial. *Nutrients* **2018**, *10*, 713. [CrossRef]
109. Di Giacomo, S.; Percaccio, E.; Gulli, M.; Romano, A.; Vitalone, A.; Mazzanti, G.; Gaetani, S.; Di Sotto, A. Recent Advances in the Neuroprotective Properties of Ferulic Acid in Alzheimer's Disease: A Narrative Review. *Nutrients* **2022**, *14*, 3709. [CrossRef] [PubMed]
110. Liu, Y.; Deng, S.; Zhang, Z.; Gu, Y.; Xia, S.; Bao, X.; Cao, X.; Xu, Y. 6-Gingerol attenuates microglia-mediated neuroinflammation and ischemic brain injuries through Akt-mTOR-STAT3 signaling pathway. *Eur. J. Pharmacol.* **2020**, *883*, 173294. [CrossRef]
111. Young, H.Y.; Luo, Y.L.; Cheng, H.Y.; Hsieh, W.C.; Liao, J.C.; Peng, W.H. Analgesic and anti-inflammatory activities of [6]-gingerol. *J. Ethnopharmacol.* **2005**, *96*, 207–210. [CrossRef]
112. Sharma, S.; Shukla, M.K.; Sharma, K.C.; Tirath; Kumar, L.; Anal, J.M.H.; Upadhyay, S.K.; Bhattacharyya, S.; Kumar, D. Revisiting the therapeutic potential of gingerols against different pharmacological activities. *Naunyn-Schmiedeberg's Arch. Pharmacol.* **2023**, *396*, 633–647. [CrossRef]
113. Konmun, J.; Danwilai, K.; Ngamphaiboon, N.; Sripanidkulchai, B.; Sookprasert, A.; Subongkot, S. A phase II randomized double-blind placebo-controlled study of 6-gingerol as an anti-emetic in solid tumor patients receiving moderately to highly emetogenic chemotherapy. *Med. Oncol.* **2017**, *34*, 69. [CrossRef]
114. Gao, F.; Lei, J.; Zhang, Z.; Yang, Y.; You, H. Curcumin alleviates LPS-induced inflammation and oxidative stress in mouse microglial BV2 cells by targeting miR-137-3p/NeuroD1. *RSC Adv.* **2019**, *9*, 38397–38406. [CrossRef] [PubMed]

115. Yu, Y.; Shen, Q.; Lai, Y.; Park, S.Y.; Ou, X.; Lin, D.; Jin, M.; Zhang, W. Anti-inflammatory Effects of Curcumin in Microglial Cells. *Front. Pharmacol.* **2018**, *9*, 386. [CrossRef]
116. Singh, A.K.; Vinayak, M. Curcumin attenuates CFA induced thermal hyperalgesia by modulation of antioxidant enzymes and down regulation of TNF-α, IL-1β and IL-6. *Neurochem. Res.* **2015**, *40*, 463–472. [CrossRef]
117. Satoskar, R.R.; Shah, S.J.; Shenoy, S.G. Evaluation of anti-inflammatory property of curcumin (diferuloyl methane) in patients with postoperative inflammation. *Int. J. Clin. Pharmacol. Ther. Toxicol.* **1986**, *24*, 651–654. [PubMed]
118. Park, S.E.; Sapkota, K.; Kim, S.; Kim, H.; Kim, S.J. Kaempferol acts through mitogen-activated protein kinases and protein kinase B/AKT to elicit protection in a model of neuroinflammation in BV2 microglial cells. *Br. J. Pharmacol.* **2011**, *164*, 1008–1025. [CrossRef]
119. Jabbari, S.; Bananej, M.; Zarei, M.; Komaki, A.; Hajikhani, R. Effects of intrathecal and intracerebroventricular microinjection of kaempferol on pain: Possible mechanisms of action. *Res. Pharm. Sci.* **2021**, *16*, 203–216. [CrossRef]
120. Alam, W.; Khan, H.; Shah, M.A.; Cauli, O.; Saso, L. Kaempferol as a Dietary Anti-Inflammatory Agent: Current Therapeutic Standing. *Molecules* **2020**, *25*, 4073. [CrossRef]
121. Ren, J.; Lu, Y.; Qian, Y.; Chen, B.; Wu, T.; Ji, G. Recent progress regarding kaempferol for the treatment of various diseases. *Exp. Ther. Med.* **2019**, *18*, 2759–2776. [CrossRef] [PubMed]
122. Akiyama, M.; Mizokami, T.; Ito, H.; Ikeda, Y. A randomized, placebo-controlled trial evaluating the safety of excessive administration of kaempferol aglycone. *Food Sci. Nutr.* **2023**. [CrossRef]
123. Kang, C.H.; Choi, Y.H.; Moon, S.K.; Kim, W.J.; Kim, G.Y. Quercetin inhibits lipopolysaccharide-induced nitric oxide production in BV2 microglial cells by suppressing the NF-κB pathway and activating the Nrf2-dependent HO-1 pathway. *Int. Immunopharmacol.* **2013**, *17*, 808–813. [CrossRef]
124. Kumar, S.; Vinayak, M. Quercetin Ameliorates CFA-Induced Chronic Inflammatory Hyperalgesia via Modulation of ROS-Mediated ERK1/2 Signaling and Inhibition of Spinal Glial Activation In Vivo. *Neuromol. Med.* **2020**, *22*, 517–533. [CrossRef] [PubMed]
125. Javadi, F.; Ahmadzadeh, A.; Eghtesadi, S.; Aryaeian, N.; Zabihiyeganeh, M.; Rahimi Foroushani, A.; Jazayeri, S. The Effect of Quercetin on Inflammatory Factors and Clinical Symptoms in Women with Rheumatoid Arthritis: A Double-Blind, Randomized Controlled Trial. *J. Am. Coll. Nutr.* **2017**, *36*, 9–15. [CrossRef] [PubMed]
126. El-Bakoush, A.; Olajide, O.A. Formononetin inhibits neuroinflammation and increases estrogen receptor beta (ERβ) protein expression in BV2 microglia. *Int. Immunopharmacol.* **2018**, *61*, 325–337. [CrossRef] [PubMed]
127. Wang, X.S.; Guan, S.Y.; Liu, A.; Yue, J.; Hu, L.N.; Zhang, K.; Yang, L.K.; Lu, L.; Tian, Z.; Zhao, M.G.; et al. Anxiolytic effects of Formononetin in an inflammatory pain mouse model. *Mol. Brain* **2019**, *12*, 36. [CrossRef] [PubMed]
128. Ong, S.K.L.; Shanmugam, M.K.; Fan, L.; Fraser, S.E.; Arfuso, F.; Ahn, K.S.; Sethi, G.; Bishayee, A. Focus on Formononetin: Anticancer Potential and Molecular Targets. *Cancers* **2019**, *11*, 611. [CrossRef]
129. Zhang, B.; Wei, Y.Z.; Wang, G.Q.; Li, D.D.; Shi, J.S.; Zhang, F. Targeting MAPK Pathways by Naringenin Modulates Microglia M1/M2 Polarization in Lipopolysaccharide-Stimulated Cultures. *Front. Cell Neurosci.* **2018**, *12*, 531. [CrossRef]
130. Pinho-Ribeiro, F.A.; Zarpelon, A.C.; Fattori, V.; Manchope, M.F.; Mizokami, S.S.; Casagrande, R.; Verri, W.A., Jr. Naringenin reduces inflammatory pain in mice. *Neuropharmacology* **2016**, *105*, 508–519. [CrossRef]
131. Rebello, C.J.; Beyl, R.A.; Lertora, J.J.L.; Greenway, F.L.; Ravussin, E.; Ribnicky, D.M.; Poulev, A.; Kennedy, B.J.; Castro, H.F.; Campagna, S.R.; et al. Safety and pharmacokinetics of naringenin: A randomized, controlled, single-ascending-dose clinical trial. *Diabetes Obes. Metab.* **2020**, *22*, 91–98. [CrossRef]
132. Cara, K.C.; Beauchesne, A.R.; Wallace, T.C.; Chung, M. Effects of 100% Orange Juice on Markers of Inflammation and Oxidation in Healthy and At-Risk Adult Populations: A Scoping Review, Systematic Review, and Meta-analysis. *Adv. Nutr.* **2022**, *13*, 116–137. [CrossRef]
133. Zhang, S.; Gao, L.; Liu, X.; Lu, T.; Xie, C.; Jia, J. Resveratrol Attenuates Microglial Activation via SIRT1-SOCS1 Pathway. *Evid. Based Complement. Alternat. Med.* **2017**, *2017*, 8791832. [CrossRef]
134. Ma, Y.; Liu, S.; Shu, H.; Crawford, J.; Xing, Y.; Tao, F. Resveratrol alleviates temporomandibular joint inflammatory pain by recovering disturbed gut microbiota. *Brain Behav. Immun.* **2020**, *87*, 455–464. [CrossRef] [PubMed]
135. Marouf, B.H.; Hussain, S.A.; Ali, Z.S.; Ahmmad, R.S. Resveratrol Supplementation Reduces Pain and Inflammation in Knee Osteoarthritis Patients Treated with Meloxicam: A Randomized Placebo-Controlled Study. *J. Med. Food* **2018**, *21*, 1253–1259. [CrossRef] [PubMed]
136. Lin, S.R.; Fu, Y.S.; Tsai, M.J.; Cheng, H.; Weng, C.F. Natural Compounds from Herbs that can Potentially Execute as Autophagy Inducers for Cancer Therapy. *Int. J. Mol. Sci.* **2017**, *18*, 1412. [CrossRef] [PubMed]
137. Rickert, U.; Cossais, F.; Heimke, M.; Arnold, P.; Preusse-Prange, A.; Wilms, H.; Lucius, R. Anti-inflammatory properties of Honokiol in activated primary microglia and astrocytes. *J. Neuroimmunol.* **2018**, *323*, 78–86. [CrossRef] [PubMed]
138. Khalid, S.; Ullah, M.Z.; Khan, A.U.; Afridi, R.; Rasheed, H.; Khan, A.; Ali, H.; Kim, Y.S.; Khan, S. Antihyperalgesic Properties of Honokiol in Inflammatory Pain Models by Targeting of NF-κB and Nrf2 Signaling. *Front. Pharmacol.* **2018**, *9*, 140. [CrossRef]
139. Eliaz, I.; Weil, E. Intravenous Honokiol in Drug-Resistant Cancer: Two Case Reports. *Integr. Cancer Ther.* **2020**, *19*, 1534735420922615. [CrossRef] [PubMed]
140. Zhu, M.D.; Zhao, L.X.; Wang, X.T.; Gao, Y.J.; Zhang, Z.J. Ligustilide inhibits microglia-mediated proinflammatory cytokines production and inflammatory pain. *Brain Res. Bull.* **2014**, *109*, 54–60. [CrossRef]

141. Wang, J.; Du, J.R.; Wang, Y.; Kuang, X.; Wang, C.Y. Z-ligustilide attenuates lipopolysaccharide-induced proinflammatory response via inhibiting NF-κB pathway in primary rat microglia. *Acta Pharmacol. Sin.* **2010**, *31*, 791–797. [CrossRef]
142. Du, J.; Yu, Y.; Ke, Y.; Wang, C.; Zhu, L.; Qian, Z.M. Ligustilide attenuates pain behavior induced by acetic acid or formalin. *J. Ethnopharmacol.* **2007**, *112*, 211–214. [CrossRef]
143. Zhang, Y.; Zhang, Y.; Han, Y.; Tian, Y.; Wu, P.; Xin, A.; Wei, X.; Shi, Y.; Zhang, Z.; Su, G.; et al. Pharmacokinetics, tissue distribution, and safety evaluation of a ligustilide derivative (LIGc). *J. Pharm. Biomed. Anal.* **2020**, *182*, 113140. [CrossRef] [PubMed]
144. Sun, X.; Zeng, H.; Wang, Q.; Yu, Q.; Wu, J.; Feng, Y.; Deng, P.; Zhang, H. Glycyrrhizin ameliorates inflammatory pain by inhibiting microglial activation-mediated inflammatory response via blockage of the HMGB1-TLR4-NF-kB pathway. *Exp. Cell Res.* **2018**, *369*, 112–119. [CrossRef]
145. Cao, Z.Y.; Liu, Y.Z.; Li, J.M.; Ruan, Y.M.; Yan, W.J.; Zhong, S.Y.; Zhang, T.; Liu, L.L.; Wu, R.; Wang, B.; et al. Glycyrrhizic acid as an adjunctive treatment for depression through anti-inflammation: A randomized placebo-controlled clinical trial. *J. Affect. Disord.* **2020**, *265*, 247–254. [CrossRef]
146. Lu, Y.; Zhao, L.X.; Cao, D.L.; Gao, Y.J. Spinal injection of docosahexaenoic acid attenuates carrageenan-induced inflammatory pain through inhibition of microglia-mediated neuroinflammation in the spinal cord. *Neuroscience* **2013**, *241*, 22–31. [CrossRef] [PubMed]
147. Hill, C.L.; March, L.M.; Aitken, D.; Lester, S.E.; Battersby, R.; Hynes, K.; Fedorova, T.; Proudman, S.M.; James, M.; Cleland, L.G.; et al. Fish oil in knee osteoarthritis: A randomised clinical trial of low dose versus high dose. *Ann. Rheum. Dis.* **2016**, *75*, 23–29. [CrossRef] [PubMed]
148. Hu, B.; Xu, G.; Zhang, X.; Xu, L.; Zhou, H.; Ma, Z.; Shen, X.; Zhu, J.; Shen, R. Paeoniflorin Attenuates Inflammatory Pain by Inhibiting Microglial Activation and Akt-NF-κB Signaling in the Central Nervous System. *Cell Physiol. Biochem.* **2018**, *47*, 842–850. [CrossRef]
149. Li, X.; Shi, F.; Zhang, R.; Sun, C.; Gong, C.; Jian, L.; Ding, L. Pharmacokinetics, Safety, and Tolerability of Amygdalin and Paeoniflorin after Single and Multiple Intravenous Infusions of Huoxue-Tongluo Lyophilized Powder for Injection in Healthy Chinese Volunteers. *Clin. Ther.* **2016**, *38*, 327–337. [CrossRef]
150. Shukla, S.M.; Sharma, S.K. Sinomenine inhibits microglial activation by Aβ and confers neuroprotection. *J. Neuroinflamm.* **2011**, *8*, 117. [CrossRef]
151. Yuan, Y.; Zhang, Y.; He, X.; Fan, S. Protective effects of sinomenine on CFA-induced inflammatory pain in rats. *Med. Sci. Monit. Int. Med. J. Exp. Clin. Res.* **2018**, *24*, 2018–2024. [CrossRef] [PubMed]
152. Liu, W.; Zhang, Y.; Zhu, W.; Ma, C.; Ruan, J.; Long, H.; Wang, Y. Sinomenine Inhibits the Progression of Rheumatoid Arthritis by Regulating the Secretion of Inflammatory Cytokines and Monocyte/Macrophage Subsets. *Front. Immunol.* **2018**, *9*, 2228. [CrossRef] [PubMed]
153. Liu, S.; Cheng, Y.; Rao, M.; Tang, M.; Dong, Z. Muscone Induces CYP1A2 and CYP3A4 Enzyme Expression in L02 Human Liver Cells and CYP1A2 and CYP3A11 Enzyme Expression in Kunming Mice. *Pharmacology* **2017**, *99*, 205–215. [CrossRef]
154. Xu, J.; Yuan, C.; Wang, G.; Luo, J.; Ma, H.; Xu, L.; Mu, Y.; Li, Y.; Seeram, N.P.; Huang, X.; et al. Urolithins Attenuate LPS-Induced Neuroinflammation in BV2Microglia via MAPK, Akt, and NF-κB Signaling Pathways. *J. Agric. Food Chem.* **2018**, *66*, 571–580. [CrossRef]
155. D'Amico, D.; Olmer, M.; Fouassier, A.M.; Valdes, P.; Andreux, P.A.; Rinsch, C.; Lotz, M. Urolithin A improves mitochondrial health, reduces cartilage degeneration, and alleviates pain in osteoarthritis. *Aging Cell* **2022**, *21*, e13662. [CrossRef] [PubMed]
156. Liu, S.; D'Amico, D.; Shankland, E.; Bhayana, S.; Garcia, J.M.; Aebischer, P.; Rinsch, C.; Singh, A.; Marcinek, D.J. Effect of Urolithin A Supplementation on Muscle Endurance and Mitochondrial Health in Older Adults: A Randomized Clinical Trial. *JAMA Netw. Open.* **2022**, *5*, e2144279. [CrossRef]
157. Niu, J.; Straubinger, R.M.; Mager, D.E. Pharmacodynamic Drug-Drug Interactions. *Clin. Pharmacol. Ther.* **2019**, *105*, 1395–1406. [CrossRef] [PubMed]
158. Liu, C.; Huang, Z.; Li, Z.; Li, J.; Li, Y. Muscone reduced the hypnotic and analgesic effect of ketamine in mice. *J. Chin. Med. Assoc.* **2020**, *83*, 148–155. [CrossRef] [PubMed]

MDPI
St. Alban-Anlage 66
4052 Basel
Switzerland
Tel. +41 61 683 77 34
Fax +41 61 302 89 18
www.mdpi.com

Pharmaceuticals Editorial Office
E-mail: pharmaceuticals@mdpi.com
www.mdpi.com/journal/pharmaceuticals

www.ingramcontent.com/pod-product-compliance
Lightning Source LLC
Chambersburg PA
CBHW061559210926
51850CB00009B/2149
9783036591896